有机光电子
双语教程

密保秀　高志强　曹大鹏　许文娟　编著

清华大学出版社
北京

内 容 简 介

本书阐述了有机光电材料中电子结构及其相关电子和光子的相关过程,解释了有机分子及其固体凝聚态的各种性质,并详细介绍了有机材料光电及热稳定性质的测试表征方法,最后对有机光电材料相关光电器件的结构、工作原理及应用进行了讨论。有机光电子学是一门发展中的交叉科学,本书融合了该领域的理论基础和前沿进展,注重深入浅出,并在每章末尾附有思考题。全书以英文为主,并对重点和难点内容辅以中文翻译。

本书可作为材料科学与工程专业或者光电信息专业的研究生、本科生、专科生的专业课教材,也可以作为从事有机光电材料及器件领域相关工作的科技工作者的参考书。

图书在版编目(CIP)数据

有机光电子双语教程:英文/密保秀等编著.—北京:清华大学出版社,2024.1
ISBN 978-7-302-64973-1

Ⅰ.①有⋯ Ⅱ.①密⋯ Ⅲ.①光电子技术-高等学校-教材-英文 Ⅳ.①TN2

中国国家版本馆 CIP 数据核字(2023)第 222428 号

责任编辑:鲁永芳
封面设计:常雪影
责任校对:欧 洋
责任印制:宋 林

出版发行:清华大学出版社
    网 址:https://www.tup.com.cn,https://www.wqxuetang.com
    地 址:北京清华大学学研大厦 A 座 **邮 编:**100084
    社 总 机:010-83470000 **邮 购:**010-62786544
    投稿与读者服务:010-62776969,c-service@tup.tsinghua.edu.cn
    质量反馈:010-62772015,zhiliang@tup.tsinghua.edu.cn
印 装 者:三河市铭诚印务有限公司
经 销:全国新华书店
开 本:185mm×260mm **印 张:**27.5 **字 数:**668 千字
版 次:2024 年 1 月第 1 版 **印 次:**2024 年 1 月第 1 次印刷
定 价:109.00 元

产品编号:095861-01

前　言

　　"有机光电子学"是一门与电子元器件相关的交叉、前沿学科,目前处于蓬勃发展阶段,国内尚未有相应的教材出版。南京邮电大学自 2006 年成立材料科学与工程学院以来,有机电子学就作为特色专业课程,成为本科生及研究生的重点教学内容之一,目的在于培养学生掌握有机光电材料在光电器件中应用的基础知识。

　　全书采取双语编排,有助于学生在掌握专业知识的同时,熟悉英文形式的科学术语,方便知识拓展及文献调研。每个章节都包括:①正文(英语,局部中文翻译),②本章小结(中文),③专业词汇(双语),④思考题(英文)。文字深入浅出,非常适合学生作为教材使用,为学生今后从事与有机光电材料相关的工作奠定基础。书中的英文大部分来自原汁原味的英文原版书籍,通过作者的二次创作及编辑加工而成,英文纯正,逻辑清晰,易于理解,有利于读者专业英语水平的提高。另外,书中含有英文重点、难点句子的中文翻译,可帮助英文相对薄弱的学生理解专业知识、提高英文水平。特别地,本书有相应配套的慕课视频及文字内容,已经在江苏省的在线开放课程平台上线。

　　本课程的教学时数建议为 32～48 学时,视学生的培养方案而定。各章的参考教学课时见如下学时分配表。

学时分配表

章　节	课　程　内　容	学　时
Chapter 1	Introduction to organic optoelectronics	2～3
Chapter 2	Bonding and electronic structures in organic materials	2～4
Chapter 3	Energy related processes in organic materials	8～12
Chapter 4	Electrical properties and related processes in organic materials	6～9
Chapter 5	Property characterization for organic materials	3～4
Chapter 6	Organic field effect transistors	3～4
Chapter 7	Solar cells containing organic materials	3～4
Chapter 8	Organic light-emitting diodes	3～4
Chapter 9	Organic sensors	2～4
课时总计		32～48

　　本书的编写参考了多本中文和英文文献(见参考文献部分),在此对这些参考文献的作者表示感谢。同时,感谢以下基金的支持:有机电子与信息显示国家重点实验室定向项目、江苏省高校优势学科建设工程(PAPD:YX03002,YX03003)、南京邮电大学卓越教师培育计划(ZYJH201402)、有机电子学慕课建设项目(2014MOOCA1)。

　　由于作者能力所限,加之双语,疏漏和不妥之处在所难免。恳请广大读者在使用过程中给予批评指正,以利于再版之际的修改和完善。与本书相关的讨论与交流,也十分欢迎。

密保秀、高志强
2023 劳动节前于南京

目　录

CHAPTER 1 INTRODUCTION TO ORGANIC OPTOELECTRONICS

Starting from historical remarks about optoelectronics，this chapter gives a brief introduction to organic optoelectronics，including history，concept，development and types/characteristics of organic materials，as well as their main optoelectronic processes and corresponding device applications.

1. 1 Optoelectronics

Optoelectronics refers to the optoelectronic science and technology not only for devices and systems of electronic in nature，yet involving light （such as the photo voltaic device and the light-emitting diode），but also for closely related areas that are essentially optical in nature but involve electronics （such as crystal light modulators）. In a broad sense，pure electronic devices，such as field effect transistor，also belong to this category. In these devices，electrons and photons are used to generate，process，transmit，and store information at unprecedented rates and with ever-decreasing power requirements[1].

Optoelectronic devices have dramatically impacted the way humans live in the twentieth and twenty-first centuries. The development of classic and quantum mechanical theories，particularly the discovery and development of classic inorganic semiconductors and subsequent optoelectronic devices such as radios，telephones，televisions，integrated circuits，and computers，has given birth to the human society of an "information age" and a "global village," where the information is processed and transferred at the level and speed of electrons and photons. In today's world，it is virtually impossible to find a piece of electrical equipment that does not employ optoelectronic devices as a basic necessity—from CD and DVD players to televisions，from automobiles and aircraft to medical diagnostic facilities in hospitals and telephones，from satellites and space-borne missions to under water exploration systems—the list is almost

1. 光电子学指的是光电子方面的科学和技术，它不仅是本质上为电子却涉及光的器件和系统（如光伏器件和发光二极管），也包括本质上为光学，但是涉及电子学的紧密相关领域（如晶体光调制器）。广义上，纯电子器件，例如场效应晶体管，也属于这个分类。在这些器件中，电子和光子被用来产生、加工、传导以及存储信息，其速率前所未有，其功率需求不断降低。

2. 在发达的现代世界，光电子学实际上存在于每一个家庭和企业办公室；存在于电话、传真机、复印机、计算机以及照明，等等。

endless. Optoelectronics is in virtually every home and business office in the developed modern world, in telephones, fax machines, photocopiers, computers, and lighting, etc[2].

As one branch of optoelectronics, electronics refers to the science and technology of how to control electric energy, energy in which the electrons have a fundamental role. Electronics deals with electrical circuit comprising active components, inert components and connection technology, and has experienced development from vacuum, solid state, micro electronics, integrated circuit (IC) and super large IC, and now is heading for molecular electronics (Figure 1.1).

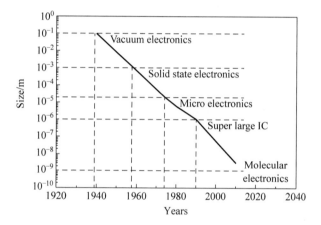

Figure 1.1　Development of electronics with times

Most of today's commercially available optoelectronic devices are inorganic material based ones, which are fabricated from inorganic semiconductors and conductors. The former includes elemental semiconductors (e.g., silicon and germanium) and compound semiconductor (e.g., gallium arsenide, zinc oxide, cadmium sulfide); Examples of latter are metals of Al, Au and Ag, and transparent oxide conductors (e.g., indium tin oxide). Inorganic materials based optoelectronics are well developed with features of fast and reliable. However, these devices bear the disadvantages of limited types and amounts in materials, consuming large amount of energy in fabrication, difficult in flexibility, incompatible to biosystem, often detrimental to environment, and hard/expensive to further miniturization, etc[3].

3. 基于无机材料的光电子学，高度发达，具有快速和可靠性好的特点。但是，这些器件存在如下缺点：材料的种类和数量少、制备时消耗大量能源、难于柔性、与生物系统不兼容、通常对环境有害，以及进一步小型化昂贵/困难，等等。

In the past several decades, research and development on organic/polymeric optoelectronic materials and devices have grown rapidly. These devices/systems have exhibited advantages such as flexibility, compatibility to biosystem, processability in solution process (e.g., spin coating, spray, roll to roll inkjet printing and doctor blade printing), and the potential for low-cost mass production, and thus being a very useful supplement to inorganic counterparts.

Devoted to organic electronics, this text book will discuss the basic science in the aspects of optoelectrical-active organic materials, and their device applications.

1.2 Organic Optoelectronics

1.2.1 Optoelectronic Organic Materials

1. Advent of Organic Materials

Materials are useful maters, which were initially divided into metals and non-metals. Later, it was found that there was a kind of non-metallic materials, which contained essential element of carbon and could only be obtained via a living organism, and hence this kind of materials were termed as organic materials. In 1828, by synthesis of small molecule urea (NH_2CONH_2), F. Wöhler demonstrated for the first time that an organic compound could be prepared in laboratory without a living organism, proving the capability for artificial synthesis of organic materials[4].

After the successful preparation of small molecules in laboratory, great progress has been made for polymer-type organic materials in 1920: German scientist H. Staudinger proposed that polymers were made of repeating molecule units regularly in long chains, forming polymer concept. Figure 1.2 schematically illustrates the formation of polymer from monomers. At that time, people thought that plastics are polymers, and unlike metals, they do not conduct electricity and can be used as insulation round the copper wires in ordinary electric cables.

4. 材料是有用的物质，最初被划分为金属和非金属。后来人们发现，有这样一种非金属材料，含有碳元素并且只能通过有机生物体才能获得；因此，这类材料就被命名为有机材料。1828 年，通过合成小分子尿素（NH_2CONH_2），F. Wöhler 首次展示了在脱离生物体的情况下，在实验室能够合成有机化合物，以此证明了人工合成有机材料的可能性。

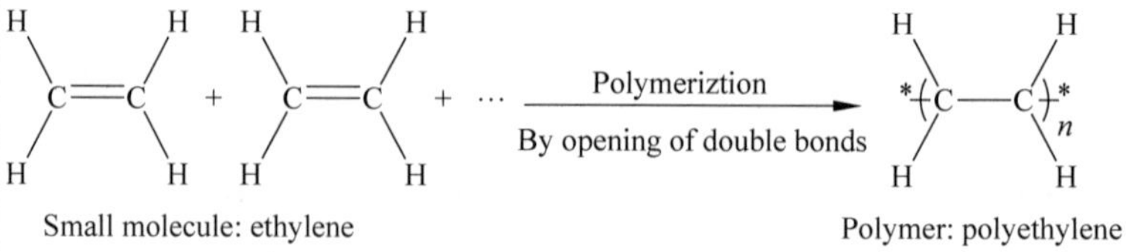

Figure 1.2 Formation of polymer from monomer

5. 随着时间的推移,有机材料的定义演变为:有机材料包含碳和氢元素,也可能包含氧、氮、卤素、硫、磷、甚至金属等;其中的碳原子相互成键,形成了小分子或者聚合物的碳骨架。

As time goes on，the concept of organic materials is evolved as such that organic materials refer to those containing carbon, hydrogen and other possible elements of oxygen，nitrogen，halogen，sulphur，phosphor，and even metal，etc，where the carbon atoms are bound to each other to form the carbon frameworks of small molecules or polymers[5]. The carbon atoms are bound together by either σ-bonds or π-bonds. Single bonds like C－H bonds are σ-bonds while double bonds consist of a hybrid orbital of a σ-bond with a π-bond. σ-bonds are strong while π-bonds are weak. Electrons involved in π-bonds are delocalized over the molecule and are responsible for most of the interesting optical properties of organic materials. On the other hand，the σ-bonds are more localized and hold the molecule together，but usually do not play a major role in optoelectronic properties.

2. Milestones in the Development of Organic Optoelectronic Materials

6. 有机材料的光导和半导体特性分别报道于 1906 年和 1950 年。自此,有机材料的基础研究稳步推进,形成了一系列重要的里程碑性历程,举例如下。

The photoconductive and semiconducting properties of organic materials were reported in 1906 and 1950，respectively. Since then，basic research has steadily continued，forming a series of important milestones. Examples are as following[6].

From 1950s，steady works on crystalline organics started，resulting in lots of crystalline organic materials with semiconductive/photoconductive properties. Figure 1.3 schematically shows some small-molecule based crystals（a），and polymer chain based crystalline and amorphous states （b）. Then，after the appearance of organic non-linear optical materials in 1970s，the first conducting polymer was prepared by A. J. Heeger，A. G. Macdiarmid and H. Shirakawa in

1977 (the reaction is shown in Figure 1.4). Due to this revolutionary work, the Nobel Price of Chemistry 2000 was given to them for the discovery and development of conductive polymers. It was clear that for a polymer to be able to conduct electric current it must consist alternatively of single and double bounds between the carbon atoms. It must also be "doped", which means that electrons can move along the molecules — it becomes electrically conductive[7]. This was of great importance for chemists as well as physicists, because: ① industrially, conductive plastics can be used as anti-static substances for photographic film, shields for computer screen against electromagnetic radiation and for "smart" windows that can exclude sunlight; ② conductive plastics also can be applied in organic optoelectronic devices, such as organic light-emitting diodes and organic solar cells[8].

7. 已经很明确,要想使聚合物导电,它必须在碳原子之间含有交替的单双键结构,还必须"掺杂",这样电子可以沿着分子移动,从而具有导电性。

8. ①工业上,导电塑料可以用作胶卷的防静电物质、计算机屏幕的电磁波辐射防护、能够去除太阳光的智能窗口。②导电塑料还可以被应用于有机光电器件中,如有机发光二极管和有机太阳能电池。

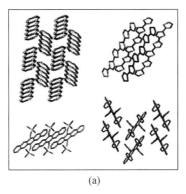
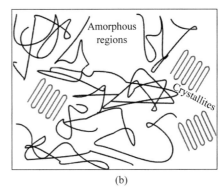

Figure 1.3 Small-molecule based crystals (a), polymer chain based crystalline and amorphous states (b)

Figure 1.4 The Noble Price work of 2000 Chemistry: Ziegler-Natta polymerization for polyacetylene (a), and the conducting mechanism for doped polyacetylene (b)

In the afterward years of 1985，1993 and 2004，semiconductor fullerene，conducting polymer PEDOT：PSS (Poly (3,4-ethylenedioxythiophene)：poly (styrenesulfonic acid)) and conductive/semiconductive graphene were reported，respectively (Figure 1.5 for their molecular structures).

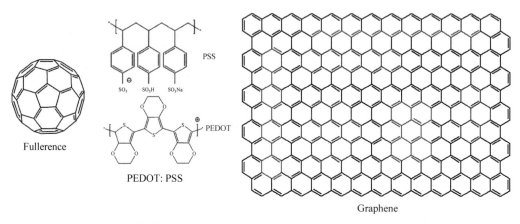

Figure 1.5　Molecular structures of fullerene，PEDOT：PSS and graphene

9. 最早的有机材料光电应用是 1980 年商业化的、作为光感器应用于电子复印机的复合有机材料,其组成是空穴传输小分子分散在绝缘体聚合物中。这个有机光感器的加工过程是涂覆,使得其成本很低。

The earliest optoelectronic application of organic materials is the composite organic materials with hole transport organic molecules dispersed in insulating polymers，which were commercialized as photoreceptors for electrophotography in 1980. The manufacturing process for this organic photoreceptor was a coating process，which contributed to the low cost of the photoreceptor[9]. Organic field effect transistor，organic light-emitting diodes (OLEDs) and organic solar cells were appeared in1986，1987，and 1986，respectively. These devices were highly efficient at that time and showed brilliant futures. OLEDs were commercialized as an automotive display in 1997，and are currently being used in high-definition OLED TVs and OLED lighting. In the future，it is expected that organic materials will be successfully applied to flexible displays，biosensors，and other devices that could not be realized with conventional inorganic semiconductors.

Currently，organic materials contribute nearly 90% of all known materials，and can exhibit the following properties：photoconductivity， electroluminescence， triboelectricity， metallic conductivity, superconductivity, photovoltaic effects, or charge storage and release, etc[10].

3. Structural and Property Features of Organic Materials

1) Bonding Characteristics

Different to other-type materials composing the smallest unit of atoms, the smallest units in organic materials are molecules that consist of atoms with covalent bondings only within a molecule. These discrete molecules are held together by weak van der Waals forces (Figure 1.6). For this reason, organic materials are also called molecular materials. Due to the weak nature of the bonding between molecules, the properties of the individual molecule are retained in the solid state to a far greater extent than would be found in solids exhibiting the other types of bonding[11].

10. 当前有机材料占所有已知材料的 90% 以上,可能具备如下性质: 光导、电致发光、摩擦电、金属导电性 超导电性、光伏效应和电荷存储与释放等。

11. 基于这个原因,有机材料也称为分子材料。由于分子间较弱的成键,相比于其他成键类型的固体,有机材料中个体分子的特性在固态时得到了极大程度的保留。

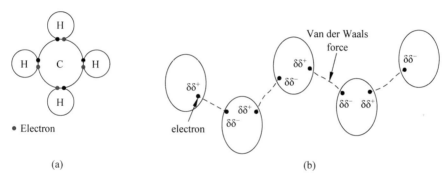

• Electron

electron

(a)   (b)

Figure 1.6   Covalent bonding within one organic molecule of methane (a), and van der Waals force between molecules (molecules are represented as circles) (b)

2) Conjugation

For use as active components in optoelectronic devices, the organic molecules must present structural feature of conjugation. The term conjugation refers to the alternating sequence of single and double bonds in between atoms (mostly are carbons) of organic molecules[12]. Figure 1.7 exemplifies four conjugated molecules and three non-conjugated molecules. As shown in Figure 1.7(b), although these molecules contain both single and double bonds, they do not have conjugation structure, and are not conjugated materials, because their double and single bonds don't present in an

12. 作为活性组分应用于光电器件,有机分子必须存在共轭结构。术语共轭指的是有机分子中的原子之间(多数情况下是碳原子)形成的单键与双键交替结构。

13. 共轭有机材料是一类非常重要的有机光电材料，因为它们表现出令人感兴趣的光电特性，是有机光电子学中主要考虑的有机材料。有机半导体就是共轭有机材料，表现出半导体特性。

alternative single-double sequence. Conjugated organic materials are one important type of organic materials，which have interesting optoelectrical properties，and are the main concerned-materials in organic optoelectronics. Organic semiconductors belong to conjugated organic materials that exhibit semiconducting properties[13].

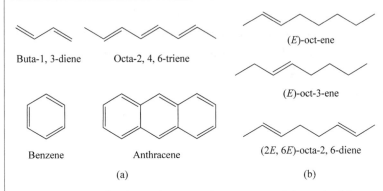

Figure 1. 7　　Molecules with conjugation structure（a），and without conjugation structure（b）

3）Large Amounts

The quantity of organic materials is large without limitation，because the molecular structures are much easier to modify compared to other type of materials. New materials can be produced just by a single modification. For example，as shown in the Figure 1. 8，merely changing a linking mode from para- to meta- in the molecule results in different materials[14].

14. 有机材料的数量是无穷无尽的，非常巨大。因为相比于其他类型的材料，有机分子结构非常容易进行分子结构修饰。任何一点点的修饰都会产生新的材料。如图1.8所示，仅将分子的一个链接模式从对位改为间位，就形成了新材料。

4. Classification of Organic Materials

There are several classification methods for organic materials.

（1）According to molecular weight，organic materials can be classified into small molecule，macromolecule and polymer. Small molecule refers to compounds possessing definite molecular formula and molecular weight. The molecular weight of small molecules is usually smaller than several thousands. Macromolecule possesses definite molecular formula with molecular weight in the range of several thousands and ten thousands. Polymer is long-chain molecule consisting of an indeterminate number of molecular repeat units, with molecular weight of several hundred thousands even several millions. Polymer is not a pure substance but a mixture[15].

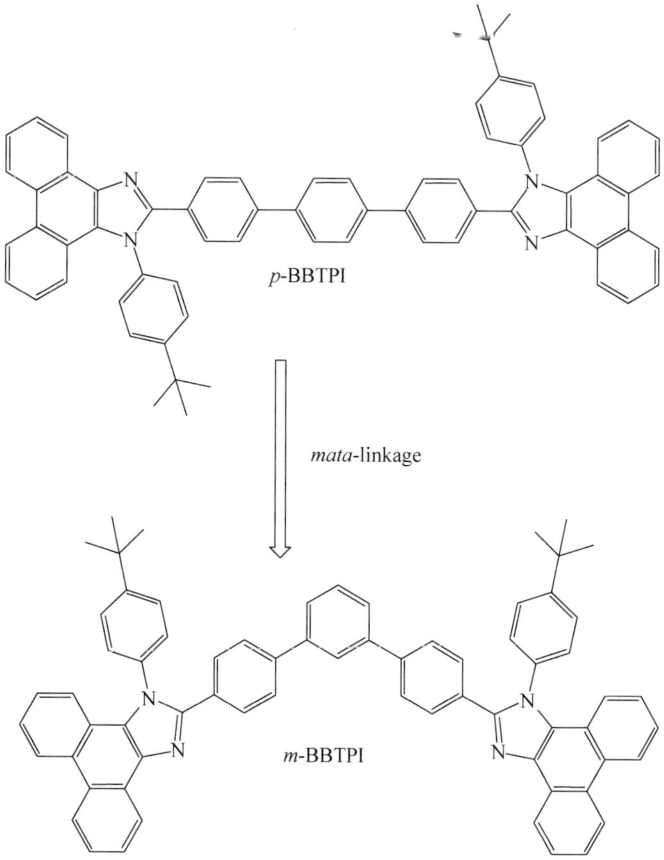

Figure 1. 8   Production of a new organic material by only changing the linking mode

15. 根据分子量,有机材料可以分为小分子,大分子和聚合物。小分子指的是具有确定分子式和分子量的化合物,其分子量通常小于几千。大分子也具有确定的分子式,其分子量在数千至数万之间。聚合物是由不确定数目的分子重复单元组成的长链分子,分子量为数十万甚至数百万。聚合物不是纯物质而是混合物。

(2) According to structure complexity, organic materials can be classified into small molecule, polymer and biomolecule (Figure 1. 9). Biomolecules are also organic molecules. Due to their less application in organic optoelectronics, they are beyond the discussion of this textbook[16].

(3) According to molecular shapes, there are disc, linear, starburst, and super branched molecules, etc (Figure 1.10).

16. 根据结构的复杂程度,有机材料可分为小分子、聚合物和生物分子(图1.9)。生物分子也是有机分子。由于它们在有机光电子学中的应用较少,本书不作讨论。

5. Differences between Organic & Inorganic Semiconductors

Semiconductors are the key components in optoelectronic

17. 光电子器件中的关键组分是半导体。有机半导体的分子特性，导致了有机半导体和无机半导体之间的一些重要物理差异。下面总结最主要的差别。

devices. The molecular nature of organic semiconductors leads to a number of significant physical differences with respect to inorganic counterpart. The most significant differences between them are summarized below[17].

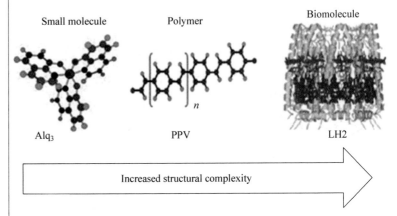

Small molecule   Polymer   Biomolecule

Alq₃   PPV   LH2

Increased structural complexity

Figure 1.9   Three types of organic materials according to structure complexity

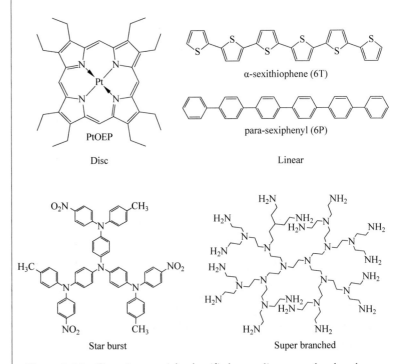

PtOEP

Disc

α-sexithiophene (6T)

para-sexiphenyl (6P)

Linear

Star burst

Super branched

Figure 1.10   Organic materials classified according to molecular shapes

1) Structural Difference

Structurally, organic and inorganic semiconductors often compose of different elements. Besides, the bondings between atoms are not the same. Organic semiconductors are molecular solids which contain weak van der Waals bonding despite of the covalent bondings within molecules，while inorganic semiconductors are made of covalently bonded atoms[18].

　　Typical van der Waals force is smaller than 42 kJ · mol^{-1}，which is much smaller than that in inorganic semiconductors，e. g.，the bond energy in silicon is 318 kJ · mol^{-1}. Morcover，the van der Waals force decreases with R^6，while the valence bond energy decreases proportionally to R^2. Figure 1.11 shows the crystal structures of silicon and a polycyclic aromatic hydrocarbon，demonstrating their differences in bonding and spacial strctching[19].

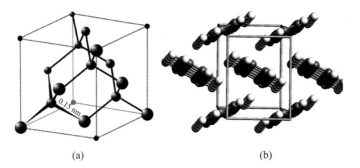

(a)　　　　　　　　　　(b)

Figure 1.11　Structural differences between organic and inorganic materials：covalently bonded crystal of silicon（a），and crystal structure of a polycyclic aromatic hydrocarbon（b）

2）Exciton Difference

　　Excitons in inorganic semiconductors belong to weakly bound Wannier − Mott excitons. They have large Bohr radii and small binding energies. For example，in GaAs the exciton Bohr radius is 140 Å and its binding energy is 4.2 meV. This is in contrast to excitons in organic materials，which are referred to as tightly bound Frenkel excitons with small Bohr radii（5～10 Å）and large binding energies（0.5～1.0 eV）. The exchange energies，i. e.，the energetic difference between singlet- and triplet-excitons in organic materials are often 0.5～1.0 eV^{20}.

3）Optoelectronic Transition Difference

　　Electronic transitions in organic materials need to be considered together with vibrational states，as opposed to

18. 在结构上，有机半导体和无机半导体通常由不同元素组成。另外，原子之间的成键也不同。有机半导体是含有弱范德瓦耳斯力的分子固体，虽然其分子内是共价键；而无机半导体材料是由以共价键相结合的原子组成。

19. 典型的范德瓦耳斯力通常小于 42 kJ · mol^{-1}，远远小于无机半导体中的作用力。例如，硅的键能为 318 kJ · mol^{-1}。更进一步，有机材料中的范德瓦耳斯力随分子间距的六次方（R^6）降低，而无机半导体中的共价键键能随原子间距的平方（R^2）降低。图 1.11 给出了 Si 晶体和有机稠环芳香烃晶体的结构示意图，展示了有机半导体和无机半导体在成键和空间伸展的不同。

20. 无机半导体中的激子属于弱结合的万尼尔-莫特激子。它们的玻尔半径较大，结合能较小。例如，GaAs 中激子的玻尔半径是 140 Å，结合能是 4.2 meV。这与有机材料中紧密结合的弗仑克尔激子不同，其玻尔半径小（5～10 Å），结合能大（0.5～1 eV）。有机材料中的交换能，即单线态和三形态激子能量差异通常在 0.5～1.0 eV。

21. 有机材料中的电子跃迁需要同时考虑振动态，与无机材料中电子跃迁和振动态可以很容易地分离不同。

22. 较小的质量使得有机材料的振动频率，即$(f/m)^{1/2}$（f为谐振子强度，m为原子质量），非常大。振动态被视为量子化简谐振动的谐振子，形成构象坐标。

23. 在无机半导体中，激子型吸收通常表现为一系列在能带之下的较强吸收峰，通常没有振动跃迁。光子可以作为束缚的电子-空穴复合的结果被发射出来，导致非常小斯托克斯位移的激子型发射。

inorganic materials, where the electronic transitions and the vibrational states can be easily separated[21]. This is because the atoms in organics are often small, e. g., hydrogen has a small mass, whereas an inorganic solid, like NaCl, has atoms with much larger mass. The smaller mass makes the vibrational frequencies, $(f/m)^{1/2}$ (f: oscillator strength, m: atomic mass), much larger in organics. The vibrational states are then treated like simple harmonic oscillators with quantized energy levels, forming configurational coordinates[22]. Optical transitions become combined electronic and vibrational transitions. Electronic transitions occur between the ground state and excited state both with vibrational quantized energy levels. In such, compared to optical absorption, the fluorescence emitted on radiative decay to the ground state may display a relatively large Stokes shift and pronounced vibronic structure. In inorganic semiconductors, excitonic absorption usually manifests itself as a series of strong absorption peaks just below the energy band gap, generally without vibrational transitions. A photon may also be emitted as a result of bound electron − hole recombination, resulting in an excitonic emission with very small Stokes shift[23].

Furthermore, in contrast to the delocalized features of excitation in inorganic materials with band structures, the excitations in organic materials are generally localized, due to the localized electron wavefunctions. For instance, the wavefunction coherence in a conjugated polymer extends only about a few repeat units (i. e., the effectively conjugated segment, ECS), but not further. Localization leads to a strong coupling between excitations and local molecular geometry, hence often leading to electron clouds and bond lengths redistribution after excitation. In the case of charged excitations, these geometric relaxations can break the local symmetry, and thus activate vibronic bands in the infrared that are symmetry-forbidden (i. e., Raman-but not IR-active) in the ground state. Consequently, electronic transition moments are strong and highly directional, parallel to the chain or molecular axis. Charged excitations in organic materials are generally more akin to the concept of the "radicalion", as familiar from solution-based chemistry, than the concept of a "polaron" in solid state physics (nevertheless, the term "hole" is commonly used for radical cations)[24].

4) Carrier and Mobility Difference

In contrast to inorganic solids where the allowed energy levels form valence and conduction bands (VB and CB), organic molecular materials are characterized by their HOMO, the highest occupied molecular orbital, and LUMO, the lowest unoccupied molecular orbital. Carrier transport occurs primarily by hopping in organics. For example, electrons and holes on polymer chains hop from one chain to another for transport. In inorganic crystalline solids, the dominant carrier transport mechanism is band-like transport, relying on the presence of a periodic crystalline structure and a well-defined density of states. As a result, the carrier mobility in organic and inorganic semiconductors is significantly different. In organic materials, the mobility of the carriers is $10^{-5} \sim 10^{-3}$ cm^2 · V^{-1} · s^{-1}. In inorganic materials, the carrier mobility is $10^2 \sim 10^4$ cm^2 · V^{-1} · s^{-1}. Figure 1.12 schematically shows the comparison between the hopping transport in organics and the bandlike transport in inorganics[25].

24. 例如,在共轭聚合物中波函数的相干仅存在于几个重复单元(即有效共轭部分,ECS),不会再多。定域性导致了激发态与局部分子几何构型的强烈耦合,因此在激发之后,电子云和键长都会重新分布。对于带电的激发,这些几何弛豫可以打破局部对称性,因此激活了基态时对称性禁阻的(拉曼而不是红外活性)红外域振动能带。结果,电子跃迁矩很强并且具有很高的方向性,平行于分子链或者分子轴。有机材料中的带电激发,相比于固体物理中极化子的概念,通常更类似于溶液化学中熟悉的自由基离子(不管怎样,"空穴"是对自由基阳离子通常使用的术语)。

25. 这导致了有机和无机半导体中载流子迁移率的显著不同。在有机材料中,载流子迁移率为 $10^{-5} \sim 10^{-3}$ cm^2 · V^{-1} · s^{-1}。在无机材料中,载流子迁移率为 $10^2 \sim 10^4$ cm^2 · V^{-1} · s^{-1}。图 1.12 给出了有机材料中跃进运输模式和无机材料中带状运输模式的比较。

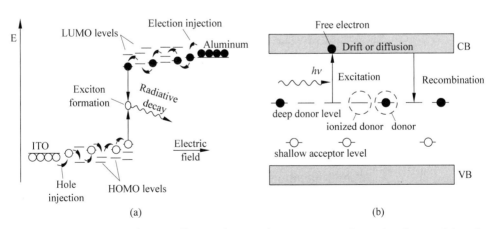

Figure 1.12 Carrier behavior difference between hopping in organic semiconductors (a) and bandlike transport in inorganic semiconductors (b)

5) Intrinsic Defect Difference

In amorphous or surface of crystalline inorganic semiconductor，e. g.，the vapor-deposited inorganic semiconductor films，the surface states and the defect-containing bulk states are not chemically saturated，hence "dangling bonds" and surface attachment of atoms，such as oxygen，exist（Figure 1.13），leading to large intrinsic defects[26].

26. 在无定形无机半导体或者晶态无机半导体的表面，例如气相沉积的无机半导体薄膜，表面态以及含有缺陷的内部态不是化学键饱和的，存在悬挂键及原子（例如氧原子）在表面的结合（图 1.13），这会导致大量的本征缺陷。

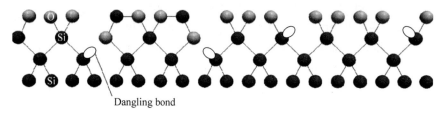

Figure 1.13　Dangling bonds in surface of crystalline inorganic semiconductor

27. 有机材料不是这样。有机材料由分子组成，其前沿轨道饱和。即使是非常薄的有机薄膜，也总是由完整的分子组成，其表面及材料内部的化学配位情况是相等的。有机材料中没有悬挂键，因而本征缺陷非常少。

Things are different in organic materials. Organic materials compose of molecules，whose frontier orbitals are saturated. Even a very thin organic film always consists of complete molecules，with the chemical coordination at the film surface equal to that in the bulk. The absent of "dangling bonds" in organic materials leads to far lower intrinsic defects[27].

These "dangling bonds" distort the band structure at the film surface from its bulk properties. In contrast，surface- and bulk-ionization potentials，and electron affinities of organic semiconductors are generally equal.

6) General Property Difference

28. 由于分子之间较弱的范德瓦耳斯力，有机半导体通常具有低熔点、高压缩系数、柔软、可燃烧、溶于有机溶剂、简单快速的制备/纯化。而无机半导体材料通常坚硬、易碎，当处于潮湿或者腐蚀环境时，相对稳定。另外，有机材料与生物系统兼容，而将无机材料移植到生物系统就比较困难。

Due to the weak van de Waals force between molecules，organic semiconductors generally possess low melting point，high compressibility，soft，combustible，soluble in organic solvents，simple and quick preparation/purification. While inorganic semiconductors are generally hard，brittle，and relatively stable in the environments of damp or corrosive. On the other hand，organic materials are compatible to biosystem，but inorganic materials are difficult to be implanted into biosystem[28].

In terms of material processing, lots of organic films can be obtained by solution process, which is not only compatible with flexible substrate, but also has the advantages of easy/cheap fabrication and large area. Figure 1.14 demonstrates two solution process methods: roll to roll injket printing and doctor blade printing[29].

29. 在材料加工方面,很多有机薄膜可以溶液法制备,这不仅仅与柔性基底兼容,还具有简便/廉价制备和大面积的优点。图 1.14 展示了 2 个溶液加工方法:卷对卷喷墨打印和刮涂打印。

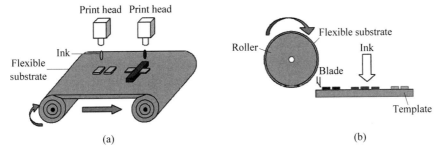

(a)   (b)

Figure 1.14   Solution processes based on roll to roll ink jet printing (a) and doctor blade printing (b)

1.2.2 Optoelectronic Processes and Device Applications of Organic Materials

Organic optoelectronics mainly deals with the optoelectronic properties of organic materials, which heavily depend on the optoelectronic processes occurring in frontier electronic levels of these materials. There are electron transition and transport, with the former generally relating to the electron transfer between low and high electronic levels, while the latter generally concerning with electron transfer among similar energy levels between molecules[30].

30. 有机光电子学主要针对有机材料的光电子性质,而这些性质强烈地依赖于材料中前沿电子能级中发生的光电过程。材料中存在电子跃迁和电子传输,前者通常与高低能级之间的电子转移有关,而后者通常考虑分子间的相似能级之间的电子转移。

Figure 1.15 shows an organic molecule with different energies. From low to high, there are molecular energy states of S_0, T_1/S_1, and T_2/S_2, representing molecules with energy of ground state, the first triplet/singlet excited state, and the second triplet/singlet excited state, respectively. Also shown in the figure, are many vibrational energy levels representing for the vibration energy of the molecules (short and fine lines) overriding on the pure electronic energy of molecules (long and coarse lines). Different energy state of molecules means that the frontier electrons of the molecule

31. 图 1.15 表示的是具有不同能量的有机分子。由小到大，分子能量依次为 S_0、T_1/S_1 和 T_2/S_2，分别代表具有基态、第一激发态的三线态/单线态和第二激发态的三线态/单线态能量的分子。图上还包含了在分子纯电子能级上（粗长线）叠加的、许多代表分振动能量的振动能级（短细线）。分子的不同能量状态意味着分子内前沿电子处于不同的电子能级上。例如，S_0 分子代表分子内最高能量的电子处于 HOMO 位置。所有种类的光电子过程可以根据图 1.15 进行描述。如图所见，在分子内存在吸收、内转换、系间窜跃、荧光和磷光电子跃迁过程。通过电子从一个分子的电子能级到另外一个分子能级的运动，可能是 Förster 能量转移和 Dexter 能量转移过程，也可能是通过跃进/俘获/脱俘获过程的电子输运。

32. 强吸收：有机材料在可见光范围内表现出良好的吸收现象，具有非常大的消光系数。因此，当在光电检测器或光伏器件中作为活性组分时，其厚度可以很薄。

33. 大的斯托克斯位移：许多荧光有机染料在吸收和发射之间表现出非常大的光谱位移，即大的斯托克斯位移。因此，有机发光二极管光发射的再吸收损失很小。

locate at different electronic levels. For example, the S_0 molecular means that electrons with the highest energy in the molecule locate at its HOMO level. The occurrence of all kinds of optoelectronic processes can be described according to Figure 1.15. As can be see, within a molecule, there are electronic transition of absorption, internal conversion, inter system crossing, fluorescence and phosphorescence. By electron moving from an electronic level of a molecule to that of another molecule, there can be electron transition processes of Förster energy transfer and Dexter energy transfer, as well as electron transport process by hopping/trapping/detrapping[31].

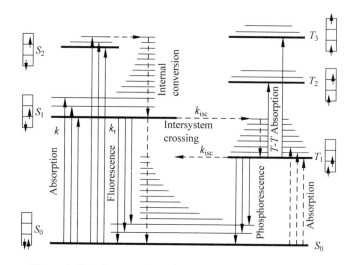

Figure 1.15　An organic molecule in different energy levels

In a common sense, the optoelectronic properties from organic materials exhibit the following features.

(1) Intensive Absorption: Organic materials exhibit good absorption in visible light region with very large extinction coefficient. Hence, when used as active components in photodetector or photovoltaic device, the active layer can be very thin[32].

(2) Large Stokes Shift: Many fluorescent organic dyes show very large spectra shift between absorption and emission, i.e., large Stokes shift. Therefore, organic light-emitting diodes have little reabsorption loss[33].

(3) Strong Fluorescence/Phosphorescence Covering Visible Range: As shown in Figure 1. 16, fluorescence is a spin-allowed radiative decay from singlet excited state to ground state; while phosphorescence refers to the radiative decay from triplet excited state to the ground state. Phosphorescence is generally spin forbidden, which can be released by introducing heavy metal in organic molecules. By molecular design, strong fluorescence or phosphorescence can be obtained with organic materials. Additionally, tailoring molecular structure can tune the emission wavelength, realizing full coverage in the visible range: approximately from 390 nm to 780 nm. Figure 1. 17 is an example for emission color-tuning[34].

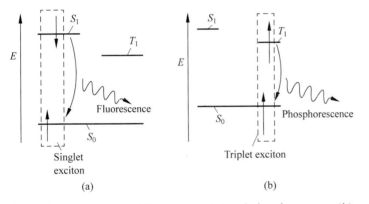

Figure 1. 16　Processes of fluorescence (a) and phosphorescence (b)

Utilizing the optoelectronic properties of organic materials, many organic optoelectronic devices can be realized, including organic field effect transistor (OFET), xerography, optical memory, liquid crystal display (LCD), organic solar cells, photo detector, organic light-emitting diodes (OLED) and organic memory. The main features of the optoelectronic devices are cheap, flexible, easy miniaturization and printable[35].

The above-mentioned optoelectronic processes/properties of organic materials, as well as their applications in devices will be discussed in detail in the later chapters.

34. 覆盖可见光范围的强荧光/磷光：如图 1.16 所示,荧光是自旋允许的单线态激发态到基态的辐射衰变过程;而磷光指的是三线态到基态的辐射衰变。磷光通常是自旋禁阻的,通过在有机分子中引入重金属,可以缓解这个禁阻。通过分子设计,可以实现有机材料的强荧光或者强磷光发射;而且,分子结构的裁剪可以调控发射波长范围,实现可见光范围的全覆盖,大约从 390 nm 到 780 nm。图 1.17 是一个发射光调控的例子。

35. 应用有机材料的光电子特性,可以实现多种有机光电子器件,包括有机场效应晶体管(OFET)、静电复印、光学存储器、液晶显示器(LCD)、有机太阳能电池、光检测器、有机发光二极管(OLED)和有机存储器。有机光电子器件的主要特点是便宜、可柔性、便于小型化和可打印。

17

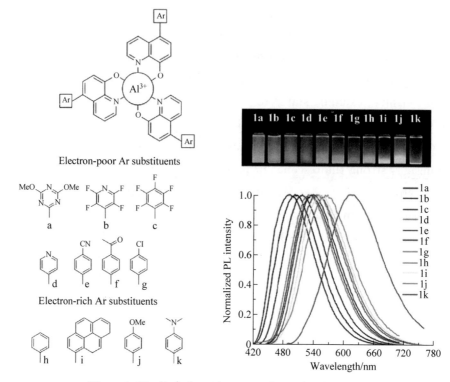

Figure 1.17　Emission color tuning by molecular design

本章小结

1. 内容概要：

本章首先从历史角度,对光电子学进行了评论,然后简明介绍了有机光电子学,包括有机材料的历史、定义、发展、类型/特征,以及它们的主要光电过程及其器件应用。

2. 基本概念：光电子学、有机材料、有机半导体、小分子、大分子、聚合物、生物分子、光电过程、器件。

3. 主要公式：无。

Appendix

1. Award Ceremony Speech（2000 Nobel Prize for Chemistry）

"for the discovery and development of conductive polymers"

Plastic that conducts electricity

We have been taught that plastics, unlike metals, do not conduct electricity. In fact plastic is used as insulation round the copper wires in ordinary electric cables. Yet this year's Nobel Laureates in Chemistry are being rewarded for their revolutionary discovery that plastic can, after modifications, be made electrically conductive. Plastics are polymers, molecules that repeat their structure regularly in long chains. For a polymer to be able to conduct electric current it must consist alternatively of single and double bounds between the carbon atoms. It must also be "doped", which means that electrons can move along the molecule—it becomes electrically conductive. Heeger, Macdiarmid and Shirakawa made their seminal findings at the end of the 1970s and have subsequently developed conductive polymers into a research field of great importance for chemists as well as physicists. The area has also yielded important practical applications. Conductive plastics are used in, or being developed industrially for, e. g. anti-static substances for photographic film, shields for computer screen against electromagnetic radiation and for "smart" windows (that can exclude sunlight). In addition, semi-conductive polymers have recently been developed in light-emitting diodes, solar cells and as displays in mobile telephones and mini-format television screens. Research on conductive polymers is also closely related to the rapid development in molecular electronics. In the future we will be able to produce transistors and other electronic components consisting of individual molecules—which we now carry around in our bags would suddenly fit inside a watch.

2. Award Ceremony Speech (2010 Nobel Prize for Chemistry)

Your Majesties, Your Royal Highnesses, Ladies and Gentlemen. This year's Nobel Laureates in Chemistry are rewarded for a method to link carbon atoms together, and this method has provided chemists with an efficient tool to create new organic molecules. The Laureates have utilized the metal palladium to couple two carbons to one another under mild conditions and with high precision.

Organic molecules contain the element carbon, where carbon atoms are bound to each other to form long chains and rings. Carbon-carbon bonds are a prerequisite for all life on earth and they are found in proteins, carbohydrates and fats. Plants and animals mainly consist of organic molecules in which carbon atoms bind to each other, and we human beings, we who have gathered here today, are to a large extent built up by carbon-carbon bonds. In living organisms, bonds between carbon atoms are created via Natur's own pathways utilizing various enzyme systems.

To create new organic molecules in an artificial manner that can be used as medicines, plastics, and various other materials, we need new efficient methods for synthesizing carbon-carbon bonds in our laboratories.

Looking back in history we find that the German chemist Kolbe synthesized the first carbon-carbon bond in 1845. Since then a number of methods for the synthesis of bonds between carbon atoms have been developed of which several have been awarded with a Nobel Prize. This year's Nobel Prize in Chemistry is the fifth that rewards the synthesis of carbon-carbon bonds.

Richard Heck's pioneering work from 1968 – 1972 laid the foundation for palladium-catalysed formation of carbon-carbon bonds. He coupled two rather unreactive molecules to one another with the aid of palladium. One of these is a molecule with a handle, e. g. bromobenzene and the other has a double bond and is called an olefin. In 1977, Ei-ichi Negishi reported a mild method to couple one of Heck's unreactive molecules to a carbon bound to zinc with the aid of palladium. Two years later, in 1979, Akira Suzuki found that the corresponding palladium-catalyzed coupling of an unreactive molecule such as bromobenzene to a carbon bound to boron could be made under very mild conditions.

Carbon is stable and carbon atoms do not easily react with one another. Earlier methods used by chemists to bind carbon atoms together were therefore based upon various techniques for rendering carbon more reactive. Such methods worked when creating simple molecules, but when synthesizing more complex molecules chemists ended up with too many unwanted by-products in their test tubes. The palladium-catalyzed cross coupling solved that problem and provided chemists with a more precise and efficient tool to work with. In the Heck reaction, Negishi reaction, and Suzuki reaction, the carbon atoms meet on a palladium atom. When the carbon atoms meet on a palladium atom, chemists do not need to activate the carbon atom to the same extent. This entails fewer by-products and a more efficient reaction.

The palladium-catalyzed cross couplings have been used for large-scale industrial manufacturing of, for example, pharmaceuticals, agricultural chemicals, and organic compounds that are used by the electronic industry.

Professors Heck, Negishi and Suzuki: You are being awarded the Nobel Prize in Chemistry for palladium-catalyzed cross couplings in organic synthesis and with these achievements you have provided organic chemists with efficient and useful methods for synthesizing compounds that were previously difficult to obtain. On behalf of the Royal Swedish Academy of Sciences, I wish to convey to you our warmest congratulations and I now ask you to step forward to receive your Nobel Prizes from the hands of His Majesty the King.

3. Award Ceremony Speech (2010 Nobel Prize for Physics)

Your Majesties, Your Royal Highnesses, Ladies and Gentlemen. The 2010 Nobel Prize in Physics is awarded for research concerning a new ultra-thin material with exceptional properties. This material is called graphene and is the first of a new class of materials. Surprisingly, it can be produced in small quantities with the help of ordinary office materials such as pencils and adhesive tape.

Carbon is probably the most important element in nature. It is the basis of all life that we know of. The most common form of pure carbon is graphite, which we find in pencils, for example. If carbon is subjected to high pressure, it can also assume the form of a diamond, which is considerably more expensive and is undoubtedly found in some of the jewelry in this hall.

Graphite consists of many one-atom-thin layers of carbon stacked on top of each other. One single such layer is what we today call graphene. A graphene layer looks something like chicken wire. We say that it has a hexagonal structure, but graphene is much thinner – actually as thin as one single atom. Each graphene layer is very strong, but in graphite these layers are only weakly connected to each other. Because of this, it is easy to split graphite, which is exactly what happens when we write with a pencil. When we drag a pencil across paper, flakes containing many layers of carbon break loose. Naturally some of the flakes are thinner than others, and it is conceivable that a tiny fraction of the flakes actually consist of single carbon layers. Most of you in this hall have thus produced graphene-like layers when you have written with a pencil.

Graphene is representative of an entire class of two-dimensional crystalline materials. By two-dimensional, we mean that atoms are added to the length and width of such a material but not to its height. This means that electrons in the material can only travel in two dimensions, but not in the third. The electrons in graphene behave in an unusual way. It actually looks as if they have no mass. In physics, this gives rise to interesting phenomena such as an unusual form of the so-called quantum Hall effect, a phenomenon which resulted in a Nobel Prize in 1985. Another example is so-called Klein tunneling. This effect was predicted by the Swedish physicist Oskar Klein in 1929. It had not been observed earlier in other systems, but in graphene it was observed last year.

Graphene also has other exceptional properties. For example, it is 100 times stronger than steel. If we imagine making a hammock out of graphene that is one square meter in size, even though it is only one atom thin it will be able to hold a newborn infant or a cat without breaking. Such a hammock would weigh about one milligram, about the same as one of the cat's whiskers. Graphene is also a good conductor of electricity, and it conducts heat 10 times better than silver. In addition, graphene is transparent, flexible and very stretchable.

It had been known for a long time that graphite consists of hexagonal carbon layers, and the behavior of electrons in graphene was calculated as early as in 1947 by Philip Wallace, but few scientists believed it would be possible to isolate graphene in a way that would enable electrical measurements of single layers. It was therefore surprising when this year's Laureates, Andre Geim and Konstantin Novoselov, together with their collaborators published their first article about graphene in October 2004. With the help of inventive methods, including the use of ordinary adhesive tape, they succeeded in isolating thin carbon layers and transferring them to a suitable surface. With the help of different microscopes, they were able to show that some of the layers were only one atom thin.

They were also able to etch the samples into a suitable shape and connect electrodes. They made electrical measurements and were able to show that the material actually had the expected properties. In subsequent research，many teams-and not least the Laureates themselves-have performed new experiments and studied a number of exciting properties of this new carbon material.

The graphene research field is still young, and it is too early to single out what applications will be the most important，but there are great hopes that the exceptional properties of graphene can be utilized in many areas. Examples of possible applications are touch screens，solar cells，fast transistors，gas sensors and lightweight super strong materials.

Glossary

Absorption	吸收
Carbohydrate	碳水化合物,糖类
Carrier	载流子
Carrier hopping	载流子跳跃/跃进
Carrier trapping	载流子俘获
Chain structure	链结构
Charge storage and release	电荷存储和释放
Compound	化合物
Crystalline	结晶度,晶相
Defect	缺陷
Definition	定义
Dexter energy transfer	Dexter 能量传递
Electric	电的
Electrical	与电相关的
Electroluminescence	电致发光
Electronic	电子的
Electronic process	电子过程
Elemental substance	基本物质(元素物质)
Excited state	激发态
Exciton	激子
Fabrication	制备,制作
Fluorescence	荧光
Förster energy transfer	Förster 能量传递
Ground state	基态
Hydrocarbon	碳氢化合物
Intensive	强烈的
Inter system crossing(ISC)	系间穿越
Internal conversion(IC)	内转换

Laureate	资金（荣誉）获得者
Living organism	活的有机体
Macromolecule	大分子
Metal	金属
Metallic conductivity	金属导电性
Mobility	迁移率
Nonmetal	非金属
Organic electronics	有机电子学
Phosphorescence	磷光
Photoconductivity	光电导性（率）
Photoconductor	光电导体
Photovoltaic effects	光生伏特效应
Polymer	聚合物
Purification	提纯
Singlet	单线态
Small molecule	小分子
Solution	溶液
Stokes shift	斯托克斯位移
Superconductivity	超导性
Triboelectricity	摩擦电
Triplet	三线态
Urea	尿素
Van de Waals force	范德瓦耳斯引力
Xerography	静电复印术

Problems

1. True or false questions：

(1) Conventional electronics is based on inorganic semiconductors.

(2) Organic molecules can contain metal atom.

(3) In organic materials there are no covalent bondings.

(4) Compared to the forces of covalent bonding in inorganic semiconductors, the van de Waals forces in organic materials are nearly ten times smaller.

(5) All organic materials have small conductivity, i.e., they are insulators.

(6) Conjugation is an essential structure in optoelectronically active organic materials.

(7) There are some differences in opto-electronics processes of organic and inorganic materials.

(8) There is only one kind of exciton.

(9) Excitons in organic materials are Wannier type with long radius.

(10) The transport modes of materials can be band-like or hopping.

2. Please state the developing stages of electronics, and briefly discuss the advantages and disadvantages of inorganic semiconductor based electronics.

3. Read the speech of the Nobel Prize in Chemistry for 2000, please answer questions:

(1) According to the speech, what are plastics?

(2) How can a polymer be conductive?

(3) Please give examples of the application or potential applications of polymers.

4. Read the speeches of the Nobel Prize in Chemistry for 2010, answer the following questions:

(1) What are organic molecules?

(2) Give examples of carbon based materials.

5. Read the speeches of the Nobel Prize in Physics for 2010, please give the difference between graphite and graphene, and state general properties of graphene. Additionally, what is the meaning of two dimensional materials?

6. Please explain the types of bonds in organic materials. Based on these bonds, can you predict some properties of organic materials (at least give 4 properties)?

7. Please tell differences between organic and inorganic semiconductors.

8. (1) Based on molecular weight, structural complexity, and molecular shape, please classify organic materials, respectively; (2) For each type, which one can be used as active component in organic electronic devices?

9. Please briefly state the optoelectronic processes in organic molecules, and give an example of electronic device application that utilizes one of the processes.

CHAPTER 2 BONDING AND ELECTRONIC STRUCTURES IN ORGANIC MATERIALS

This chapter gives a comprehensive introduction to the knowledge of bonding and electronic structures in organic materials, which is the basic for understanding their optoelectronic property. After briefly presenting four kinds of bonds in solids and their corresponding features, in the first part, focusing on carbon atom, which is the most essential element in organic materials, hybrid orbital and covalent bonding are discussed. In the second part, various intermolecular forces in organic materials are studied. In the third part, after brief introduction of different bonding theories (e.g., molecular orbital theory, crystal/ligand field theory and band theory), electronic structure of organic materials are discussed in detail. Finally, material properties related to some particular bonding and electronic structures are demonstrated.

2.1 Bonding in Organic Materials

2.1.1 Bonding in Solids

Ordinary solid is an aggregate of atoms or molecules, holding together by bonds. A molecule, in turn, can be defined as a group of a limited number of atoms which are strongly bonded together but whose bonds with the atoms of other adjacent molecules are relatively weak. Therefore, in solids, bonds or forces of attraction, which are responsible for the aggregation of atoms and molecules into ordinary solids, can be divided into primary bonds and secondary bonds; the former refers to chemical bonds between atoms, which arc classified into ionic bonds, metallic bonds and covalent/atomic bonds; and the latter is intermolecular attraction forces being frequently called van der Waals forces[1]. Figure 2.1 presents these four types bonding in solids.

1. 因此固体中,使原子和分子聚集形成普通物质的键或者吸引力,可以分为主键和次级键;前者是指原子之间的化学键,分为离子键、金属键和共价/原子键;后者是分子间引力,通常被称为范德瓦耳斯力。

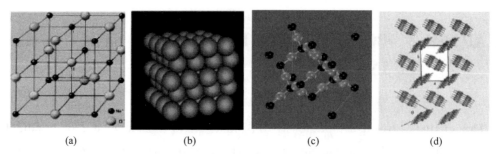

Figure 2. 1　Bonding in solids: ionic（a），metallic（b）and covalent（c）bondings; and molecular bonding between molecules（d）

1. Ionic Bond

Ionic bond is formed by the essentially mutual electrostatic attraction between positive and negative charges, which are derived from the neutral atoms by the loss or gain of electrons. Atoms with incomplete outer shells will tend to gain or lose electrons in order to attain stable configurations with filled outer shells or subshells, becoming negative or positive ions in the process[2]. The ease with which atoms form ions depends on the electronic configurations of atoms. Taking sodium chloride as an example: when sodium atom and chlorine atom close to each other, sodium atom loses an electron and has 8 electrons in its outer subshell, simultaneously chlorine atom gets an electron from sodium atom and forms 8 electrons in its outer shell. Then the positive sodium ions and the negative chloride ions attract each other to form sodium chloride. The interaction between sodium cation and chloride anion is ionic bond. Ionic bond is essentially Coulombic attraction, and without saturation/directionality[3].

A crystal consisting of ions bound together by their electrostatic attraction is ionic crystal. This means that there is no molecule presents in ionic crystal. Generally, ionic solids exhibit such characters: ① they are hard and brittle; ② exhibit relatively high melting and boiling points, e. g., sodium chloride melts at 801 ℃ and boils at 1413 ℃; ③ they are not good electrical conductors at low temperatures but exhibit ionic conductivity at temperature close to and above

2. 离子键本质上是带正电和带负电的离子之间的静电引力，其中正负离子通过中性原子失去或获得电子形成。未满外壳层的原子倾向于获得或者失去电子来达到外壳层或者亚外壳层充满的稳定排布，这个过程中它们变为负的或者正的离子。

3. 离子键本质上是库仑吸引，没有饱和性/方向性。

the melting point；additionally they can conduct electricity in aqueous solution；④they are more soluble in water than other solvents[4].

2. Metallic Bond

The metallic bond is also essentially Coulombic force，but the negatively charged species are electrons，which are delocalized and frcc to move throughout the solid. Once the valence electrons detach from their original atomic owners and float around in the sea，the metal atoms become positive ions. The result is an orderly structure of positive metal ions surrounded by a sea of negative electrons that hold the ions together. Similar to ionic bonds，metallic bonds also exhibit non-saturation and without directionality. Metals have characteristic properties such as：① high thermal and electrical conductivity，② luster and high reflectivity，malleability and ductility，they can be beaten or shaped without fracture；③ variability of mechanical strengths （ranging from soft alkali metals to Tungsten，which is hard）；④melting point is high，though it tends to be lower than that of the ionic solids[5].

3. Covalent Bond

Covalent bond refers to the interatomic linkage resulting from the sharing of electrons between two neighboring atoms；by such，both atoms attain a full outer shell and hence a stable electronic configuration. A covalent bond forms when the bonded atoms have a lower total energy than that of widely separated atoms[6]. Taking the formation of a hydrogen molecule as an example：one hydrogen atom possesses one electron in the outer shell that does not reach a stable electronic configuration，once another hydrogen atom is bound to it by sharing the two electrons，both hydrogen atoms reach a stable configuration and hydrogen molecule is formed. The nature of the covalent bond is overlapping of electron cloud，the larger the overlapping，the more stable the molecule. Covalent bond has directionality due to different shape of s，p，d and f atomic orbitals. This in turn gives molecules definite shapes，as in the angular （bent）

4. 通常地,离子固体具有这样的特征：①坚硬且脆；②表现出相对较高的熔沸点，如氯化钠在 801 ℃ 熔化,在 1413 ℃ 沸腾；③它们低温下不导电,但是在接近或高于熔点时,表现出离子导电性,另外它们的水溶液也可以导电；④相对于其他溶剂,它们更容易溶于水。

5. 金属有如下特质：①高的导热和导电性；②具有金属光泽和高反射率,具有可锻造性和延展性,它们可以无断裂地被敲打及塑型；③机械强度各异（碱金属较柔软,而钨很硬）；④熔点高,但倾向于低于离子固体的熔点。

6. 共价键是指两个相邻原子通过共用电子而形成的原子间的作用力；这样,两个原子都可以达到全满外壳层,形成稳定的电子构型。共价键形成后,参与成键的两个原子的总能量,要比分开很远的原子总能量低。

structure of the H_2O molecule. In addition, since the number of bonds of an atom is related to its valence electron number, the maximum covalent bond that can be formed in an atom is fixed, i. e., covalent bond has saturation state. Covalent bonds between identical atoms (as in H_2) are nonpolar—i. e., electrically uniform—while those between unlike atoms are polar—i. e., one atom is slightly negatively charged and the other is slightly positively charged. Crystals formed from atom aggregation by covalent bonding are atomic/covalent crystals, which often present properties of high melting/boiling point, hard and brittle[7].

4. Molecular Bond

Molecular bond is intermolecular attraction, and will be discussed in the next section. Solids containing molecular bonds are molecular solids. In molecular solids, except for monatomic molecules, there are two kinds of bonds, covalent bond and molecular bond: the former strongly holds atoms together to form a molecule; while the latter renders the stacking of molecules to solid. Generally, molecular solids exhibit characters of: ① soft; ② relatively low melting/boiling point; ③ soluble in organic solvents; ④ they are not good electrical conductors. Because of the weak nature of the bonding between molecules in a molecular solid, it is to be expected that the properties of the individual molecules are retained in the solid state to a far greater extent than would be found in solids exhibiting the other types of bonding[8]. Organic materials are a kind of molecular substance.

2.1.2 Orbital Hybrid

Molecules are made of atoms, between which are bonds to hold them together, as discussed in previous section. When bonds are formed, in some cases of organic materials, orbital hybrid may occur. Since the essential component in organic materials is carbon. Here, we shall talk about the orbital hybrid in carbon atom.

The electron configuration of a carbon is $1s^2\,2s^2\,2p^2$ (Figure 2.2(a)), possessing two unpaired electrons in the outer shell ($2p$ orbital). In the old days, people are confused

7. 原子之间通过共价键聚集形成的晶体是原子/共价晶体,它们通常具有较高的熔/沸点,坚硬且脆。

8. 分子固体通常具有以下特征:①柔软;②熔/沸点相对低;③溶于有机溶剂;④不导电。由于分子固体中分子之间较弱的键合力,可以预期其孤立分子的特性,能够在固体时得以保持,这远远大于任何其他键连接的固体中可以保持(其构成单元特性)的程度。

by the fact that, for example, in CH_4, there are only two unpaired electrons in a carbon atom, but it can form four bonds with other atoms. To explain this, L. Pauling proposed the hybrid orbital theory: when forming a molecule, atomic orbitals with similar energy in an atom combine and redistribute, leading to hybrid atomic orbitals with same shape and energy; meanwhile the number of the hybrid orbitals is equal to the original ones. Hybrid orbital theory is very useful in the explanation of molecular geometry[9].

9. 为了解释这个, L. Pauling 提出杂化轨道理论: 在形成分子时, 一个原子中能量相近的原子轨道发生组合并重新分布, 形成能量和形状都相同的杂化原子轨道; 同时, 杂化轨道数目与杂化前的原子轨道数目相同。杂化轨道理论可以很好地解释分子形状。

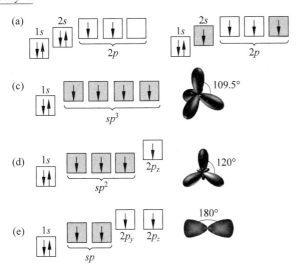

Figure 2.2   Electron configuration of carbon atom (a), Jumping of one $2s$ electron to empty $2p$ orbital (b), sp^3 hybridization (c), sp^2 hybridization (d), sp hybridization (e)

To produce a hybridization among the outer shell orbitals ($2s$ and $2p$) in carbon atoms, firstly, one $2s$ electron need to jump to the empty orbital of $2p$, forming 4 unpaired orbitals, i.e., one $2s$ and three $2p$, for hybridization (Figure 2.2 (b)). As shown in Figure 2.2(c) to (e), there are three patterns for orbital hybrid in carbon atoms: sp^3, sp^2 and sp hybridization. If the unpaired $2s$ orbital hybridizes with three unpaired $2p$ orbitals, four equivalent orbitals with tetrahedron distribution are formed. The separation angle between the two sp^3 orbitals is 109.5°. If the unpaired $2s$ orbital hybridizes with two unpaired $2p$ orbitals, three equivalent orbitals with planar triangle shape are formed.

10. 如果未成对的 $2s$ 轨道与三个未成对的 $2p$ 轨道杂化，形成具有四面体分布的四个等效轨道。两个 sp^3 轨道之间夹角是 109.5°。如果一个未配对的 $2s$ 轨道与两个未配对的 $2p$ 轨道杂化，形成具有平面三角形形状的三个等效轨道。两个 sp^2 轨道之间的夹角是 120°。剩下的 $2p$ 轨道（$2p_z$）和杂化轨道垂直。如果未配对的 $2s$ 轨道与一个 $2p$ 轨道杂化，则将形成直线形状的两个等效轨道。两个 sp 轨道之间夹角是 180°。剩下的两个 $2p$ 轨道（$2p_y$ 和 $2p_z$）相互垂直，同时它们垂直于 2 个 sp 杂化轨道。

11. 当两个电子轨道以头对头重叠时，形成的键称为 σ 键（图 2.3(a)）。当重叠采用肩并肩方式时，该键称为 π 键（图 2.3(b)）。

The separation angle between two sp^2 orbitals is 120°. The remained $2p$ orbital ($2p_z$) is perpendicular to the hybrid orbitals. If the unpaired $2s$ orbital hybridizes with one $2p$ orbital, two equivalent orbitals with linear shape will be formed. The separation angle between the two sp hybrid orbitals is 180°. The remained two $2p$ orbitals ($2p_y$ and $2p_z$) are mutually perpendicular, and concomitantly, they are perpendicular to the two sp hybrid orbitals[10].

2.1.3　Covalent Bondings within a Molecule

As previously discussed, organic materials are made of molecules consisting covalently bound atoms. In terms of binding pattern, covalent bonding in organic materials can be classified into σ bond and π bond. When two electron orbitals overlap in head-to-head manner, the formed bond is called σ bond (Figure 2.3(a)). When the overlapping adopts a side-by-side way, the bond is called π bond (Figure 2.3(b))[11].

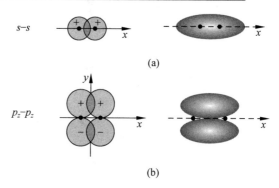

Figure 2.3　Formation of σ bond (a) and π bond (b)

The head-to-head combination in σ bonds results in an axial symmetry and the most electron cloud locates around the axis. Therefore σ bond is strong. In contrast, the side-by-side overlap in π bond leads to a nodal plane at the central of the two atoms and small overlap of their electron clouds. Therefore, π bond exhibits plane symmetry and not as strong as the σ bond. Due to the unstable nature compared to σ bonds, a π bond can be broken by forming a new σ bond. In this sense, it can be said that π bonds are unsaturated, while σ bonds are saturated bonds. Unsaturated bond is not the maximum bonding situation which has potential for chemical reaction and the atom is more active. In saturated atoms, only σ bonds exist, thus it will be relatively chemical stable.

A feature of covalent bonding in organic materials is that the sharing of electron pairs between two atoms can be one, two and three, corresponding to single, double and triple bonds. When carbon forms single bond with neighbour atoms, it's in sp^3 hybridization, the bond is σ bond. When double bonds are formed, carbon is in sp^2 hybridization, and there are one σ bond and one π bond. To form triple bonds between atoms, carbon is in sp hybridization, and there are one σ bond and two π bonds[12]. As shown in the Table 2.1, there are differences in bond length and bond energy among single, double and triple bonds.

Table 2.1　Differences among single, double and triple bonds

	Bond length	Bond energy
C—C	1.54 Å	~3.6 eV
C=C	1.33 Å	~6.4 eV
C≡C	1.21 Å	~8.4 eV

According to the type of covalent bonding in organic materials, the frontier electrons (i.e., valence electrons) can be classified into σ, π, and n electrons. Namely, σ and π electrons exist in σ bond and π bond, respectively; while the n electrons refer to the lone-pair electrons in the outer shell of an atom, which do not form bond, such as found in atoms of O, N, S, P, F, Cl, Br, and I[13]. The order of electron energy is as follows: $\sigma < \pi < n < \pi^* < \sigma^*$ (π^* and σ^* are electrons in antibonding orbital).

2.2　Intermolecular Forces in Organic Materials

Intermolecular forces include intermolecular attraction and intermolecular repulsion, the former applies under long distance while the latter under short distance. The intermolecular attractive force is frequently called van der Waals force, which is, by nature, the electrostatic attraction between electrons in one molecule and the nuclei of another[14].

12. 当碳原子与邻位原子形成单键时，是 sp^3 杂化，该键为 σ 键。当形成双键时，碳原子采用 sp^2 杂化，该双键含有一个 σ 键和一个 π 键。原子之间形成三键时，碳原子采用 sp 杂化，该三键含有一个 σ 键和两个 π 键。

13. 顾名思义，σ 电子和 π 电子分别存在于 σ 键和 π 键中；而 n 电子指的是原子最外层没有参与成键的孤对电子，如在 O、N、S、P、F、Cl、Br 原子和 I 原子中存在的孤对电子。

14. 分子间作用力包括分子间引力和分子间斥力，前者作用于长距离之间，后者作用于短距离之间。分子间引力通常被称为范德瓦耳斯力，其本质上是一个分子中的电子和其他分子的原子核之间的静电引力。

2.2.1 Intermolecular Attractive Forces

Attractive forces between molecules can be divided into three types: dipole-dipole interaction (Keesom force), inductive force (Debye force), dispersive force (London force).

1. Dipole-Dipole Interaction

Molecular attraction between two molecules that are both permanently polarized, such as between two HCl gas molecules, are referred to dipole-dipole interaction, which is also called Keesom force. In concept, an electrical dipole is a separation of positive and negative charges. The measure of the electrical polarity of a dipole is electrical dipole moment, which can be expressed as P (electric dipole moment) = qd, where q is electrical charge, d is a vector directed from the negative towards the positive charge (Figure 2.4(a))[15].

15. 定义上,电偶极子是指被分开的正、负电荷。偶极子的极性是用电偶极矩(P)来表示,即 $P = qd$,q 是单位电荷电量,d 是正负电荷中心之间的向量,从负电指向正电(图 2.4(a))。

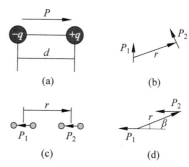

(a)   (b)

(c)   (d)

Figure 2.4   A schematic diagram of a dipole (a), and interaction between two dipoles: (b) general case, (c) a parallel orientation of the two dipole axes with the vector connecting them, (d) two parallel dipoles at a distance of r with a given angle of β

In a general case (Figure 2.4(b)), the potential energy of the interaction between two dipoles is expressed as

$$E_{\Psi} = \frac{1}{4\pi\varepsilon_0} \frac{P_1 P_2 - 3(P_1 e_r)(P_2 e_r)}{r^3} \qquad (2\text{-}1)$$

where, $e_r = r/r$, is the unit vector connecting to the two dipoles, r is distance between the two dipoles. P_1 and P_2 are the dipole moments of the two dipoles with distance of d_1 and d_2, ε_0 is vacuum dielectric constant. For two parallel dipoles with distance vector in the same direction (Figure 2.4(c)), the potential energy is

$$E_\Psi = -\frac{1}{4\pi\varepsilon_0}\frac{2P_1P_2}{r^3} \qquad (2\text{-}2)$$

In this case，the force between the two dipoles is attractive，with the value of

$$f_{\text{force}} = -\frac{\mathrm{d}E_\Psi}{\mathrm{d}r} = -\frac{1}{4\pi\varepsilon_0}\frac{6P_1P_2}{r^4} \qquad (2\text{-}3)$$

For situation shown in Figure 2.4(d)，the potential energy and the force between the two parallel dipoles are

$$E_\Psi = -\frac{1}{4\pi\varepsilon_0}\frac{P_1P_2}{r^3}(3\cos^2\beta - 1) \qquad (2\text{-}4)$$

$$f_{\text{force}} = -\frac{\mathrm{d}E_\varphi}{\mathrm{d}_r} = -\frac{1}{4\pi\varepsilon_0}\frac{3P_1P_2}{r^4}(3\cos^2\beta - 1) \qquad (2\text{-}5)$$

Clearly，the strength of dipole-dipole interaction is reciprocal to the distance to the power of four，and relates to the orientation of the two dipoles.

2. Inductive Force

When an induced dipole (that is, a dipole that is induced in what is otherwise a non-polar atom or molecule) interacts with a molecule that has a permanent dipole moment，this interaction is termed as an inductive force or a Debye force[16]. As shown in Figure 2.5，a point dipole P_1 can induce a dipole moment P_{ind} in a nonpolar molecule with a polarisability of α，which satisfies $P_{\text{ind}} = \alpha F$，with F presenting the electric field at the location of the second molecule

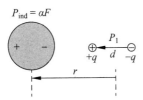

Figure 2.5　Interaction between one dipole and an induced dipole

According to Coulomb's law，the field strength F at a distance r from a point charge q is $F = \frac{1}{4\pi\varepsilon_0}\frac{q}{r^2}e_r$. Therefore，the field strength at distance r，due to the first dipole P_1 is equal to

$$F = \frac{1}{4\pi\varepsilon_0}\left[\frac{+q}{\left(r-\frac{d}{2}\right)^2} + \frac{-q}{\left(r+\frac{d}{2}\right)^2}\right]e_r \qquad (2\text{-}6)$$

16. 当一个诱导偶极子（即一个中性原子或分子被诱导形成的偶极子）与一个具有永久偶矩的分子相互作用时，这种相互作用称为诱导力或德拜力。

When $d \ll r$, $F = \dfrac{1}{4\pi\varepsilon_0}\dfrac{2P_1}{r^3}$. The induced dipole moment of a neutral molecule in a dipole field can be expressed as

$$P_{\text{ind}} = \alpha F = \alpha\,\frac{1}{4\pi\varepsilon_0}\frac{2P_1}{r^3} \qquad (2\text{-}7)$$

Thus，combine Equation（2-3）and Equation（2-7），the attractive force between the molecule with a permanent dipole moment P_1 and the polarizable molecule is

$$f_{\text{force}} = -\frac{1}{4\pi\varepsilon_0}\frac{6P_1P_{\text{ind}}}{r^4} = -\frac{1}{(4\pi\varepsilon_0)^2}\frac{12P_1^2\alpha}{r^7} \qquad (2\text{-}8)$$

The potential energy between them：

$$E_{\Psi} = -\int_{\infty}^{r} f_{\text{force}}\,\mathrm{d}r = -\frac{1}{(4\pi\varepsilon_0)^2}\frac{2P_1^2\alpha}{r^6} \qquad (2\text{-}9)$$

The inductive force is reciprocal to the distance to the power of seven，and proportional to the polarisability of the neutral molecule.

3. Dispersive Force

In a strict sense，dispersive force is an attractive force originated from two instantaneous dipoles，which is produced by electron movement in atoms，i. e. positive and negative charge-centers instantaneously depart from overlapping in a neutral molecule. Dispersive force is also termed as London force and exists in all molecules[17]. Figure 2. 6 shows the formation of fluctuating dipole moment （（1）in Figure 2.6） and its inducing of adjacent molecules to produce instantaneous inductive-dipoles （（2）in Figure 2.6）.

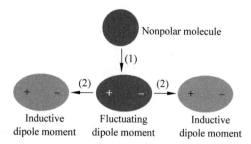

Figure 2. 6　Formation of fluctuating dipole moment in a nonpolar molecule （1），inducing of instant dipoles （2）

The force between the temporally fluctuating dipole of one molecule and its inductive dipole from another molecule is attractive，with value of

17. 严格地讲,色散力是指两个瞬时偶极子之间的吸引力,它产生于原子中的电子运动,即中性分子中的正负电荷中心瞬间偏离了重叠。色散力也称为伦敦力,存在于所有分子中。

$$f_{force} = \frac{1}{(4\pi\varepsilon_0)^2} \frac{A'\alpha^2}{r^7} \qquad (2\text{-}10)$$

where α is polarisability, r is the distance, A' is a factor whose magnitude is specific to a particular molecule. Like the inductive forces, dispersive forces are of very short range owing to their reciprocally r^7 dependence, and they are proportional to the square of the molecular polarisability. Thus, for larger organic molecules, especially for aromatic molecules like anthracene with their strongly polarisable π electrons, inductive forces are relatively strong.

4. Hydrogen Bonding

A hydrogen bonding is an electrostatic attraction between a hydrogen atom which is chemically bound to an atom (X) with large electronegativity (x) and small radius, such as nitrogen ($x = 3.07$), oxygen ($x = 3.61$), and fluorine ($x = 4.19$), and another atom (Y) with large electronegativity and small radius, forming X II—⋯—Y type attraction between two molecules or within a molecule (X and Y can be equal). Hydrogen bonding between molecules is a kind of van de Waals force (dipole-dipole interaction, Figure 2.7(a)), and it is usually larger compared to other van de Waals forces. Hydrogen bonding can also be formed within a molecule, as shown in Figure 2.7(b). In this case, the hydrogen bonding can not be classified as van de Waals force[18].

18. 氢键是一个与电负性(x)大且半径小的原子(X),诸如氮($x = 3.07$)、氧($x = 3.61$)和氟($x = 4.19$),以化学键结合的氢原子,与另外一个电负性大且半径小的原子(Y)之间,形成的静电吸引力,在两个分子之间或者一个分子内形成 X—H—⋯—Y 型吸引力(X 和 Y 可以相同)。分子之间的氢键是一种范德瓦耳斯力(偶极-偶极作用,图 2.7(a)),它通常比其他范德瓦耳斯力大。氢键也可能存在于分子内,如图 2.7(b) 所示。这时的氢键不是范德瓦耳斯力。

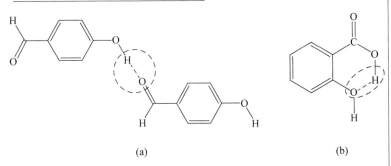

(a)　　　　(b)

Figure 2.7　Hydrogen bonding between two molecules (a) and within a molecule (b)

2.2.2　Repulsive Force and Overall Force between Molecules

The repulsive forces between atoms/molecules are based

on Coulomb repulsion. These effects become important only at very small distances and increase very rapidly with further decreasing distance. For organic materials, the simplest approximation of a rectangular potential was adopted for the repulsion force: r (distance)$>r_0$ (equilibrium distance), $E_\Psi = 0$ (no repulsion energy); $r \leqslant r_0$, $E_\Psi = \infty$ (repulsive force is infinite). For refinement, the expression for repulsive force is written as[19]

$$E_\Psi = \frac{C^n}{r^n}, \quad n \text{ is usually taked as } 12 \qquad (2\text{-}11)$$

Considering attraction and repulsion force together, the overall potential energy can be obtained (Figure 2.8 and Equation (2-12) and Equation (2-13)).

19. 对于有机材料,排斥力采用最简化的矩形电势:r（距离）$>r_0$（平衡距离），$E_\Psi = 0$（没有排斥力）；$r \leqslant r_0$，$E_\Psi = \infty$（排斥力为无穷大）。为了细化,排斥力表达为

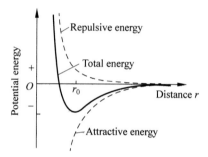

Figure 2.8　Interaction energy considering attraction force and the repulsion force. The interaction energy has its minimum at the equilibrium distance r_0

1) Lennard-Jones Potential Curve

$$\Psi = \frac{R}{r^{12}} - \frac{A}{r^6}, \quad \left.\frac{\mathrm{d}\Psi}{\mathrm{d}r}\right|_{r=r_0} = 0 \qquad (2\text{-}12)$$

where R and A are empirical constants, representing repulsion and attraction parameters, r is distance between atoms/molecules.

2) Buckingham Potential Curve

$$\Psi = -\frac{A}{r^6} + B\mathrm{e}^{-ar} \qquad (2\text{-}13)$$

where the repulsion potential is in the Born-Meyer form: $\Psi = B\mathrm{e}^{-ar}$, A, B and α are parameters, r is distance between atoms/molecules.

2.2.3 Material Properties Relating to Intermolecular Forces

1. Melting/boiling Points

Melting and boiling points ($m.p.$ and $b.p.$) of organic materials are strongly influenced by molecular forces, because in some sense, they indicate the starting temperature for breaking the forces between molecules with different initial states. As indicated in Table 2.2, the organic materials in the first three rows are long-chain alkanes with different chain-length; with the increase of molecular weight, both $m.p.$ and $b.p.$ increase. This is because that among these non-polar analogues, the increase of molecular weight is an indicative of stronger dispersive force, originating from the increased amount of electrons in the material for higher polarisibility. Comparing between n-hexane and benzene, it is clear that although the molecular weight of benzene is smaller, its $m.p.$ and $b.p.$ are greatly larger. This is because that in benzene, the conjugation allows π electron delocalization, greatly improving material polarisibility, hence the molecular force is larger, leading to significantly larger $m.p.$ and $b.p.$[20].

Table 2.2　Melting/boiling point comparison

Material	Molecular weight	Melting point /℃	Boiling point /℃
$CH_3(CH_2)CH_3$	44	−190	−42
$CH_3(CH_2)_6CH_3$	114	−57	125
$CH_3(CH_2)_{18}CH_3$	282	37	343
n-hexane	86.2	−95	69
Benzene	78.1	5.5	80.1
Anthracene	178.2	216	342
Pentacene	278.4	370 (sublimated)	—

2. Packing Density

Molecular forces also have effect on crystal formation of organic materials, because to form stable crystal, lattice energy should be as much as possible, this can be achieved by either close packing of molecules or large molecular forces.

20. 有机材料的熔点和沸点($m.p.$ 和 $b.p.$)受分子间作用力的强烈影响,因为在某种程度上,它们表示了不同初始状态的有机分子之间的作用力开始遭到破坏的温度。如表 2.2 所示,前三行的有机材料是不同链长的长链饱和烃;随着分子量的增加,其 $m.p.$ 和 $b.p.$ 都增加。这是因为,在这些非极性同系物中,分子量增加,表明色散力增加,因为分子量增加使得材料中电子增多,因而极化率提高。比较正己烷和苯,可以明显看出,虽然苯分子量较小,其 $m.p.$ 和 $b.p.$ 都高很多。这是因为在苯中,共轭使得 π 电子可以离域,极大地提高了材料的极化率,因此分子间作用力增强,显著地提高了 $m.p.$ 和 $b.p.$。

As pointed by Kitaigorodskii, the packing density of molecules in a crystal can be characterized by a packing coefficient K:

$$K = ZV_0/V \qquad (2\text{-}14)$$

where V is the volume of a crystallographic unit cell in which each molecule has the volume V_0, which can be calculated from the known atomic radii and the dimensions of the molecules. Z is the number of molecules in a unit cell. The K for polar materials is generally smaller than nonpolar materials. For example, K for aromatic hydrocarbon is in the range of 0.68 (benzene)~ 0.8 (perylene), while K for ice is 0.38. The dominated binding force in ice is hydrogen bonding, which is the strongest among all inter molecular forces in organic materials. Therefore in ice, due to the strong intermolecular interaction, large lattice energy can be obtained without dense packing, leading to small packing coefficient[21].

21. 冰中起主导作用的键合力是氢键,它是有机材料中所有分子间作用力中最强的。因此冰中,由于很强的分子间作用力,不需要密堆积就可以获得较大的晶格能,导致较小的堆叠系数。

2.3　Electronic Structures in Organic Materials

In terms of optoelectrical property, the main microprocesses involved are the transition and transportation of electrons locating in frontier orbitals of a material. Therefore, it is important to understand the electronic structures, both for material-property explanation and prediction. This section discusses electronic structure related issues for organic materials.

2.3.1　Theories

In atoms, there are various quantities of electrons moving surrounding the nuclei. An electron of an atom has no traditional trajectory, but possesses wave property, and obeys the Heisenberg uncertainty principle. It is impossible to tell the exact position of an electron; the best that one can do is to predict the probability of finding the electron within a given volume element, which is closely related to statistic. The probabilities of finding an electron in the whole space are termed as electron cloud (also called electron shape). To describe an electron in an atom，electron orbital is used，which consists of the orbital shape (e. g.，electron cloud) and the orbital energy (i. e.，electron energy). A wavefunction of a single electron (ψ) are assigned to represent the electron's moving status，which tells information of both electron energy and electron cloud.

If atoms aggregate to form molecules/solids，outer shell electrons in different atoms can form bonds；and the energy distribution of these electrons，i. e.，the electronic structure，can be described by several theories，such as molecular orbital theory，crystal/ligand field theory and band theory. By such the frontier electrons，which are the main component for optoelectronic processes in materials，can be theoretically predicted.

1. Molecular Orbital Theory

In molecular orbital (MO) theory，electrons are not assigned to individual bonds between atoms，but are treated as moving species under the influence of the nuclei in the whole molecule.

1) Main Points of MO Theory

Definition of molecular orbital：In a molecule，the movement and energy of an electron is associated with the average potential of the nuclei and other electrons. By introducing a wavefunction (ψ)，i.e. molecular orbital，both the energy and movement of an electron in the molecule can be described[22].

LCAO：Molecular orbital can be obtained by linear combination of atomic orbitals (LCAO)：

$$\Psi = c_1\varphi_1 + c_2\varphi_2 + \cdots + c_n\varphi_n \qquad (2\text{-}15)$$

where Ψ and φ_n are molecular and atomic orbitals，respectively. The coefficient c_n in the linear combination shows the extent to which each atomic orbital (φ) contributes to the molecular orbital (Ψ)：the greater the value，the higher the contribution of that atomic orbital to the molecular orbital. Criteria for the effective LCAO：① Matched symmetry. To form a molecular orbital，two adjacent atomic orbitals must be matching in symmetry. ② Similar energy. Only atomic orbitals with similar energies can be effectively

22. 分子中，一个电子的运动及其能量都与原子核以及其余电子的平均势场相关联。通过引入波函数(ψ)，即分子轨道，分子中电子的能量和运动可以被描述出来。

23. 有效 LCAO 要求：①对称性匹配。为了形成分子轨道，两个相邻原子轨道必须是对称匹配的。②能量相近。只有能量相近的原子轨道才能有效地结合成分子轨道。通常要求原子轨道能级差小于 15 eV。③最大轨道重叠。取决于适当的核间距离和重叠方向，原来的原子轨道之间的重叠倾向于最大化。

24. 因此在这些分子轨道中有成键轨道，即被占据 MOs，和反键轨道，即空置 MOs。成键轨道的能量低于原来的原子轨道，而反键轨道的能量高于原来的原子轨道。

25. 键级描述的是分子的稳定性，它等于分子中成键轨道中电子数和反键轨道中电子数之差的一半。一对成键轨道电子创建一个键，而一对反键轨道电子否定一个键。

combined into molecular orbitals. The energy level difference is usually required to be less than 15 eV. ③ Maximum overlap. Depending on appropriate inter-nuclear distance and overlapping direction of the electron cloud, maximum overlapping between the two original atomic orbitals is preferred[23].

Orbital conservation：It was approved by quantum mechanics that the number of molecular orbitals is equal to the number of atomic orbitals that are involved in bond formation.

Electron filling pattern：Electrons in a molecule prefer to fill lower energy MOs, and the filling of electrons on MOs obeys the Pauli exclusive principle and the Hund's rule. The former implies that each molecular orbital can be occupied by up to two electrons and if two electrons are present, their spins must be paired; while the latter tells that, for degenerate orbitals with equal energy, electrons tend to occupy different orbitals with the same spin before they pair up. As a result, among these MOs, there are bonding orbitals, i.e., the occupied MOs, and anti-bonding orbitals, i.e., the unoccupied MOs. The energies of the bonding orbitals are lower than those of the original atomic orbitals; while the energies of the anti-bonding orbitals are higher than those of the original atomic orbitals[24].

HOMO and LUMO：HOMO refers to the highest occupied molecular orbital (HOMO); and LUMO is the lowest unoccupied molecular orbital that is the next higher molecular orbital to HOMO. HOMO and LUMO are frontier orbitals of a molecule, the electrons in HOMO and LUMO are frontier electrons, which are relatively active. When the molecule is excited, electrons in the HOMO will jump to LUMO or molecular orbital with even higher energy.

Bond order of a molecule：Bond order describes the stability of the molecular; it can be determined by half of the difference between the number of electrons in bonding orbitals and the number of electrons in antibonding molecular orbitals. A pair of electrons in a bonding orbital create a bond, whereas a pair of electrons in an antibonding orbital negate a bond[25]. A bond order of greater than zero suggests a stable molecule. The higher the bond order is, the more stable the molecule.

Figure 2.9 shows the configuration of electrons in H_2, H_2^+, He_2 and He_2^+, their bond order can be calculated as 1, 0.5, 0 and 0.5, respectively. Bond orders of H_2, H_2^+ and He_2^+ are greater than zero, hence they all exist stably. He_2 has a bond order of zero and this is the reason that the He_2 molecule is not observed[26].

26. 图 2.9 给出了 H_2、H_2^+、He_2 和 He_2^+ 的电子排布,经过计算它们的键级分别为 1、0.5、0 和 0.5。H_2、H_2^+ 和 He_2^+ 的键级大于 0,因此它们都可以稳定存在。He_2 的键级为 0,这是 He_2 不能被观测到的原因。

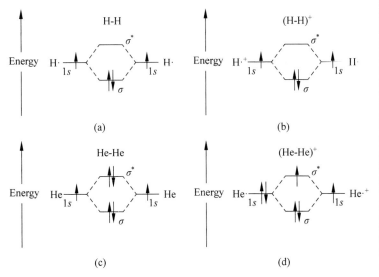

Figure 2.9   Configuration of electrons in H_2(a), H_2^+(b), He_2(c) and He_2^+(d)

2) Examples for MO Theory

Figure 2.10(a) depicts the formation of MO with two homonuclear atoms of hydrogen. When the two $1s$ atomic orbitals of the two hydrogen (H) atoms overlap to form hydrogen molecule (H_2), two molecular orbitals with different energy are formed. Since only the MO with lower energy (i. e., bonging orbital) is filled with two paired electrons, the one with higher energy (i. e., antibonding orbital) is empty. Compared to electrons in H atoms, the electron energy in H_2 is lowered by 431 kJ · mol^{-1}. For the formation of molecular orbital from two different atoms with different electron energies, the energy of the bonding orbital will close to that of the atomic orbital with lower energy, and the energy of the anti-bonding orbital will close to that of the

27. 对于具有不同电子能量的两个不同原子之间形成的分子轨道而言,成键轨道的能量将接近较低的原子轨道能量,而反键轨道的能量将接近较高的原子轨道能量。

atomic orbital with higher energy[27]. As can be seen in Figure 2.10(b), when one $1s$-orbital of H with higher energy and three $2p$-orbital of F with lower energy combine to form HF, three bonding MOs (3σ and 1π) close to the lower $2p$-orbital and one antibonding MO close to the higher $1s$-orbital are formed.

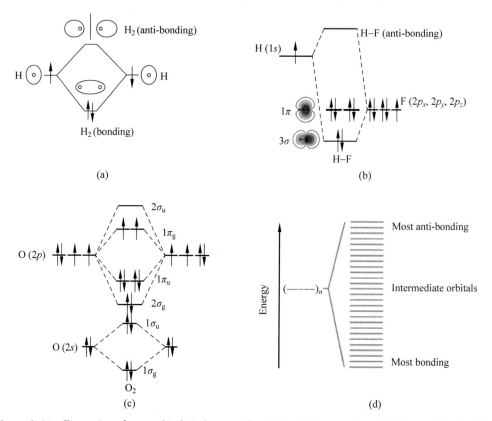

(a)

(b)

(c)

(d)

Figure 2.10　Formation of new orbitals in homonuclear (a) and heteronuclear (b) diatomic molecules. Electron configuration of O_2 (c), and energy bands in extended solids (d)

The molecular orbital diagram and electron configuration of O_2 are depicted in Figure 2.10(c). The notation σ signifies that the orbital has cylindrical symmetry; atomic orbitals that can form σ orbitals include the $2s$ and $2p_z$ orbitals on the two atoms. As indicated in Figure 2.10(c), four σ molecular orbitals can be constructed, i.e., $1\sigma_g$, $1\sigma_u$, $2\sigma_g$, and $2\sigma_u$, two of which arise predominantly from interaction of the $2s$ orbitals of the two O atoms, and two from interaction of the $2p_z$ orbitals of two O atoms. The remaining two $2p$ orbitals on each O atom, which contain three electrons and have a nodal plane containing the z-axis, overlap to give four π orbitals: $1\pi_u$ and $1\pi_g$[28]. Bonding and antibonding π orbitals can be formed from the mutual overlap of the two $2p_x$ orbitals, and also from the mutual overlap of the two $2p_y$ orbitals. This pattern of overlap is termed as resonance and gives rise to the two pairs of doubly degenerate energy levels of the $1\pi_u$ and $1\pi_g$. Electrons occupy the MO orbitals is the order of increasing energy. According to Hund's rule, each $1\pi_g^*$ orbital has one electron occupation, and the electrons in the half-filled orbitals adopt parallel spins. Therefore, O_2 presents paramagnetism property.

As shown in Figure 2.10(d), extension to solids with many molecules, due to the interaction among molecules, the originally same level MOs split in energy. Since the difference among the energy-splitted orbitals is very small, semi-continuous band is supposed to form[29].

2. Crystal/ligand Field Theories

Complexes are substances composed of two or more components capable of an independent existence. A coordination complex is one in which a central atom or ion is joined to one or more ligands through what is called a coordinate covalent bond in which both of the bonding electrons are supplied by the ligand. In such a complex, the central atom acts as an electron-pair acceptor and the ligand as an electron-pair donor; and the coordinate covalent bonds are generally much weaker than ordinary covalent bonds. A coordination complex whose center is a metal atom (or ion) is called a metal complex with a common formula of $M(L)_n$, where M is a central metal atom or ion, accepting electron pair from covalently bonded ligand of L[30].

Metal complexes are a very important class of materials in organic electronic devices. Due to the easy oxidation or reduction, some metal complexes possess good electrical conductivity, i.e., Alq_3 (8-hydroxyquinolinato aluminum) (Figure 2.11(a)) is one of the state-of-the-art electron transport materials and is widely used in OLEDs. In heavy metal involved complexes, spin-orbital coupling can

28. 如图2.10(c)所示，（两个氧原子结合）形成4个σ分子轨道，即$1\sigma_g$、$1\sigma_u$、$2\sigma_g$和$2\sigma_u$，其中的两个，来自于两个氧原子中$2s$轨道的相互作用，另外两个是氧原子中$2p_z$轨道的相互作用。每个氧原子中剩余的两个$2p$轨道包含3个电子，并且存在一个包含z-轴的节面，它们相互重叠形成4个π轨道：$1\pi_u$和$1\pi_g$。

29. 如图2.10(d)所示，扩展到含有很多分子的固体，由于分子之间的相互作用，原来能量相等的分子轨道产生能量分裂。由于能量分裂轨道之间的差异非常小，形成了半连续的能带。

30. 复合物是由两种或两种以上可以独立存在的成分组成。配位复合物是一个中心原子或者离子与一个或者多个配体通过称为配位共价键结合的复合物，这里的共价键中的成键电子都来自于配体。在这样的复合物中，中心原子作为电子对受体，配体是电子对给体，配位共价键通常比普通共价键要弱很多。中心是金属原子（或者离子）的配位复合物称为金属配合物，其通用表达式为$M(L)_n$，M是中心金属原子或离子，它从共价键合的配体L中获得电子对。

effectively occur, rendering significant radiative decay from triplet state. The emission of triplet state is becoming more and more important, one model of such a molecule is Irppy$_3$ (Figure 2.11(b)).

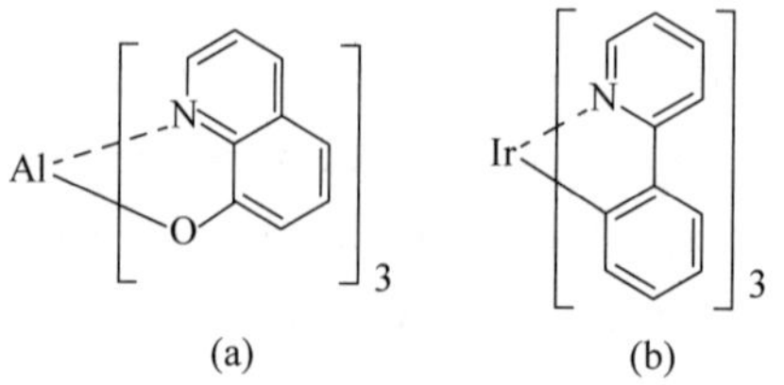

Figure 2.11　Molecular structure of Alq$_3$(a) and Irppy$_3$(b)

There are two widely used models describing the electronic structure of d-metal complexes. One ("crystal-field theory") emerged from an analysis of the spectra of d-metal ions in solids; the other ("ligand-field theory") arose from an application of molecular orbital theory. Crystal-field theory (CFT) is more primitive, and strictly speaking it applies only to ions in crystals; however, it can be used to capture the essence of the electronic structure of complexes in a straightforward manner. Ligand-field theory (LFT) builds on crystal-field theory: it gives a more complete description of the electronic structure of complexes and accounts for a wider range of properties. We shall discuss them in detail here.

1) Crystal Field Theory

In CFT, complex formation is assumed to be due to electrostatic interactions between a central metal ion and a set of negatively charged ligands or ligand dipoles arranged around the metal ion. Depending on the arrangement of the ligands, the d orbitals split into sets of orbitals with different energies. Stronger ligand field results in larger d orbital splitting[31]. CFT successfully accounts for some magnetic properties, optical spectra (colors), hydration enthalpies, and spinel structures of transition metal complexes.

Generally, metal complexes exhibit various molecular structures, they can be octahedral, tetrahedral or planar, depending the orientation of the ligands. Due to different orientation of ligands in metal complexes, electrons from the ligand will be closer to some of the d orbitals and farther

31. 在晶体场理论中,配合物的形成被认为是中心金属离子和一组带负电的配体或者配体偶极子之间的静电相互作用。取决于配体的排列方式,d 轨道分裂为几组具有不同能量的轨道。配体场越强,d 轨道的分裂越大。

away from others, causing different degree of interaction between the d orbitals and the electrons from the ligand, and hence the degeneracy of the d orbitals is broken. In an octahedral complex (Figure 2.12(a)), due to the head-to-head approaching from the ligand, the $d_{x^2-y^2}$ and d_{z^2} orbitals receive larger repulsion than d_{xy}, d_{yz} and d_{xz} orbitals, which locate at the interstitial positions of the ligands. As a result, the energy increment in $d_{x^2-y^2}$ and the d_{z^2} orbitals are larger than the rest three orbitals, with splitting energy of Δ_O (the subscript "O" signifies an octahedral crystal field)[32].

32. 在八面体配合物中（图 2.12(a)），由于配体头碰头的接近方式，$d_{x^2-y^2}$ 和 d_{z^2} 轨道受到的排斥力比位于配体间隙位置的 d_{xy}、d_{yz} 和 d_{xz} 大。因此相比于其他三个轨道，$d_{x^2-y^2}$ 和 d_{z^2} 轨道能量增加较大，分裂能为 Δ_O（下标"O"表示八面体晶体场）。

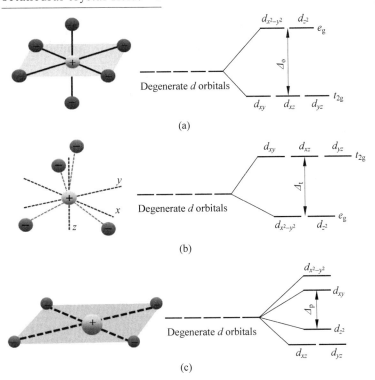

Figure 2.12   Geometry and energies of the d orbitals in an octahedral (a), tetrahedral (b), and planar (c) crystal fields

In a tetrahedral complex (Figure 2.12(b)), none of the five d orbitals points directly to or between the ligands, but the d orbitals still split into two groups, with an energy difference of Δ_t. Because the d_{xy}, d_{xz}, and d_{yz} orbitals (the t_{2g} orbitals) interact more strongly with the ligands than do the $d_{x^2-y^2}$ and d_{z^2} orbitals (the e_g orbitals), the order of orbital energies in a tetrahedral complex is the opposite of the order in an octahedral complex[33]. Furthermore, since the

33. 在四面体配合物中（图 2.12(b)），五个 d 轨道都不直接指向配体或者处于配体之间，但 d 轨道仍然分成两组，轨道能量差为 Δ_t。因为 d_{xy}、d_{xz} 和 d_{yz} 轨道（t_{2g} 轨道）与配体作用比 $d_{x^2-y^2}$ 和 d_{z^2} 轨道（e_g 轨道）更强烈，四面体配合物轨道能量的顺序和八面体配合物是相反的。

ligand electrons in tetrahedral symmetry are not oriented directly towards the d orbitals, the energy splitting will be lower than in the octahedral case.

In square-planar complexes (Figure 2.12(c)), four ligands coordinate to the central metal ion and approach $\pm x$, $\pm y$ axis. Due to the head-to-head approaching of the ligand, the $d_{x^2-y^2}$ orbital receives the largest repulsion, d_{xy} points between the ligands and receives repulsion followed by $d_{x^2-y^2}$, d_{xz} and d_{yz} orbitals receive the smallest repulsion. The energy order of d orbitals splitting in planar complexes is $d_{x^2-y^2} > d_{xy} > d_{z^2} > d_{xz}$ and d_{yz}[34]. Many other geometries can also be described by CFT, such as pentagonal bipyramidal, square pyramidal, trigonal bipyramidal. Here we would not discuss them further.

The first property that can be interpreted by crystal-field theory is the absorption spectrum of metal complexes, which, as explained by CFT, can be assigned to the transition across the ligand-filed splitting, Δ. The ligands can be arranged in a spectrochemical series, in which the members are arranged in order of increasing energy of transitions that occur when they are present in a complex: $I^- < Br^- < S^{2-} < SCN^- < Cl^- < NO_3^- < N_3^- < F^- < OH^- < C_2O_4^{2-} \approx H_2O < NCS^- < CH_3CN < py < NH_3 < en < bipy < phen < NO_2^- < PPh_3 < CN^- \approx CO$ (Spectrochemical series). The series indicates that, for the same metal, the optical absorption of the cyano (CN) complex will occur at higher energy than that of the corresponding chloride complex. A ligand that gives rise to a high energy transition (such as CO) is referred as a strong-field ligand, whereas one that gives rise to a low-energy transition (such as Br^-) is referred as a weak-field ligand[35].

The ligand-field strength also depends on the identity of the central metal ion, the order being approximately: $Mn^{2+} < Ni^{2+} < Co^{2+} < Fe^{2+} < V^{2+} < Fe^{3+} < Co^{3+} < Mn^{4+} < Mo^{3+} < Rh^{3+} < Ru^{3+} < Pd^{4+} < Ir^{3+} < Pt^{4+}$. The value of Δ increases with increasing oxidation state of the central metal ion (compare the two entries for Fe and Co) and also increases down a group (compare, for instance, the locations of Co, Rh, and Ir). The variation with oxidation state

34. 由于配体的头碰头接近,$d_{x^2-y^2}$ 轨道受到的斥力最大,d_{xy} 轨道指向配体之间,受到的斥力次之,d_{xz} 和 d_{yz} 轨道受到的斥力最小。在平面型配合物中 d 轨道分裂后的能量大小顺序为 $d_{x^2-y^2} > d_{xy} > d_{z^2} > d_{xz}$ 和 d_{yz}。

35. 这个序列表明,对于同样的金属,氰基(CN)复合物的光学吸收能量比相应的氯基复合物高。产生高能量跃迁的配体(如 CO),称为强场配体,而产生低能量跃迁的配体(如 Br^-)称为弱场配体。

reflects the smaller size of more highly charged ions and the consequently shorter metal-ligand distances and stronger interaction energies. The increase down a group reflects the larger size of the $4d$ and $5d$ orbitals compared with the compact $3d$ orbitals and the consequent stronger interactions with the ligands.

Figure 2. 13（a） shows such an example of ［Ti (OH$_2$)$_6$］$^{3+}$, that is a metal complex with six coordinated ligands of water. The octahedral orientation of the ligands results in the splitting of d-orbital as three t_{2g} orbitals $<$ two e_g orbitals. Ti has an outer configuration of $4s^2 3d^2$, so that Ti^{3+} will be a $4s^0 3d^1$ ion. In its ground state, one electron will occupy the lower group of d orbitals （t_{2g}）, and the upper group （e_g） will be empty. The d-orbital splitting in this case is 240 kJ per mole which corresponds to light of blue-green color （peaked at 493 nm）; absorption of this light promotes the electron to the upper set of d orbitals, which represents the exited state of the complex. If we illuminate a solution of ［Ti（OH$_2$）$_6$］$^{3+}$ with white light, the blue-green light is absorbed and the solution appears violet in color[36]. Different d orbital splitting patterns occur in square planar

36. 图 2.13(a)展示了这样一个例子[Ti(OH$_2$)$_6$]$^{3+}$,即一个含有六个水配体的金属配合物。八面体方位的配体产生了 3 个 t_{2g} 轨道<2 个 e_g 轨道的 d 轨道分裂。Ti 的外层电子排布为 $4s^2 3d^2$,因此 Ti^{3+} 为 $4s^0 3d^1$ 离子。在其基态,一个电子占据低能量组的 d 轨道(t_{2g}),上面一组(e_g)是空置的。这时 d 轨道的裂分为 240 kJ/mol,相当于蓝绿光(峰位 493 nm);该光的吸收促使电子进入代表配合物激发态的高能量 d 轨道。如果用白光照射 [Ti(OH$_2$)$_6$]$^{3+}$ 溶液,蓝绿色光被吸收,该溶液呈现紫色。

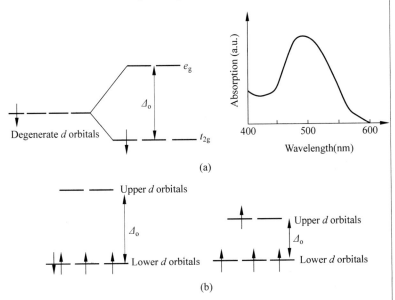

(a)

(b)

Figure 2. 13   The outer electron configuration and absorption of ［Ti（OH$_2$）$_6$］$^{3+}$ （a）, and the reason for high and low spin complexes （b）

and tetrahedral coordination geometries, so a very large number of arrangements are possible. In most complexes the value of Δ corresponds to the absorption of visible light, accounting for the colored nature of many such compounds in solution and in solids such as $CuSO_4 \cdot 5H_2O$ (blue).

Another successfulness of CFT is the explanation of the magnetism in metal complexes. Typically, unpaired electrons in the central ion of a metal complex significantly influence the magnetism of the complex. Due to the degree of d orbital splitting, complexes with a same metal ion coordinating to different ligands may be low-spin if the crystal field is strong but high-spin if the field is weak; this is true in complexes with metal ion of $3d^4$, $3d^5$, $3d^6$, and $3d^7$; while, the ground-state electron configurations of $3d^1$, $3d^2$, $3d^3$, $3d^8$, $3d^9$, and $3d^{10}$ complexes are unambiguous[37]. Taking $3d^4$ complexes with octahedral coordination as an example, when the ligand field is strong (left in Figure 2.13(b)), the d-splitting is large, therefore, the fourth electron from $3d^4$ may enter one of the t_{2g} orbitals and pair with the electron already there; the resulting configuration of t_{2g}^4 leads to low-spin metal complex. Alternatively, weak ligand field renders small d splitting (right in Figure 2.13(b)), and the fourth electron from $3d^4$ of the central ion in the metal complex may occupy one of the e_g orbitals, giving the configuration $t_{2g}^3 e_g^1$ and leading to high-spin complex.

2) Ligand Field Theory (LFT)

As previously talked, CFT is a model that describes the breaking of degeneracies of electron orbital states, usually d or f orbitals, due to a static electric field. However, the theory is defective because it treats ligands as point charges or dipoles and does not take into account the overlap of ligand and metal-atom orbitals. One consequence of this over simplification is that crystal-field theory cannot exactly account for the ligand spectrochemical series. Combining CFT and molecular orbital theory leads to ligand field theory (LFT), which can better explain some properties in metal complexes, such as the spectrochemical series.

Two dominant contents for LFT is the building up of σ type and π type molecular orbitals between the central metal

37. CFT另一个成功之处是对金属配合物磁性的解释。一般地，中心离子的非成对电子会极大地影响配合物的磁性。缘于d轨道的分裂程度，同一金属离子与不同配体配位形成的配合物，在强晶体场下是低自旋，但在弱场下是高自旋；这对金属离子为$3d^4$、$3d^5$、$3d^6$和$3d^7$的配合物是事实，而基态电子排布为$3d^1$、$3d^2$、$3d^3$、$3d^8$、$3d^9$和$3d^{10}$的配合物，（其自旋）是唯一的。

ion and the periheral ligands. In molecular orbital theory, symmetry-adapted linear combinations (SALC) are adopted to build molecular orbitals of a given symmetry. SALC refers to the specific combinations of atomic orbitals. For metal complexes, the SALCs originate from specific combinations of the molecular orbitals of the lone pairs that are involved in metal-lignad bonding. For examples, in an octahedral complex, as shown in Table 2. 3, there are six SALCs accounting for σ coordination bonding with metal ions. Molecular orbitals are formed by combining SALCs and metal-atom orbitals of the same symmetry. Because there is no combination of lone-electron-pair orbitals in the ligand that has the symmetry of the metal t_{2g} orbitals, so the latter do not participate in σ coordination bonding[38].

38. 例如,在八面体配合物中,如表 2.3 所示,存在六个 SALCs 与金属离子形成 σ 配位键。分子轨道是由相同对称性的 SALCs 和金属原子轨道组合而成。因为配体中不存在与金属 t_{2g} 轨道相同对称性的孤对电子轨道的组合,因此后者不参与 σ 配位键的形成。

Table 2. 3 Symmetry of atomic orbitals and symmetry-adapted linear combinations (SALCs) orbitals

Symmetry	Atomic orbital	Orbital of ligand	MO between metal and ligands		
			Bonding	Anti-bonding	Non-bonding
A_{1g}	s	$\sigma_1 + \sigma_2 + \sigma_3 + \sigma_4 + \sigma_5 + \sigma_6$	a_{1g}	a_{1g} *	
T_{1u}	p_z	$\sigma_1 - \sigma_6$	t_{1u}	t_{1u} *	
	p_x	$\sigma_3 - \sigma_5$			
	p_y	$\sigma_2 - \sigma_4$			
E_g	d_z^2	$2\sigma_1 + 2\sigma_6 - \sigma_2 - \sigma_3 - \sigma_4 - \sigma_5$	e_g	e_g *	
	$d_x^2 - d_y^2$	$\sigma_2 - \sigma_3 + \sigma_4 - \sigma_5$			
T_{2g}	d_{xy}, d_{xz}, d_{yz}				t_{2g}

Figure 2. 14 shows molecular orbitals of an octahedral metal complex based on ligand field theory. Firstly, according to crystal field theory, the five $3d$ orbitals from metal ion (left hand in Figure 2.14) split into two sets under the repulsive force from ligands. The splitting energy is noted as Δ_C. Then based on ligand field theory, by combining SALCs and metal-atom orbitals of the same symmetry, molecular orbitals between the metal and the ligand can be obtained. The three triply degenerate metal t_{2g} orbitals

remain nonbonding and fully localized on the metal atom. The doubly degenerate metal e_g orbitals and the ligand e_g SALCs overlap to give four molecular orbitals (two degenerate bonding, two degenerate antibonding). Similarly, we can give a_{1g} and t_{1u} orbitals (bonding and antibonding). It is obviously that the splitting energy Δ_L is larger than Δ_C in this situation.

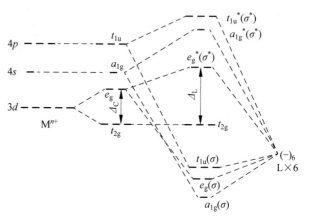

Figure 2.14 Molecular orbital energy levels of a typical octahedral complex according to LFT

39. 如上所讨论的，t_{2g} 轨道是非键轨道。如果一个配合物的配体具有相对于 M-L 轴 π 对称性轨道，那么这些轨道就可以与金属轨道组合产生 π 成键轨道和 π 反键轨道。对于一个八面体配合物，可以形成这种键的配体 π 轨道，是具有 t_{2g} 对称性的 SALCs。

40. 对于配体是 π 给体的情形，包括 Cl^-、Br^-、OH^-、O^{2-} 甚至 H_2O，充满的 π 给体配体轨道通常比部分填充的金属 d 轨道能量低，当它们与金属 t_{2g} 轨道形成分子轨道时，成键的组合比配体轨道能量低，反键组合比原来金属原子 d 轨道能量高（图 2.15(a)）。

As discussed above, t_{2g} orbitals are non-bonding orbitals. If the ligands in a complex have orbitals with local π symmetry with respect to the M-L axis, they may form bonding and anti-bonding π orbitals with the metal orbitals. For an octahedral complex, the combinations that can be formed from the ligand π orbitals are SALCs of t_{2g} symmetry[39]. These ligand combinations have net overlap with the metal t_{2g} orbitals, which are therefore no longer purely nonbonding on the metal atom.

In the case of a π-donor ligand, including Cl^-, Br^-, OH^-, O^{2-}, and even H_2O, the full π orbitals of π-donor ligands normally lie lower in energy than the partially filled d orbitals of the metal, when they form molecular orbitals with the metal t_{2g} orbitals, the bonding combination lies lower than the ligand orbitals and the antibonding combination lies above the energy of the d orbitals of the original metal atom (Figure 2.15(a))[40]. The electrons supplied by the ligand π orbitals occupy and fill the bonding combinations, leaving the electrons originally in the d orbitals of the central metal atom to occupy the antibonding t_{2g} orbitals. The net effect is that the previously nonbonding metal t_{2g} orbitals become antibonding and hence are raised closer in energy to the antibonding e_g orbitals. It follows that π-donor ligands decrease Δ_O[41].

41. 最终的结果是,原来的金属非键 t_{2g} 轨道变成了反键轨道,因此能量升高了,更加接近于 e_g 反键轨道。由此可见,π 给体配体会降低 Δ_O。

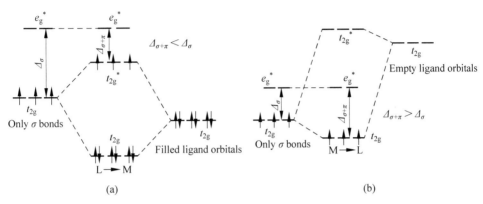

Figure 2.15　The effect of π bonding on the ligand-field splitting parameter in octahedral ligands: (a) Ligands that act as π donors decrease Δ_O. (b) Ligands that act as π acceptors increase Δ_O

In the case of a π-acceptor ligand, the ligand has empty π orbitals, e.g., in CO and N_2. The two π^* orbitals of CO, for instance, have their largest amplitude on the C atom and have the correct symmetry for overlap with the metal t_{2g} orbitals, so CO can act as a π-acceptor ligand. Phosphines (PR_3) are also able to accept π-electron density and also act as π acceptors. Because the π-acceptor orbitals on most ligands are higher in energy than the metal d orbitals, they form molecular orbitals in which the bonding t_{2g} combinations are largely of metal d orbital character (Figure 2.15(b)). These bonding combinations lie lower in energy than the d orbitals themselves. The net result is that π-acceptor ligands increase Δ_O[42].

42. 由于大多数配体的 π 受体轨道的能量要比金属 d 轨道要高,它们形成分子轨道时,t_{2g} 成键轨道具有更多的金属 d 轨道特征(图 2.15 (b))。成键轨道处在 d 轨道下方。总的结果是,π 受体配体使 Δ_O 增加。

Based on LFT, the spectrochemical series can be well interpreted. The order of ligands in the spectrochemical series is partly that of the strengths with which they can participate in M-L σ bonding. For example, both CH_3^- and H^- are high in the spectrochemical series (similar to NCS^-)

because they are very strong σ donors. However, when π bonding is significant, it has a stronger influence on Δ_O: π-donor ligands decrease Δ_O and π-acceptor ligands increase Δ_O. This effect is responsible for CO (a strong π acceptor) being high on the spectrochemical series and for OH^- (a strong π donor) being low in the series. The overall order of the spectrochemical series may be interpreted in broad terms as dominated by π effects (with a few important exceptions), and in general the series can be interpreted as follows: Δ_O increases in the order of π donor $<$ weak π donor $<$ no π effect $<$ π acceptor.

3. Band Theory

MO, CFT and LFT describe states of electrons in molecules, however, when the molecules aggregate in solid, the effects of other molecules on electrons are not taken into account. Band theory describes state and motion of electrons in solid (mainly inorganic materials), including metal, insulator and semiconductor. Band theory has been successfully used to explain many physical properties of solids, such as electrical resistivity and optical absorption, and forms the foundation of the understanding of many solid-state devices (transistors, solar cells, etc.).

Band theory includes the following points.

1) Electron Sharing

When a large quantity of atoms forms regularly-stacked crystal, there exists periodical potential field in the crystal. The valence electrons will delocalize among the lattices and are shared by all lattice atoms, no longer belonging to the atom they originate. The degree of delocalization depends on the degree of the overlap of the valence orbitals.

2) Band Formation

When atoms form aggregates, since each electron suffers Coulombic interactions from nuclei and other electrons, the electron energies that are initially identical, split into a series of different levels. These new energy levels are almost continuous due to the tremendous quantity and the small energy intervals, forming energy band[43]. By quantum mechanical approach, the band formation can be calculated.

43. 当原子形成聚集态时,因为每个电子都受到原子核和其他电子的库仑作用,最初能量相同的轨道,分裂成一系列不同的能级。由于巨大的数量和很小的能量间隔,这些新的能级几乎是连续的,从而形成能带。

Taking Si, whose electron configuration is $1s^2 2s^2 2p^6 3s^2 3p^2$, as an example, after quantum mechanical calculation for solid Si with N atoms of Si, the band structure can be obtained (Figure 2.16). In a silicon atom, 10 of the 14 electrons in a silicon atom occupy deep-lying energy levels that are tightly bound to the nucleus; the four outer shell electrons ($3s^2 3p^2$) belong to valence electrons and are relatively weakly bound, which, are the electrons involved in chemical reactions. As shown in Figure 2.16, with the decrease of interatomic distance (r), the $3s$ and $3p$ states interact and overlap; at the equilibrium interatomic distance (a_0; for Si, $a_0 = 2.35$ Å), the bands have again split, but now four quantum states per atom are in the lower band and four quantum states per atom are in the upper band. At absolute zero degrees, electrons are in the lowest energy state, so that all states in the lower band (the valence band) will be full and all states in the upper band (the conduction band) will be empty. The bandgap energy E_g between the top of the valence band and the bottom of the conduction band is the width of the forbidden energy band. For Si, $E_g = 1.1$ eV at 200 K[44].

44. 如图 2.16 所示,随着原子间距(r)的减小,$3s$ 和 $3p$ 态相互作用并重叠;在平衡原子间距(a_0;对于硅,$a_0 = 2.35$ Å),能带再次分裂,但现在,每个原子的四个量子态处在低能带,四个量子态处在高能态。在绝对零度,电子的能量最低,所以低能带(价带)全部被电子占据,高能带(导带)将是空置的。价带顶和导带底之间的带隙 E_g 是禁带宽度。对于硅,在 200 K,$E_g = 1.1$ eV。

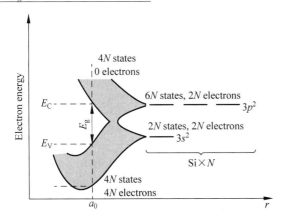

Figure 2.16   The splitting of the $3s$ and $3p$ states of silicon into the allowed and forbidden energy bands

The width of energy band is influenced by many factors. The first one is the atomic distance-the smaller the distance, the wider the energy band; the second factor is related to the states of electrons. The inner electrons are in close vicinity to their nucleus, thus slightly influenced by the adjacent atoms.

Hence the width of inner bands is narrow; while the valence electrons in the outer shell split strongly, due to the similar degree of influence from its own nucleus and the adjacent atoms, leading to wide energy band.

3) Terminologies

Based on band theory, a set of terminologies are created, which are summarized here: ① allowed band refers to the region of an energy band, where electrons are allowed; ②forbidden band is the region in between two energy bands, where no electron is allowed; ③filled band implies that all the energy states in a band are filled; ④empty band refers to the unfilled energy bands under unexcited situation; ⑤valence band indicates the band that accommodates valence electrons, which is generally the highest band with electrons, and may be either fully occupied or partially occupied; ⑥conduction band usually refers to the first empty band in semiconductors and insulators, or else some partially occupied energy bands (e. g., valence band) in some metals; ⑦band gap refers to the energy difference between the valence band and the conduction band.

4) Electrical Conduction Mechanism

In band theory, conductivity can be explained as such that electrical conduction in solids originates from the drifting of electrons in the conduction band or hole in the valence band under external electric field[45]. For a conductor (Figure 2.17(a)), which possesses ambiguous band gap e. g., the valance band and conduction band in metals come very closer to each other and may even overlap. This means electrons can move freely in the band, leading to good conductivity. A typical example is sodium with half-filled valence band, which allows electrons to flow in valence band, making sodium a good conductor. A semiconductor has fully occupied valence band and empty conduction band (Figure 2.17(b)), but its bandgap is relatively small (about $0.1 \sim 3$ eV); hence, by optical or thermal excitation, electrons in valence band can jump to conduction band to produce free electrons (in conduction band) and holes (in valence band) for electrical conduction. Besides, electrons or holes can also be obtained by doping, realizing electrical conduction[46]. For an

45. 在能带理论中导电解释如下：固体导电是由于导带电子或者价带空穴在外电场下的迁移。

46. 半导体拥有全满的价带和空置的导带(图2.17(b))，但是其带隙比较小(0.1～3 eV)；因此，通过光学或者热激发，价带中电子可以跳跃到导带，产生自由电子(在导带)和空穴(在价带)用以导电。另外，电子或者空穴也可能通过掺杂获得，从而实现导电。

insulator, although the electronic structures are similar to semiconductors, its band gap is very large (about 3.0 eV $<$ $E_g <$ 6 eV); hence valence electrons cannot jump to conduction band for free moving, hence the conductivity of an insulator is poor (Figure 2.17(c)).

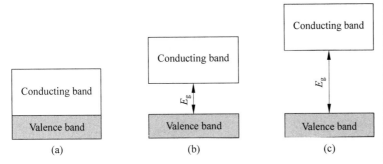

Figure 2.17　Schematic views for conductor (a), semiconductor (b), and insulator (c) base on band theory

The conductivity of organic materials can be interpreted by band theory. In saturated organic materials, atoms are bound together by σ bonds. Due to the stable nature of σ bonds, the gap between σ and σ^* is very high, leading to large energy gaps similar to inorganic semiconductors. On the other hand, electrons in σ bonds are localized, prohibiting flowing. Hence σ electrons have no contribution to conductivity; and saturated organic materials are insulators. In unsaturated organic materials, the parallel p_z orbitals can form π orbitals and π^* orbitals. Electrons are filled in the π orbitals and the π^* orbitals are empty. Due to the relatively unstable nature of π compared to σ bonds, π orbitals lie higher than σ orbitals, hence gap between π and the π^* orbitals is small, with value similar to those in inorganic semiconductors[47]. Conjugated polymers belong to unsaturated organic materials, and generally show semiconductor property. One of typical examples of conjugated polymers is polyacetylene (Figure 2.18(a)), which is a long molecular chain repeating the pattern C_2H_2 a great many times and displaying an alternating pattern of single and double bonds. Due to the large conjugation in polyacetylene, the gap between the π and the π^* orbitals is very small, e.g., 1.5 eV. In solid state, the interaction between the molecular chains

47. 在不饱和有机材料中,平行 p_z 轨道可以形成 π 轨道和 π^* 轨道。电子填在 π 轨道,π^* 轨道空置。由于 π 键比 σ 键的稳定性相对差,π 轨道的位置比 σ 轨道高,因此 π 与 π^* 之间的能隙小,其值与无机半导体的类似。

48. 在固态,分子链之间的相互作用使得在孤立状态下视为简并的 π 及 π* 轨道裂分;由于 π 及 π* 轨道的数量巨大且能量间隔小,这些轨道不再分立,而是形成准连续的 HOMO 和 LUMO 能带,如图 2.18(b)所示。这两个能带分别相当于无机半导体的价带和导带。

results in the splitting of the π as well as the π* orbitals, which are considered degenerate in isolated state; because the quantities of the π and the π* orbitals are huge, and the energy intervals are small, these orbitals are no longer discrete, but instead form quasi-continuous bands of HOMO and LUMO as shown in Figure 2.18(b). These two bands are, respectively, the equivalent of a valence band and a conduction band in inorganic semiconductors[48]. Due to the small bandgap, the π-electrons can jump to π* orbitals, leading to free carriers for electrical conduction.

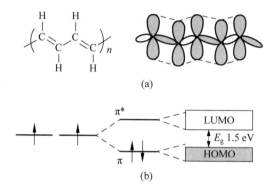

Figure 2.18   Molecular structure of polyacetylene (a), and formation of bands in organic conjugated materials (b)

2.3.2   Electronic Structures in Organic Materials

Combining molecular orbital theory, ligand field theory and band theory, the electronic structure of isolated organic molecule and that for organic crystal can be obtained, as schematically shown in Figure 2.19.

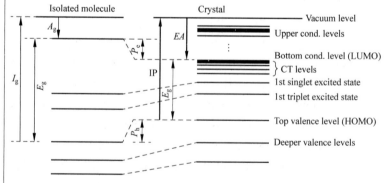

Figure 2.19   The electronic structure of isolated organic molecule and organic crystal

1. Vacuum Level

Vacuum level refers to the energy level of a free stationary electron that is outside of any material, i. e., a place where the electron possesses zero energy. There are two situations in practical application, i. e., surface vacuum level, $E_{vac}(s)$ and infinite vacuum level, $E_{vac}(\infty)$, as shown in Figure 2.20(a). Surface vacuum level is used at times in describing the electron energetics of surfaces in solid state devices, and it is defined as the energy of an electron at rest just outside the surface of the solid. In the case of metal, this distance is also the one where there is no effect of image forces, i. e., where the image potential is essentially zero. This vacuum level is a characteristic of the surface and depends sensitively, through the electronic surface dipole, on the atomic, chemical, and electronic structures of the outer atomic layers of the solid[49]. The vacuum level at infinity is defined as the energy of an electron at rest at infinite distance from the surface. That energy level is invariant, but this is a reference level that is experimentally not accessible.

49. 在实际应用中, 有两种情况, 即表面真空能级($E_{vac}(s)$)和无穷远真空能级($E_{vac}(\infty)$), 如图 2.20(a)所示。表面真空能级用于描述固体器件表面的电子能量特性, 是指离开固体表面的静止电子的能量。对于金属, 这个距离也是指没有镜像力的位置, 即镜像势能实际为零的位置。这个真空能级是表面的特性, 并敏感地依赖于表面电偶极子, 依赖于固体外层原子层的原子、化学以及电子结构。

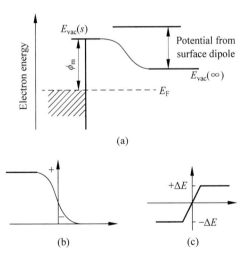

(a)

(b)   (c)

Figure 2.20 (a) Vacuum levels of a metal surface $E_{vac}(s)$, and the infinite distance, $E_{vac}(\infty)$. (b) Electron density distribution and (c) electron potential inside and outside metal surface

The energy difference between these two vacuum levels stems from the contributions of surface dipoles and, in the case of non-metallic surfaces also from extra charges on the solid surface, to the work function. The surface dipole on metals normally raises the vacuum level, as it is caused by spilling of the electronic charge density out of the surface leaving a positive charge inside (Figure 2.20(b)). Note that, according to the scale of solid state, the total electron energy increases upward (Figure 2.20(c)). As the electron crosses the surface dipole on its way out of the solid, its potential energy is raised by an amount equal to the dipole energy barrier; while, as the electron moves away from the surface, the dipole field and the potential energy decrease, leading to the decreased infinite vacuum level as depicted in Figure 2.20(a).

2. HOMO, LUMO and Charge Transfer (CT) Levels

HOMO refers to the highest occupied molecular orbital. LUMO refers to the lowest unoccupied molecular orbital. Energy levels under HOMO are generally occupied. In band theory terms, HOMO corresponds to top valence level, while LUMO corresponds to bottom conducting level. Charge transfer (CT) levels refer to those that beneath LUMO, the states formed due to the columbic interaction between an excited state of one molecule and a ground state of an adjacent molecule, as such, the excited state is stabilized and lowered to CT level[50].

3. Excited Singlet State and Triple State

When a ground state electron is excited to LUMO, the excited electron can also be stabilized by columbic interaction with the hole leaved in HOMO. Singlet state refers to the excited state that the excited electron possesses opposite spin as the unpaired HOMO electron. Triplet state refers to the excited state that the excited electron possesses the same spin as the unpaired HOMO electron. Singlet state shows higher energy than corresponding triplet state[51].

4. Fermi Level and Work Function

Fermi level is the energy level where the probability of finding an electron is 0.5. As shown in Figure 2.21, Fermi level of metals refers to the orbital where the electrons

50. 电子转移(CT)能级指的是,在 LUMO 下方,一个激发态分子和相邻基态分子之间的库仑相互作用而形成的状态,这样激发态得到稳定并降低到 CT 能级。

51. 当一个基态电子被激发到 LUMO 上时,这个激发态电子还可以与 HOMO 中剩下的空穴产生库仑作用,从而得到稳定。单线态是指激发态电子与 HOMO 上未配对电子自旋方向相反。三线态是指激发态电子与 HOMO 上未配对电子自旋方向相同。单线态能量比相应的三线态高。

possess the highest energy; for insulator and intrinsic semiconductor, Fermi levels locate at the half of the bandgap; in *p*- and *n*-semiconductors, Fermi levels are close to the top of the valence band and the bottom of the conducting band, respectively. Organic materials can be regarded as insulator or intrinsic semiconductor, hence their Fermi levels locate at the half of the bandgap[52]. Work function (referred as ϕ in Figure 2.21) is the energy needed to remove an electron from Fermi level to vacuum, i.e., the energy difference between Fermi level and vacuum level.

52. 如图 2.21 所示，金属费米能级指的是电子拥有最高能量的轨道；绝缘体和本征半导体的费米能级位于带隙的一半处。在 *p*-和 *n*-半导体中，费米能级分别接近于价带顶部和导带底部。有机材料可以被认为是绝缘体或本征半导体，因此其费米能级处于带隙的一半位置。

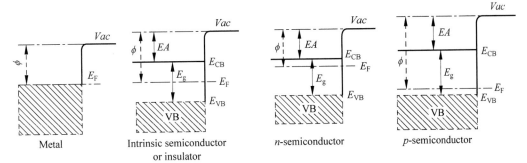

Figure 2.21　Fermi levels (E_F) and work functions (ϕ) of metal, intrinsic semiconductor, *n*- and
p- semiconductors

5. EA, IP and Bandgap

Electron affinity (EA) refers to the energy released when put a vacuum electron to LUMO level. Ionization potential (IP) refers to the energy needed to bring an electron from HOMO to the vacuum level.

In molecular orbital theory, bandgap is energy difference between HOMO and LUMO, while in band theory, bandgap is energy difference between VB and CB. With respect to bandgap, there are many confused concepts: optical gap, optical bandgap, and electronic bandgap. As shown in Figure 2.22, optical gap is energy difference between 1st singlet excited state (S_1) and the ground state, it refers to the optical transition from the ground state to the S_1 state. Optical bandgap is energy difference between HOMO (VB) and LUMO (CB), it refers to the formation of free electrons in LUMO by optical excitation from ground state, i.e., there is no interaction between LUMO electron and HOMO hole.

53. 如图 2.22 所示，光隙是第一激发态单线态（S_1）与基态之间的能量差，指的是从基态到 S_1 态的光学跃迁。光学带隙是 HOMO（VB）和 LUMO（CB）之间的能量差，指的是通过光激发基态分子而形成自由的 LUMO 电子，即 LUMO 电子和 HOMO 空穴之间没有相互作用。电学带隙指的是通过电激发而形成自由的 LUMO 电子。与光隙类似，由于电子离域能的存在，电学带隙也小于光学带隙。这里离域能的大小依赖于带宽和热激活能。

Electrical bandgap refers to the formation of free electrons in LUMO by electrical excitation. Similar to optical gap, electrical bandgap is also smaller than that of optical bandgap due to delocalization energy of electrons. Here delocalization energy is dependent on the width of the band and the thermal activation energy[53].

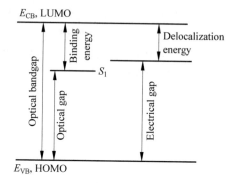

Figure 2.22　Optical gap, optical bandgap and electronic bandgap

Adiabatic bandgap（E_g^{ad}）and vertical bandgap（E_g^{opt}, also called optical gap）are two other concepts in application. As shown in Figure 2.23, vertical bandgap refers to the gap involved in vertical electronic transition; while adiabatic bandgap refers to the gap involved in 0 to 0 electronic transition.

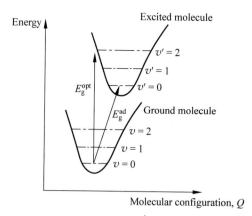

Figure 2.23　Adiabatic bandgap（E_g^{ad}）and vertical bandgap（E_g^{opt}）

2.3.3　Electronic Differences between Isolated Molecules and Crystals

1. Polarization among Molecules

When isolated molecules approach and aggregate,

electrons in HOMO repel each other，increasing electron energy，i.e.，HOMO level rises. The lifted energy is called hole polarization energy. The approach of unoccupied energy levels，e.g.，LUMO，results in band broadening and energy level lowering. The dropped energy is electron polarization energy. Generally，hole polarization energy and electron polarization energy are equal but opposite. Compared to isolated molecules，crystal shows higher HOMO，lower LUMO and lower bandgap[54].

2. Energy Level and Absorption Spectrum Differences

As compared to isolated molecules（Figure 2.24(a)），in crystal aggregates，besides the lowering of LUMO and the rising of HOMO due to polarization energy，the periodical interaction between molecules in the crystalline solid makes the degenerated electron orbitals split，forming energy band，with typical width smaller than 100 meV（ΔE）. In going from molecular crystals to disordered organic solids，one also has to consider locally varying polarization energies due to different molecular environments which lead to a Gaussian density of states for electron orbitals，with mean square deviation（σ）of $80 \sim 120$ meV.

As shown in Figure 2.24(b)，in isolated molecules，such a molecular gas，since its energy levels are discrete，its absorption spectrum is in line shape. In dilute solution，molecules are supposed to have no interaction；however，due to the interaction between molecule and solvent，its absorption spectrum presents as continuous line with redshift of Δ_1. In crystal，owing to the weak electronic delocalization and hence the minor energy-level broadening，to first order approximation，the optical absorption spectrum of organic molecular crystal is very similar to that of the isolated molecule，with slightly line-broadening；and compared to solution，the spectrum has redshift of Δ_2. In particular，intramolecular vibrations play an important role in solid state spectra and often these vibronic modes can be resolved even at room temperature. Nevertheless，solid state spectra can differ in detail with respect to selection rules，oscillator strength and energetic position often lead to a pronounced

54. 当孤立分子靠近并聚集时，HOMO 电子会相互排斥，造成电子能量的增加，即 HOMO 能级升高。而未占据轨道如 LUMO 轨道相互靠近，导致能带展宽和能级下降。下降的能量是电子极化能。一般来说，空穴极化能和电子极化能是大小相等、方向相反的。相对于孤立分子，晶体的 HOMO 上升，LUMO 下降，带隙减小。

anisotropy. Additionally disordered organic solids usually show a considerable spectral broadening. Detail optical property of organic materials can be referred to Chapter 3.

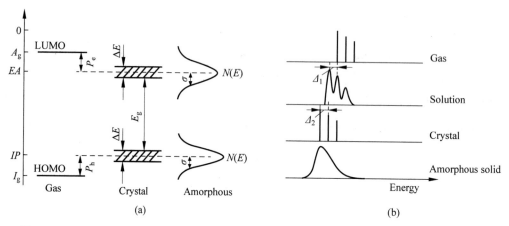

Figure 2.24　Energy levels (a) and absorption spectra (b) of a material in different situations

2.4　Effects of Bonding/Electronic-structures on Property of Organic Materials

2.4.1　sp^3 Bonding versus sp^2/sp Bonding

Different types of bonding result in different material property; this point can be well understood by comparison between pure carbon materials of diamond, graphite and fullerene[55].

55. 不同键型导致了不同的材料性质,这点可以通过比较金刚石、石墨和富勒烯这三个碳材料,得到很好的理解。

As shown in Figure 2.25(a), in diamond, the outer shell electrons in carbon atoms are of sp^3 hybridization with tetrahedron shape. Atoms stack and stretch to 3-dimension space along tetrahedron pattern, forming macromolecules. In a whole, diamond consists of solo σ bonds totally based on sp^3 orbitals and belongs to cubic crystal; it is very precious and extremely hard with super thermal stability and being an insulator. Graphite (Figure 2.25(b)) is made of carbon atoms with sp^2 hybridization. It exhibits layered structure. Within a layer, carbon atoms are coplanar and stretch to space along sp^2 orbitals via σ bonds; meanwhile, the unhybridized $2p_z$ orbitals, which are perpendicular to the carbon plane, can overlap, forming π orbitals. Between two layers, due to regular arrangement and close distance, strong π-π interaction exists. Compared to diamond, due to the existence of π system, graphite is less stable; on the other hand, the big π system within a layer and the strong π-π interaction between layers enable electron delocalization, leading to good electrical conductivity, hence graphite belongs to conductors. Fullerene (Figure 2.25(c)) exhibits a ball shape and is composed of sp^2 hybridized carbon atoms, possessing π bonds. Different to graphite, in fullerene, due to the requirement of spherical shape, the sp^2 orbitals in fullerene are not coplanar, but require some degree of bending. Therefore, besides the worst thermal stable nature compared to diamond and graphite, the bending shape of fullerene limits π electron delocalization, hence fullerene is less electrically conductive compared to graphite and is a semiconductor[56].

56. 与石墨不同,由于球形的需要,富勒烯中的 sp^2 轨道不是共平面的,而是必须有一定程度的弯曲。因此,除了与金刚石和石墨比较其热稳定性最差,富勒烯弯曲的形状限制了 π 电子的离域,因此富勒烯的导电性比石墨的差,是半导体。

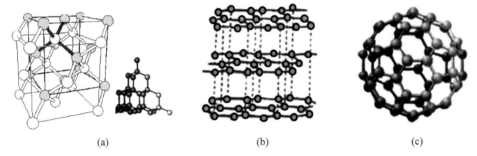

(a)   (b)   (c)

Figure 2.25　Structures of diamond (a), graphite (b) and fullerene (c)

2.4.2　Conjugation

As briefly introduced in Chapter 1, conjugation is a special structural feature of organic materials, which refers to the alternating sequence of single and double/triple bonds in molecules. It should be noted that although conjugation in molecular formulas has been written in the way of alternative sequence of single and double bonds in the literature for many years, it was known that all the bond lengths between the carbon atoms in the conjugation structure were essentially equal. Conjugation is the most interest feature in both small molecules and polymers for optoelectronic applications.

57. 图 2.26 展示了最小的共轭分子 1,3-丁二烯,同时展示了电子云分布和前沿能级。在共轭分子中,比如 1,3-丁二烯,σ 键将原子结合在一起,π 键提供了电子的离域。因为 π 电子可以在共轭结构内移动,它们有时也称为移动电子。

Figure 2.26 shows the smallest conjugated molecule,1,3-budadiene, with the presentation of electron cloud distribution and the frontier energy levels. In a conjugated molecule, e. g., 1, 3-budadiene, the σ bonds hold atoms together；the π bonds offer electron delocalization. Since the π electrons can move within conjugation，they are sometime called moving electrons[57].

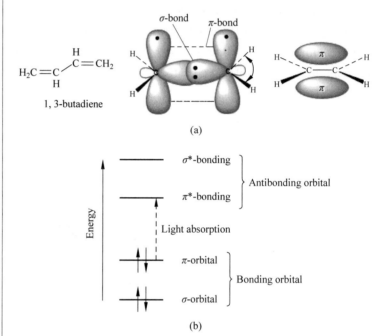

Figure 2.26　（a）Molecular structure of 1,3-budadiene and example for electron distribution of σ and π bonds in it. （b）Schematic view for energy levels of the σ and π bonds

The frontier orbitals of conjugated molecules belong to π orbitals. Generally π bond is much weaker than σ bond，and π electron possesses higher energy. Hence the electrons locating at HOMO levels are π electrons and LUMO is π^* orbital. This means that in organic conjugated molecules，the lowest optical transition generally occurs with π electrons；the transition is π-π^* transition. Due to the narrower splitting between π and π^* orbitals compared to that between σ and σ^* orbitals，bandgaps of conjugated materials are narrower compared to non-conjugated materials. In a word，organic conjugated materials generally present more active HOMO-electrons，their absorption/emission locates in ultraviolet， visible, and near infrared regions, e. g., bandgap is around 1.5～3.0 eV; with the increase of conjugation length, material bandgap decreases and absorption/emission redshifts[58].

In terms of electricity, due to the delocalization of π electrons, conjugated materials present some degree of conductivity. Depending on the scale of effective conjugation, and the stacking mode between molecules, the conductivity of organic conjugated materials can be quite different: they can be conductor, semiconductor, or even insulator.

2.4.3 Aromatic Rings

Aromatic structure refers to the cyclic planar conjugated unit, a special conjugation type in organic materials. The term aromatic derives from the characteristic odour of the natural compounds in which these materials are found. Criteria for aromaticity: molecule must be cyclic and planar; every atom in the ring must have p orbital perpendicular to molecular-plane, which allows overlapping between those on either side atom（completely conjugated）; it must satisfy Hückel rule: the number of π electrons $= 4n + 2$, where n is the number of atoms in the ring[59]. 8-annulene（cycloocta-1, 3,5,7-tetraene, COT）shown in Figure 2.27（a）is a non-aromatic compound, because the eight carbon atoms in COT are not coplanar. The cyclic molecule shown in Figure 2.27（b）（cyclopenta-1,3-diene）also presents no aromaticity though all the carbon atoms are coplanar. This is because C-5 adopts sp^3 hybrid and does not contribute to p-electron, leading to the broken of conjugation in the planar cyclic ring.

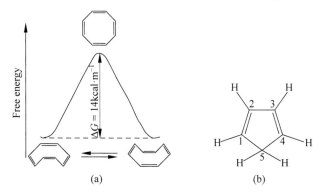

Figure 2.27　（a）Structure and energy schematic of COT with different spatial configuration.（b）Structure for a cyclic molecule without aromaticity

58. 共轭分子的前沿轨道属于 π 轨道。通常地，π 键比 σ 键弱很多，π 电子拥有较高的能量。因此，HOMO 能级上的电子是 π 电子了，LUMO 是 π^* 轨道。这意味着在有机共轭分子中，最低光学跃迁通常发生在 π 电子上；跃迁是 π-π^* 跃迁，由于 π 和 π^* 轨道之间的裂分比 σ 和 σ^* 之间的小，共轭材料的能隙比非共轭材料的能隙窄。总之，有机共轭材料通常表现出较活跃的 HOMO 电子，它们的吸收/发射在紫外、可见以及红外区域，如能隙为1.5～3.0 eV；随着共轭长度的增加，材料的带隙减小，同时吸收/发射红移。

59. 芳香结构指的是环状平面共轭单元，是有机材料中的一种特殊共轭。术语芳香衍生于天然化合物中存在的材料的气味特点。芳香性的要求：分子必须是环状且平面的，环上的每一个原子必须拥有垂直于分子平面的 p 轨道，用来与两边原子的 p 轨道重叠（完全共轭）；它必须满足 Hückel 规则：π 电子数量 $=4n+2$，其中 n 是环上原子的个数。

60. 苯分子由 6 个 sp^2 碳原子和 6 个氢原子组成。每一个碳原子与一个氢原子和两个碳原子通过 σ 键链接形成六元六边形分子平面,伴随有 6 个未杂化的 $2p_z$ 轨道垂直于该平面,它们相互重叠形成大 π 键。因为填充的 π 轨道能量低于环上孤立 p_z 轨道,因此 π 键的形成能够降低体系能量,稳定分子体系。这个大 π 共轭体系含有 6 个 π 电子,它们是离域而不是定域在键上,并且在 π 共轭体系上移动。

Benzene ring is a well-known aromatic ring, its molecular structure and energy levels are shown in Figure 2.28. Benzene molecule is made of six sp^2 carbon atoms and six hydrogen atoms. Each carbon attaches to one hydrogen and two carbons by σ bond, forming a six-member and hexagonal molecular plane with six unhybrid $2p_z$ orbitals perpendicular to the plane, which overlap mutually to form a big π bond. Since the energy of the filled π orbitals is lower than that of isolated p_z orbital in ring, thus the formation of the π bond can lower system energy and stabilize molecular system. The big π conjugated system contains 6 π electrons, which are delocalized rather than being localized in bonds and move in the π conjugation system[60]. Similar to previously talked conjugated materials, frontier orbitals in benzene are π and π^* orbitals.

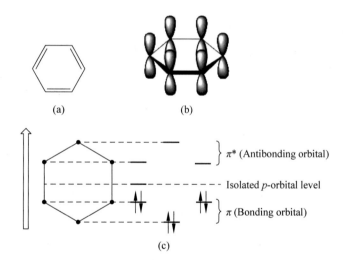

Figure 2.28　Molecular structure (a), big π orbital (b) and frontier energy levels (c) of benzene molecule

Table 2.4 summarises most of the typically used polycyclic aromatic hydrocarbons (PAH). These materials generally exhibit strong emission in ultraviolet to near infrared range, and due to the rigidity, the emission spectra of PAHs are usually mirror images to their absorption spectra; and with the increase of effective conjugation length, material bandgap decreases and absorption/emission redshifts.

Besides the aromatic structures formed by pure carbon and hydrogen atoms，there are lots of heterocyclic aromatic materials（Table 2.5）containing non-carbon atoms that generally possess non-bonding lone-pair electrons；and when forming aromatic ring via sp^2 hybrid orbitals, the lone-pair electrons locate either in the molecular plane or in the π conjugation system that perpendicular to the molecular plane[61].

61. 除了由纯碳氢构成的芳香结构，存在很多含有非碳原子的杂环芳香材料（表2.5），这些非碳原子通常拥有非键的孤对电子；在它们通过 sp^2 杂化轨道形成芳环时，孤对电子可以在分子平面上，也可以位于垂直于分子平面的 π 共轭体系中。

Table 2.4　Molecular structures and their optical absorption peaks in some typical PAHs

Structure	English name	Chinese name	Absorption /nm
	Benzene	苯	254
	Naphthalene	萘	311
	Anthracene	蒽	375
	Tetracene	并四苯	471
	Pentacene	并五苯	582
	Pyrene	芘	352
	Perylene	苝	434
	Fluoranthene	荧蒽	356
	2,3-Dimethyl-naphthalene	2,3-二甲基萘	315
	Fluorene	芴	303, 260

As shown in Figure 2.29(a), in pyridine molecule with one heteroatom of nitrogen, the lone pair electrons belong to the aromatic plane, taking up one sp^2 orbital, and there is one unpaired p_z electron from the N that contributes to the aromatic π system; but in the case of pyrrole molecule (Figure 2.29(b)), which also possesses one heteroatom of nitrogen, all of the three in-plane sp^2 orbitals in the N contain a unpaired electron for each, and involve in the σ bonding, two with carbon atom, and one with hydrogen atom; therefore, the p_z orbital of the N contains the lone-pair electrons contributing to the π system. For heteroatom of oxygen and sulphur that possesses two lone-pair electrons, like furran (Figure 2.29(c)), the lone pair electrons locate in the molecular plane and the p_z orbital of the O atom, which means that the electron contribution from O to the π system is one lone-pair electrons. It should be noted that when the lone-pair electrons are in the aromatic molecular plane, they will be easy to give out for metal coordination or for itself oxidation; if the lone-pair electrons of the heterocyclic aromatic compound locate at the π system, they will be not easy to give out for coordination/oxidation reaction.

Due to the polar nature caused by the heteroatoms (i.e., non-carbon atoms), heterocyclic aromatic materials/moieties often exhibit preferred carrier transport. For instance (Table 2.5), compounds with dominant building block of 1,10-phenanthroline, phthalazine, silole, 1,3,4-oxidiazole, quinoline, or 1,3,5-triazine, often present electron transport, while those with dominant building block of isoindole, carbazole, or thiophene, often exhibit hole transport property[62]. Compared to pure PAHs, the emission efficiency of heterocyclic aromatic compounds is generally lower due to the forbidden nature of electron transition from lone-pair electron in the HOMO to the LUMO orbital of π^*.

62. 由于杂原子(非碳原子)引起的极性,杂环芳香材料/基团,通常表现出选择性的电荷输运。例如(表2.5),以1,10-菲咯啉、酞嗪、硅杂环戊二烯、1,3,4-噁二唑、喹啉或者三嗪为主导构成单元的化合物,通常表现出电子传输特性,而那些以异吲哚、咔唑或者噻吩为主要构成单元的化合物,通常表现出空穴传输特性。

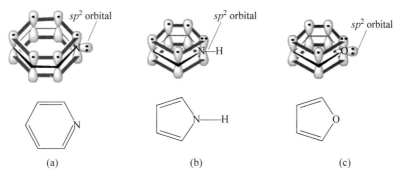

Figure 2.29　Heterocyclic aromatic molecules of pyridine (a), pyrrole (b) and furran (c)

Table 2.5　Typical heterocyclic aromatic hydrogen carbons

Structure	English name	Chinese name	Structure	English name	Chinese name
	1,10-Phenanthroline	1,10-菲咯啉		Thiophene	噻吩
	Isoindole	异吲哚		Benzothiophene	硫茚(苯并噻吩)
	1,3,4-Oxidiazole	1,3,4-噁二唑		Pyridine	吡啶
	Phthalazine	酞嗪		Acridine	吖啶
	Quinoline	喹啉		Pyrazine	吡嗪
	1,3,5-Triazine	三嗪		Quinoxaline	喹喔啉
	Pyrrole	吡咯		Pyrazole	吡唑
	Carbazole	咔唑		3,4-Ethylene-dioxithiophene	3,4-次乙基-二氧噻吩
	Furane	呋喃		Dibenzothiophene	硫芴(二苯并噻吩)
	Benzofurane	苯并呋喃		Silole	硅杂环戊二烯

本章小结

1. 内容概要：

本章全面介绍了有机材料中的成键和电子结构，这是理解有机材料光电性质的基础。第一部分，在简要地介绍了固体中的四种键及其特点后，聚焦于有机材料中最重要的碳原子，讨论了轨道杂化及共价键。第二部分，学习了有机材料中的各种分子间作用力。第三部分，在简要地介绍了有助于理解有机材料电子结构的不同成键理论后（如分子轨道理论、晶体/配位场理论和能带理论），详细讨论了有机材料的电子结构。最后的部分，说明了材料性质与特定成键方式及电子结构的关系。

2. 基本概念：金属键、离子键、共价键、分子键、范德瓦耳斯力、诱导力、色散力、杂化轨道、分子轨道、成键轨道、反键轨道、HOMO、LUMO、d 轨道能级分裂、空带、满带、禁带、导带、价带、带隙、单线态、三线态、真空能级、电子/空穴极化能、费米能级、功函数、电荷转移态、解离能、亲和能、吸收、发射。

3. 主要公式：

（1）平行且在一条直线上的偶极-偶极作用力 $f_{\text{force}} = -\dfrac{\mathrm{d}E_{\Psi}}{\mathrm{d}r} = -\dfrac{1}{4\pi\varepsilon_0}\dfrac{6P_1P_2}{r^4}$

（2）诱导力 $f_{\text{force}} = -\dfrac{1}{4\pi\varepsilon_0}\dfrac{6P_1P_{\text{ind}}}{r^4} = -\dfrac{1}{(4\pi\varepsilon_0)^2}\dfrac{12P_1{}^2\alpha}{r^7}$

（3）色散力 $f_{\text{force}} = \dfrac{1}{(4\pi\varepsilon_0)^2}\dfrac{A'\alpha^2}{r^7}$

（4）堆叠系数 $K = ZV_0/V$

Glossary

Aggregate	聚集体，聚集
Amorphous	非结晶的
Anti-bonding orbital	反键轨道
Aromatic	芳香族的
Atom	原子
Attractive force	引力
Band theory	能带理论
Bandgap	带隙
Binding energy	结合能
Bonding	成键
Charge transfer	电荷传输
Conducting band	导带
Conductor	导体
Conjugation	共轭

Conservation	守恒
Coordination field theory	配位场理论
Covalent bond	共价键
Crystal field theory	晶体场理论
Debye interactions	德拜相互作用
Delocalization	离域
Dense stacking	密堆积
Diamond	金刚石,钻石
Dipole-dipole interaction	偶极-偶极相互作用
Double bond	双键
Electron affinity	电子亲和力,电子亲和势
Electron cloud	电子云
Electron sharing	电子共享
Electronic structure	电子结构
Empty band	空带
Energy degenerate	简并能级
Energy splitting	能级分裂
Exciton	激子
Fermi level	费米能级
Filled band	满带
Fluorine	氟气,氟元素
Fullerene	富勒烯
Graphene	石墨烯
Graphite	石墨
Hybrid orbital	杂化轨道
Hydrogen bonding	氢键
Inductive force	诱导力
Insulator	绝缘体
Interaction	相互作用
Interatomic spacing	原子间距
Intermolecular forces	分子间力
Intramolecular	分子内的
Intrinsic semiconductor	本征半导体
Ionic bond	离子键
Ionization potential	电离势
Isolated	隔离的,单独的,分立的
Lattice energy	晶格能
Ligand field theory	配位场理论
Malleability	展性,韧性

Metal complex	金属络合物
Metallic bond	金属键
Metastable state	亚稳态
Molecular bond	分子键
Molecule	分子
Nucleus	原子核
Occupy	占据
Orbital	轨道
Packing coefficient	堆积系数
Pauli exclusion principle	泡利不相容原理
Point charge	点电荷
Polarisability	极化率
Repulsive force	斥力
Saturated bond	饱和键
Semiconductor	半导体
Single bond	单键
Singlet	单线态
Subouter shall	亚外壳层
Triple bond	三键
Triplet	三线态
Unsaturated bond	不饱和键
Vacuum level	真空能级
Valence band	价带
Van der Waals bond	分子键（范德瓦耳斯键）
Work function	功函数

Problems

1. True or false questions:

(1) An ionic crystal consists of cations and anions, therefore, the ionic crystal can conduct electricity.

(2) Conjugation size of organic molecular partially determines the carrier transportation property.

(3) Covalent bonding only exists in organic materials.

(4) σ electron can easily delocalize.

(5) If there is only one covalent bond between two atoms, it is σ bond.

(6) Hydrogen bonding belongs to permanent dipole-dipole interaction.

(7) Compared to Debye force, London force is more polarisability-dependent.

(8) Atomic orbital describes orbital energy only.

(9) Generally, π electron shows higher energy than that of σ electron, so π^* electron also shows higher energy than that of σ^* electron.

(10) A molecular orbital describes the moving status of an electron under the average potential of the whole molecule.

(11) If the energy level difference is very large, the two atomic orbitals cannot effectively combine into molecular orbitals.

(12) Both HOMO and LUMO are frontier orbitals.

(13) According to Hund's rule and crystal field theory, electrons will occupy different d orbitals before they pair up.

(14) Compared to ligand field theory, spectrochemical series can be better explained by crystal field theory.

(15) In octahedral complex, regarding π bonding, t_{2g} orbitals are non-bonding orbitals.

(16) Ligand field theory can be regard as the combination of molecular theory and crystal field theory.

(17) Compared to crystal field theory, d orbital splitting in ligand field theory energy generally changes.

(18) When π bonds are formed between ligand and metal, d orbital splitting energy always increases.

(19) Ionization potential determines the ability of hole injection to a molecule.

(20) Electron affinity is the ability of a molecule to attract electron.

(21) Fully filled band can conduct electricity.

(22) Conduction band is fully filled with electrons.

(23) In terms of a fixed semiconductor, the inner band is narrower than outer band.

(24) Hole polarization energy is equal to electron polarization energy numerically.

(25) Phosphorescence refers to the radiative decay from singlet excited state to the ground state.

(26) Energy transfer in organic materials refers to electron or hole hopping from one molecule to another molecule.

(27) Band gap of a crystal is narrower than that of the corresponding isolated molecule.

(28) LUMO is the lowest unoccupied molecular orbital, so 1st single state is higher than LUMO.

2. Please explain electron orbital and electron cloud.

3. What are saturated and unsaturated bonds? Please point out their differences.

4. Please explain valence, σ, π and n electrons. Please put them in ascending order of energy.

5. Please describe the three types of carbon hybridization, and point out their shapes in space.

6. Why, among analogues, the melting point and the boiling point generally increase with their molecular weight?

7. Why the packing coefficient of ice (0.38) is far smaller than those of aromatic hydrocarbons (0.68 to 0.8)?

8. Dipole-dipole force is reciprocal to R^4, while dipole-induced dipole force is reciprocal to R^7, please give necessary deducing processes.

9. At room temperature, Ar is gaseous, while at temperature below 84 K ($m.p.$ of Ar), it is solid, why? In Ar and He, which one possesses higher melting point?

10. Please describe the main points of molecular orbital theory.

11. Please schematically draw MO levds of O_2, and the electron configuration.

12. According to MO theory, please explain why He_2 does not exist, but He_2^+ can exist.

13. Please describe the five d orbitals in an atom using their names and compare their energy in the case of no disturbing. According to ligand field theory, please briefly deduce the d orbital splitting in octahedral metal complex with help of schematic drawing, and please give the order of d orbital energy.

14. Why are metal complexes often highly colored?

15. Please explain crystal field theory and tell reason for d orbital splitting.

16. In octahedral metal complexes, on σ bond basis, please schematically draw metal-to-ligand π bond, and point out the change of splitting energy.

17. Please briefly state the main points of the ligand field theory.

18. Please schematically draw the HOMO and LUMO level distributions of (1)gaseous molecules, (2)crystals, and (3)amorphous solids.

19. Discussion and explanations according to the energy level diagram shown in Figure 2.19.

(1) Please explain the levels between HOMO and LUMO in isolated molecules, and give the reason for their formation.

(2) Why there exist energy-level changes when isolated molecules aggregate to crystal?

(3) Compared to the energy levels in isolated molecule, please explain the origination of the charge transfer state in crystal.

20. Please outline the main points of band theory.

21. Please explain HOMO and LUMO, and point out their relationship to ionization potential, electron affinity, conducting band and valence band, if any.

22. Please give the concept of Fermi level and explain it in metal, insulator, intrinsic-semiconductor, n- and p-semiconductors.

23. Please simply explain why filled and empty bands can not conduct electricity? In term of an intrinsic semiconductor, conduction band is usually empty, while valence band is usually fully filled, how to make intrinsic semiconductor conductive while keeping its

intrinsic nature?

24. Conjugation and doping are two most effective strategies to increase the conductivity of an organic material. Please explain this phenomenon according to band theory.

25. Please briefly explain optical bandgap, optical gap and electronical bandgap.

CHAPTER 3   ENERGY RELATED PROCESSES IN ORGANIC MATERIALS

This chapter focuses on energy related processes in organic materials, which are the main concern for their optoelectronic applications. After a brief introduction, some basic concepts relating to energy processes are discussed. Then, electronic transitions within a molecule are studied, including optical absorption to obtain excited state, and the decay processes of excited states. In order to better understand the optical transition, selection rules are presented comprehensively. Then topics are moved to energy processes in organic aggregates, with the discussion on Davydov splitting, excimer/exciplex, charge-transfer/surface exciton, and fusion/fission. Energy transfer mechanism is the next topic, including cascade energy transfer, Förster energy transfer and Dexter energy transfer. Finally, various dynamic processes of excitons are summarized.

3.1   Introduction

1. 考虑材料的光电性质,通常只有最活跃的电子即前沿电子参与;并且取决于前沿电子的能量,有机材料中的分子可以处于基态(S_0)、第一或者更高单线态激发态(S_1, S_2, ⋯)、第一或者更高三线态激发态(T_1, T_2, ⋯)。

Concerning optoelectronic properties of materials, generally, only the most energetic electrons, i. e., the frontier electrons, are involved; and depending on the energy of their frontier electrons, the molecules in organic materials can be in ground state (S_0), the first or higher singlet excited state (S_1, S_2, ⋯), the first or higher triplet state (T_1, T_2, ⋯)[1]. As can be seen in Figure 3.1, there are lots of electronic processes within a molecule and between molecules, such as light-absorption (up solid arrow), light emission (down solid arrow), non-irradiate excitation (up wave arrow), non-radiative decay (down wave arrow), fission (down dash line arrow), and fusion (up dash line arrow), etc. The occurrence of these processes involves electron transition and accompanies with energy absorption or release. All of these are the focuses of this chapter.

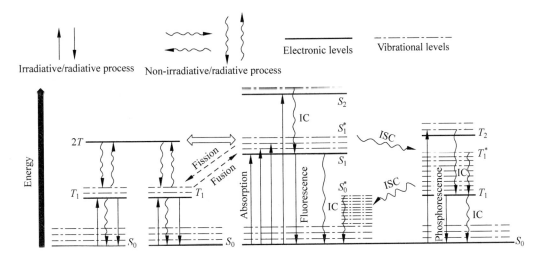

Figure 3.1　Molecules with different energy and various electronic transition processes in organic materials ($2T$ is the energy of two triplet state, here, it is roughly equal to the energy of S_1^*; IC and ISC are internal conversion and inter system conversion, respectively). Note that this energy diagram and electronic transitions in a molecule also refers as Perrin-Jablonski diagram

3.2　Concepts for Ground/Excited State and Exciton

3.2.1　Ground and Excited States

Electronic transitions in organic materials generally involve frontier orbitals of the highest occupied molecular orbital (HOMO) and the lowest unoccupied molecular orbital (LUMO). When the electrons of the highest energy in a molecule locate at HOMO levels, the molecule is in its ground state; when they locate at higher level than HOMO, and in the meantime, these electrons are Coulombically bound to HOMO hole, it is an excited state molecule. Subjecting to the Coulombic attraction from nucleus, the energy level of the excited electron (e.g., S_1 and T_1 states) is lowered compared to LUMO level, and the excited electron can be stabilized in this way[2].

Excited states are classified into singlet and triplet excited states according to multiplicity (M), which is determined by $M = 2S + 1$, here S is the total spin number of the molecule. When $S = 0$, $2S + 1 = 1$, the excited state is

2. 当分子中最高能量的电子处于 HOMO 能级时,分子是基态分子;当这些电子的能级比 HOMO 高,同时它们与 HOMO 的空穴通过库仑引力结合,分子是激发态分子。受控于原子核的库仑吸引力,激发态的电子能级(如 S_1 和 T_1)比 LUMO 能级低,激发态电子得到稳定。

singlet; when $S = 1$, $2S + 1 = 3$, the molecule is in triplet state. As shown in Figure 3.2(a), a ground state molecule generally possesses paired electrons in HOMO orbital, it can be referred as singlet state, i.e., S_0; while, singlet and triplet excited states are those, in which the higher electron possesses the opposite and the same spin to the unpaired HOMO electron, respectively (Figures 3.2(b) and (c)). According to Hund's rule, when electrons occupy different orbitals, system energy is the minimum when their spins are the same. Therefore, singlet state is higher than the corresponding triplet state. One should be noted that if the excited electron in the molecule isn't attracted by the HOMO hole, and the lowest level for its occupation is LUMO, the molecule becomes negative polaron (Figure 3.2(d)); on the other hand, if the excited electron in a molecule is autoionized (lost from the molecule), the molecule becomes positive polaron (Figure 3.2(e)). Detail for polaron can be found in Figure 4.3, Figure 4.13 and Figure 4.14 and the corresponding contexts.

From the perspective of quantum chemistry, the spin orbitals of singlet (S) and triplet (T) can be expressed as

$$S = \frac{1}{\sqrt{2}}(|{\uparrow\downarrow}\rangle - |{\uparrow\downarrow}\rangle) \tag{3-1}$$

$$T = |{\uparrow\uparrow}\rangle, \text{ or} = |{\downarrow\downarrow}\rangle, \text{ or} = \frac{1}{\sqrt{2}}(|{\uparrow\downarrow}\rangle + |{\uparrow\downarrow}\rangle) \tag{3-2}$$

Singlet and triplet states in organic materials have features of: ① Singlet state is spin-asymmetric with only one wavefunction, and its multiplicity is 1; triplet state is spin-symmetric with three wavefunctions, and its multiplicity is 3. ② Within the same electronic state, singlet energy is higher than triplet one. ③ The lifetime of singlet state is smaller than that of triplet state (ns versus μs). ④ Optical transition from triplet is generally forbidden, while that from singlet is allowed. ⑤ Singlet state molecule is diamagnetic, triplet state molecule is paramagnetic. ⑥ While employing absorption and emission spectra, both singlet and triplet can be investigated, the feature of the triplet state can also be studied by electron spin resonance spectrum (ESR)[3].

3. 有机材料中的单线态和三线态有如下特点：①单线态是自旋非对称的,只有一个波函数,其多重性是1；三线态是自旋对称的,具有三个波函数,其多重性是3。②在一个电子能级内,单线态比三线态能级高。③单线态的寿命比三线态的短(纳秒相对微秒)。④三线态的光学跃迁通常是禁阻的,而单线态是允许的。⑤单线态是抗磁的,三线态是顺磁的。⑥单线态和三线态都可以通过吸收,以及发射光谱来研究,此外三线态还可以通过电子自旋共振谱(ESR)研究。

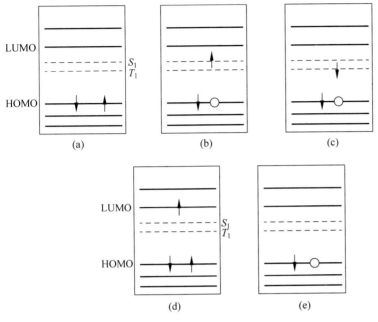

Figure 3. 2　Frontier electron configuration in a molecule of ground state, S_0 (a), singlet excited state, S_1 (b), triplet excited state, T_1 (c), and that in a negative polaron (d) and a positive polaron (e). The empty circle indicates hole in HOMO level

3.2.2　Exciton

1. Concept

In a material, electron in the excited state and the mutually attracted positive polaron (also refers as hole) is termed as exciton. Simply, exciton is also expressed as Coulombically bound electron-hole pair[4]. Different electronic structure endows different type of exciton. In inorganic semiconductors, sine the higher-energy electrons in conduction band (CB) are able to delocalize, the separation between the Coulombically bound CB electron and VB (valence band) hole, i.e., the exciton radium, is long (40~100 Å), and hence, the binding force is weak (about 10 meV). In contrast, in organic materials, sine the higher-energy electrons are generally localized within a molecule, exciton radium is small (about 10 Å) and the binding force is strong (0.3 ~ 1 eV). Excitons found in inorganic

4. 材料中,激发态电子和与其相互吸引的正极化子(也称为空穴)组成了激子。简单来说,激子就是库仑力束缚的电子-空穴对。

semiconductors are generally Wannier exciton, and those in organic materials are often Frenkel type. Besides the Wannier and Frenkel excitons, in organic materials, there exists another type excition, i. e., the charge transfer (CT) exciton, which is referred to the Coulombic binding between higher-energy electron in one molecule and the lower-level hole in adjacent molecule. Both the exciton radius and the binding force of CT excitons locate in between those of Wannier and Frenkel excitons, with values of $10\sim15$ Å and $0.01\sim0.3$ eV, respectively. Figure 3.3 schematically illustrates the space orientation of these three types of excitons. Similar to excited state, exciton is also classified into singlet and triplet excitons. As shown in previous Figure 3.2（b）and（c）for the discussion of excited states, the combination of the higher-energy electron and the hole in HOMO that possesses the unpaired electron with different and same spin can be referred as singlet exciton and triplet exciton, respectively[5].

5. 如图 3.2（b）和（c）所讨论的激发态时所表示的,高能量电子与 HOMO 空穴的结合,HOMO 中的非成对电子与高能电子自旋相反和相同的两种情形,分别对应于单线态和三线态激子。

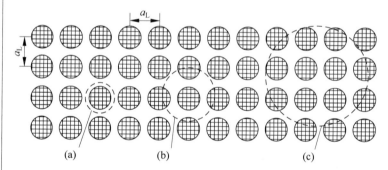

Figure 3.3　Frenkel（a）, CT（b）and Wannier（c）excitons

2. Exciton Generation

Generally, exciton can be produced by direct/indirect light excitation, carrier injection followed by recombination, or energy transfer.

As shown in Figure 3.4（a）, a ground state molecule can absorb light with proper wavelength and jump to excited state S_1, forming Coulombically bound electron and hole pair, i. e., exciton. Due to the spin conservation rule and singlet nature of the ground state, optical generation of exciton is generally in the form of singlet. Figure 3.4（b）schematically demonstrates the exciton formation by carrier injection: holes and electrons can be injected into HOMO and LUMO of molecules from anode and cathode, respectively; then after transport oppositely, electron and hole are approaching each other, some of which can be eventually bound by Coulombic force, forming exciton[6]. Depending on multiplicity, exciton formed by electrical pumping can be either singlet or triplet; and due to the ratio of 1 : 3 for singlet and triplet orbitals, statistically, electrical pumping of exciton often contains 25% for singlet and 75% for triplet. Equations（3-3）to（3-5）describe the formation of optically pumped singlet exciton, and electrically pumped singlet and triplet excitons.

$$S_0 + h\nu \rightarrow S_1 \tag{3-3}$$

$$p^+ + n^- + S_0 \rightarrow S_n^* \rightarrow S_1 \tag{3-4}$$

$$p^+ + n^- + S_0 \rightarrow T_n^* \rightarrow T_1 \tag{3-5}$$

where, S_0, S_1/T_1, S_n^*/T_n^*, $h\nu$, p^+/n^- represent ground state, the first singlet/triplet excited state, the nth singlet/triplet excited state overridden with vibrational energy, the excitation light and the hole/electron, respectively. Exciton also can be formed by energy transfer, which will be discussed in section 3.6.

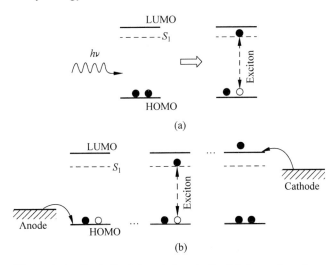

Figure 3.4　Optically（a）and electrically（b）formed excitons

3.3　Electronic Transitions within a Molecule

Electronic transition refers to the up energy-level or

6. 如图 3.4(a)所示,处在基态的分子可以吸收合适波长的光,跳到激发态 S_1,形成库仑结合的电子-空穴对,即激子。基于自旋守恒和基态的单线态特点,光生激子通常是单线态形式。图 3.4(b)示意地展示了载流子注入到产生激子的情形:空穴和电子可以分别从阳极和阴极注入到分子的 HOMO 和 LUMO,通过相向输运,逐渐接近,其中的一些终将通过库仑力结合形成激子。

down energy-level jumping of electron in material，the former is also called electron excitation，and the latter is electron decay process. Electron in ground state of a molecule can obtain energy either by optical or thermal/electrical method，becoming a higher energy-level electron. In contrast，electrons in excited states possess higher energy hence aren't stable，they tend to decay to the ground state within a molecule or in between molecules. There are several types of electronic transition，which is the focus of this section.

3.3.1　Absorption

Absorption refers to the process that electrons in materials take up electromagnetic energy with the result of electron excitation，i.e.，light absorption. Depending on the energy of the incident light，the result of light absorption can be the excitation of ground state electron to the first excited state or state with even higher energy；when the light energy is high enough，electrons in materials can emit out of the material they belong to. If the absorption process is carried out under irradiation light in optical range (i.e.，ultraviolet to visible light)，it is referred as optical absorption. In terms of organic materials，especially those with conjugation structure，the bandgap generally lies in energy of optical range，they are hence generally optical-absorption active. Therefore，UV-visible absorption is a very useful tool to probe electronic behaviours in organic materials[7]. Optical absorption of organic materials corresponds to the electron transition from ground state to first excited state，e.g.，the first singlet and triplet excited states. For a pure organic material such as fluorescent material，triplet absorption is generally forbidden，hence only singlet exciton is produced；while for a phosphorescent material，due to the strong spin-orbital coupling，triplet absorption is partially allowed，hence both singlet and triplet excitons can be produced by optical absorption (Figure 3.5).

An example of absorption spectrum from an organic material is shown in Figure 3.6，which is simply a plot of absorbance versus wavelength. The data can also be expressed

7. 如果吸收过程是在光学范围（即紫外光和可见光）的光辐射下进行的，就是光学吸收。对于有机材料，尤其是包含共轭结构的，其带隙通常在光学范围，它们通常具有光学吸收活性。因此，紫外-可见光吸收是研究研究材料电子行为非常有用的手段。

as a graphical plot of molar extinction coefficient（ε）versus wavelength. The use of ε has the advantage that all intensity

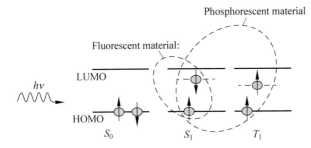

Figure 3.5　Excited states produced by light absorption in a fluorescence and a phosphorescence materials

values refer to the same number of absorbing molecules, which is convenient to compare the absorption between materials. Usually, absorption spectra of organic materials are broad, with width spans up to 1 eV, and some time in a material with rigid molecular structure, fine structure representing vibrational levels can be observed[8].

8. 通常地，有机材料的吸收光谱很宽，可以达到1 eV，有时在刚性结构的材料中，代表振动能级的精细结构可以被观测到。

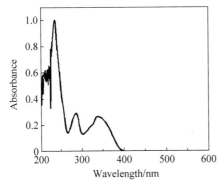

Figure 3.6　Absorbance-wavelength curve of an organic material

By recording UV-visible absorption spectra of dilute solution and thin film, some photophysical parameters can be obtained. Generally, for dilute solution, Lambert-Beer's law （historically more correct: the Bouguer-Lambert-Beer Law） is applicable:

$$A = \lg\left(\frac{I_0}{I}\right) = \lg\left(\frac{1}{T}\right) = -\lg T = \varepsilon \cdot c \cdot l \tag{3-6}$$

where A is absorbance at a concerned wavelength λ, also called optical density （OD）, indicating the capacity of a substance to absorb radiation; I_0 and I are the intensity of the light beam before and after passing through the sample, respectively; T is transmittance, ε is molar extinction coefficient in unit of $L \cdot mol^{-1} \cdot cm^{-1}$, c is sample concentration, and l is length of optical path.

In film, the absorbance in absorption spectrum complies with:

$$A = OD = \alpha d \qquad (3\text{-}7)$$

where α is the absorbance constant, in unit of cm^{-1}, d is the thickness of film. According to Equations (3-6) and (3-7), with the optical absorption spectrum, when the concentration, the cuvette length and film-thickness are known, $\alpha(\lambda)$ and $\varepsilon(\lambda)$ can be obtained. Organic materials often exhibit high absorption intensity. For a pure organic material without heavy metal, its molar extinction coefficient, absorbance constant and oscillation strength from $S_0 - S_1$ transition can be high as $10^4 \ M^{-1} \cdot cm^{-1}$, $10^5 \ cm^{-1}$ and 10^{-3} to 2, respectively[9].

3.3.2 Energy Decay Processes

Electron decay processes in organic materials include radiative decay and nonradiative decay. Radiative decay refers to the electron transition from higher energy level to lower energy level with light emission, while nonradiative decay refers to the electron transition from higher energy level to lower energy level without light emission, such as internal conversion, inter system crossing, etc[10]. Both radiative decay and nonradiative decay conserve multiplicity. For examples, $S_1 \rightarrow S_0$ and $T_2 \rightarrow T_1$ are allowed transitions; while $T_1 \rightarrow S_0$ and $S_1 \rightarrow T_1$ are forbidden. In excited state molecules, the nonradiative decay processes always compete with the radiative decay processes. Thus nonradiative decay processes are an important class of processes when considering energy transfer and light emission.

1. Internal Conversion

Internal conversion (IC) refers to the nonradiative decay processes that the electronic energy coverts to vibrational energy, which generally occurs between the same multiplicity (e.g., $S_2 \rightarrow S_1$, $T_2 \rightarrow T_1$). Internal conversion can occur

9. 有机材料通常具有较高的吸收强度。对于没有重金属的纯有机材料，其 S_0 到 S_1 跃迁的摩尔消光系数、吸收常数以及谐振子强度可以分别高达 $10^4 \ M^{-1} \cdot cm^{-1}$、$10^5 \ cm^{-1}$ 和 10^{-3} 到 2。

10. 辐射衰减是指从较高能级到较低能级的电子跃迁并伴随光的发射，而非辐射衰减是指从较高能级到较低能级且无光发射的电子跃迁过程，就像内转换和系间窜越等。

within the same electronic level as well as between different electronic levels, and the rate of internal conversion is very fast, usually within picoseconds (10^{-12} s)[11].

When a ground state molecule absorbs light and changes to excited state molecule, its orbital energy usually contains the electronic energy and the vibrating energy. The overlap between the electronic and the vibrating energy level results in the excited state with higher energy level. Via vibrational relaxation, the electron loses its vibrating energy quickly and returns to the energy state with only electronic energy. This is internal conversion process within the same electronic state. If the excited state couples with the vibration vigorously, it can decay to ground state by nonradiative decay process. This is the internal conversion process between different electronic levels.

2. Inter System Crossing

Inter system crossing (ISC) refers to the nonradiative decay process between energy levels with different multiplicities (e. g., $S_1 \rightarrow T_1$, $T_1 \rightarrow S_0$). Generally, ISC processes are forbidden. Spin-orbital coupling can release the forbidden feature[12]. When the energies of the singlet state and the triplet states are comparable, and they are in close vicinity, coupling can occur between the spin states of these two orbitals, resulting in the blurring of the triplet state. This process is termed as spin-orbital coupling. Spin-orbital coupling renders triplet state partially loses its feature and exhibits some singlet characteristic, which can result in inter system crossing, optical excitation of triplet state and phosphorescent emission. Due to the ISC process, in pure organic material, the absorbance constant of $S_0 \rightarrow T_1$ is not zero but is around $10^{-5} \sim 10^{-4}$ cm^{-1}, even though it is forbidden process. In complexes containing heavy metals, the combination of metal and ligand leads to metal-ligand charge transfer (MLCT) state, which often produces spin orbital coupling in the molecule. As shown in Figure 3.7, spin orbital coupling between singlet and triplet states of MLCT (^{1}MLCT and ^{3}MLCT) may bring the population of ^{3}MLCT by optical absorption, as well as phosphorescence from

11. 内转换（IC）是指电子能量转化为振动能量的非辐射衰变过程，一般发生在相同多重性之间（如 $S_2 \rightarrow S_1$，$T_2 \rightarrow T_1$）。内转换可以发生在同一个电子能级内，也可以发生在不同电子能级之间，并且内转换的速率非常快，通常在皮秒内（10^{-12} s）。

12. 系间窜越（ISC）是指具有不同多重态的能级之间的非辐射衰变过程（如 $S_1 \rightarrow T_1$，$T_1 \rightarrow S_0$）。一般来说，系间窜越是被禁阻的。自旋轨道耦合可以缓解该禁阻特性。

13. 对于纯有机物而言，尽管 $S_0 \rightarrow T_1$ 是禁阻过程，由于存在系间窜越，$S_0 \rightarrow T_1$ 的吸收常数不是零，而是可达 $10^{-5} \sim 10^{-4}$ cm^{-1}。在含重金属的金属配合物中，由于金属与配体的结合导致了金属-配体电荷转移（MLCT）态，分子内通常产生自旋轨道耦合。如图 3.7 所示，MLCT 的单线态和三线态之间的自旋轨道耦合（^{1}MLCT 和 ^{3}MLCT）可以产生基于光吸收的 ^{3}MLCT 电子，以及 ^{3}MLCT 磷光；^{1}MLCT 与基于配体的三线态（^{3}LC）之间的自旋耦合可以产生光激发的 ^{3}LC 电子，以及该态的磷光。

^{3}MLCT; spin orbital coupling between ^{1}MLCT and ligand-centered triplet state（^{3}LC）may bring the population of ^{3}LC by optical absorption, as well as phosphorescence from ^{3}LC[13]. As a result from ISC and spin orbital couple, in dilute solution of a metal complex, the absorption extinction coefficient of $S_0 \rightarrow T_1$ is in the range of 10^3 $\mathrm{M}^{-1} \cdot \mathrm{cm}^{-1}$ (equivalent to absorbance constant of 10^4 cm^{-1}, which is far larger than that for pure organic materials（10^{-4} cm^{-1}）; at room temperature, the photoluminescence quantum yield（PLQY）of metal complexes can be up to 100%, while phosphorescence from pure organic material is generally negligible.

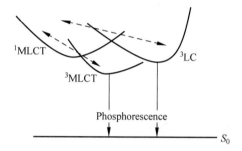

Figure 3.7　The diagram of ISC between ^{1}MLCT and ^{3}MLCT, and between ^{1}MLCT and ^{3}LC in metal complexes, leading to phosphorescence emission

3. Emission

Contrarily to absorption, the process that the excited state molecule decays to its ground state accompanying with radiation is termed as emission, which is also known as luminescence. Generally speaking, emission from organic materials is related to the first excited state and the ground state. This can be explained by the lifetime of the excited states. Although there are many excited states above the ground state, only the first excited states（S_1 and T_1）possess relatively long lifetime（about nanosecond to microsecond, even several millisecond）, which permits the radiative decay process to occur; while the lifetime of the energy states higher than the first excited states is very short（within picosecond）, hence before radiative decay, electrons in these states decay to the first excited state by internal conversion and other processes[14].

As shown in the following Figure 3.8(a), emission can be classified into fluorescence and phosphorescence, according to its origination. Fluorescence refers to the radiative decay from S_1 to S_0, while phosphorescence is the radiative decay from T_1 to S_0; typically, the former is an allowed process and the latter is a forbidden process. Lifetime of phosphorescence ($10^{-6} \sim 20$ s) is generally longer than that of fluorescence ($10^{-9} \sim 10^{-6}$ s). Besides the prompt fluorescence with lifetime at nanosecond scale, delayed fluorescence exhibiting lifetime comparable to phosphorescence has been observed, which is primarily demonstrated in materials with energy level of S_1 almost being double higher than that of T_1, with such, by triplet-triplet annealing (TTA), a singlet of S_1 can be produced to afford fluorescence (Figure 3.8(b)). Because lifetime of this fluorescence is longer than that from direct singlet state, it is called delayed fluorescence (DF), and it is also known as p-type DF; another delayed fluorescence was found afterward, which involved the thermal activation of the triplet state. In an organic molecule, when the energy-level difference between the levels T_1 and S_1 is small, the endothermic RISC (reversed inter system crossing) process of $T_1 \rightarrow S_1$ can take place by the thermal motions of the molecule atoms, causing the population of the singlet state and the afterword fluorescence. This is an extension of DF, and can be termed as thermal activated delayed fluorescence (TADF) and is also known as E-type DF (Figure 3.8(c))[15].

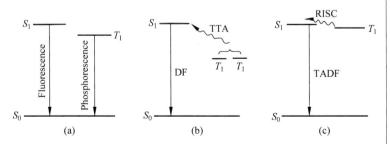

Figure 3.8 Fluorescence and phosphorescence (a), delayed fluorescence (b), thermal activated delayed fluorescence (c)

14. 一般来说,发射与第一激发态和基态有关。这可以用激发态的寿命来解释。虽然基态之上,有许多激发态能级,但只有第一激发态(S_1 和 T_1)具有相对较长的寿命(大约是纳秒到微秒,甚至几毫秒),允许辐射衰变过程发生;而高于第一激发态的能态寿命很短(在皮秒内),因此在辐射衰变之前,这些能态的电子通过内转换和其他过程衰变到第一激发态。

15. 除了纳秒寿命的即时荧光,与磷光寿命相仿的延迟荧光也被观测到,这个最初是在 S_1 能级几乎高于 T_1 两倍的材料中发现的,如此,通过三线态-三线态湮灭(TTA)可以产生一个单线态 S_1,由此产生荧光(图 3.8(b))。由于该荧光的寿命比直接单线态荧光寿命长,所以被称为延迟荧光(DF),也被称为 p-型延迟荧光。后来发现了另外一种延迟荧光,它涉及三线态的热激活过程。在有机材料中,当 T_1 和 S_1 间的能级差较小时,通过分子中原子的热运动就能产生吸热的 RISC 过程(反系间窜跃过程),由此产生单线态电子,以及后续的荧光。这个过程是 DF 过程的扩展,可以称为热激活延迟荧光,也称为 E-型延迟荧光(图 3.8(c))。

3.4 Optical Transition Rules

3.4.1 Quantum Formula

According to quantum mechanics, in principle, all the macroscopic properties of a system can be deduced if the wavefunction of the system is known. Optical transition, which refers to the process of electron transition mediated with optical light, can also be handled by quantum method. The wavefunction for an electron, Ψ, can be obtained by solving the Schrödinger equation:

$$H\Psi = E\Psi \tag{3-8}$$

where H is the Hamilton operator, which is a sum of several energy operator terms of the system. Ψ is the wavefunction, E is the corresponding energy. For electrons in a molecule, it holds,

$$\left[-\sum_A \frac{1}{2M_A} \nabla_A^2 - \sum_P \frac{1}{2} \nabla_P^2 + \sum_{p<q} \frac{1}{r_{pq}} - \right.$$
$$\left. \sum_A \sum_P \frac{ZA}{r_{PA}} + \sum_{A<B} \frac{Z_A Z_B}{R_{AB}} \right] \Psi = E_T \Psi \tag{3-9}$$

Here, the Hamilton operator is presented in bracket, which includes five parts: ① kinetic energy of nucleus, ②kinetic energy of electrons, ③ repulsion energy among electrons, ④attraction energy between electron and nucleus, and ⑤repulsion energy between nuclei[16]. This equation is too complicated to be solved, and an approximation known as Born-Oppenheimer (B-O) approximation was proposed based on the fact that the movement of the electron is three orders magnitude faster than the nucleus, the nuclei may be assumed to remain stationary during the electronic transition. Therefore, some items in the operator can be omitted, and the wavefunction was divided into electron (Ψ) and nucleus (Φ) parts:

$$-\sum_A \frac{1}{2M_A} \nabla_A^2 \Phi + E(R)\Phi = E\Phi \quad \text{(nucleus part)} \tag{3-10}$$

$$-\frac{1}{2} \sum_p \nabla_p^2 \Psi + V(R,r)\Psi = E(R)\varphi \quad \text{(electron part)}$$

$$\tag{3-11}$$

Under B-O approximation, only electronic movement is

16. 这里,括号中的是哈密顿算符,它包括五个部分:①核动能、②电子动能、③电子之间的排斥能、④电子与核之间的吸引能以及⑤核之间的排斥能。

considered and the nuclei are supposed to be motionless. The total wavefunction (Ψ_t) of a molecule can then be approximated by the product of the electronic wavefunction $\Psi_e \Psi_s$, the vibrational wavefunction χ_v and the rotational wavefunction Ψ_r, be expressed as

$$\Psi_t = \Psi_e \Psi_s \chi_v \Psi_r \qquad (3\text{-}12)$$

The electron energy can be expressed as

$$E_t = E_e + E_v + E_r \qquad (3\text{-}13)$$

where the e, v and r represent electronic, vibrational and rotational components, respectively; and the electronic wavefunction ($\Psi_e \Psi_s$) contains two information of two noninteracting terms, the electron part (Ψ_e) and the spin part (Ψ_s). As revealed by both experimental and theoretical studies, the electronic, vibrational and rotational energies are generally of the order of 10^0 eV ($\sim 10^4$ cm^{-1}), 10^{-1} eV (~ 1000 cm^{-1}, the near infrared region), and 10^{-2} eV (far infrared region, ~ 10 cm^{-1}), respectively. In spectra of absorption and emission, vibrational levels some time are observable as fine structure, while the rotational energies only present as spectrum broadening and cannot be recognized[17]. Ignored the rotational wavefunction Ψ_r, the total wavefunction of a molecule can then be approximated as,

$$\Psi_t = \Psi_e \Psi_s \chi_v \qquad (3\text{-}14)$$

The energy of a given electronic state is

$$E_t = E_e + E_v \qquad (3\text{-}15)$$

Having the wavefunction of an electron in hand, we now try to describe the process of optical transition by quantum manner. As schematically shown in Figure 3.9, upon light absorption, the occurred process is electron wavefunction changes from Ψ to Ψ^*, meaning that the probability of finding this electron is changed and the electron energy changes from E to E^*.

According to Equation (3-15), the electron energy includes two parts: electronic part (E_e) and vibrational part (E_v). For a given energy level, vibrational levels can be expressed as

$$E_v^* = \frac{1}{2}(2m+1)E_v \qquad (3\text{-}16)$$

17. 实验和理论都揭示了电子、振动和转动能量的大小,通常分别在 10^0 eV($\sim 10^4$ cm^{-1})、 10^{-1} eV(~ 1000 cm^{-1},近红外区)、以及 10^{-2} eV(~ 10 cm^{-1},远红外区)。在光吸收和发射光谱中,振动能级有时可被观察到,表现为精细结构;而转动能量则仅表现为光谱扩展,不能被识别。

where E_v is the zero energy of vibration, m is the vibrational quantum number. The degree of vibrational energy is mainly determined by temperature, and according to Boltzmann statistics, the probability (f) of finding an electron in an energy level is

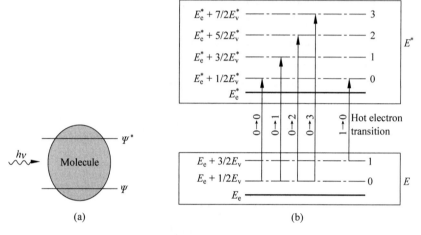

Figure 3.9 Optical transition from Ψ to Ψ^* (a) with electron-energy change from E to E^* (b)

$$f = \exp\left(-\frac{E}{kT}\right) \tag{3-17}$$

where, k is Boltzmann constant, T is absolute temperature, and E is energy. Thus, the electron ratio of zero vibrational level ($m = 0$) to nonzero vibrational level (m) is

$$r = \frac{f_m}{f_0} = \frac{\exp(-(2m+1)E_v/2kT)}{\exp(-E_v/2kT)} = \exp(-mE_v/kT)$$

$$\tag{3-18}$$

18. 假设 $E_v \approx 1000$ cm$^{-1} \approx$ 1/8 eV，且当 $m = 1$，$f_1/f_0 \approx$ 6.7×10^{-3}。这意味着在一个能级，电子存在于 $m = 1$ 振动能级的概率是在零振动能级的 6.7×10^{-3}，可忽略不计。因此，可以粗略地认为电子位于零振动能级，光吸收主要来自 $0 \rightarrow n$ 跃迁。

Suppose $E_v \approx 1000$ cm$^{-1} \approx 1/8$ eV, when $m = 1$, $f_1/f_0 \approx 6.7 \times 10^{-3}$. This means that in an energy level, the probability for electron locating at vibrational level of $m = 1$ is 6.7×10^{-3} times smaller than zero vibrational level, which is negligible. Thus, electrons can be roughly considered locating at zero vibrational level, and optical absorption mainly originates from $0 \rightarrow n$ transitions[18]. In some cases, hot electron transition, which mean nonzero vibrational transition, also exists, e.g., the $1 \rightarrow 0$ transition in Figure 3.9(b).

3.4.2 Selection Rules

Optical absorption involves the interaction between electromagnetic field of light and the static/dynamic electric field of molecules. Since the influence of magnetic field from the light electromagnetic field is very small, it can be neglected. Therefore, the probability of exciting a molecule (R_{lu}^2) is mainly determined by the transition dipole moment of the two states[19]:

$$R_{lu}^2 \propto |\langle \Psi_u | \boldsymbol{M} | \Psi_l \rangle|^2 \qquad (3\text{-}19)$$

where the subscripts refer to the initial (l for lower) and final (u for upper) states, $\boldsymbol{M}$ is the dipole moment vector. Neglecting rotation and under Born-Oppenheimer approximation, one get

$$R_{lu}^2 \propto |\langle \Psi_{el} | \boldsymbol{M} | \Psi_{eu} \rangle|^2 |\langle \chi_{el} | \chi_{eu} \rangle|^2 |\langle \Psi_{sl} | \Psi_{su} \rangle|^2$$
$$(3\text{-}20)$$

Here Ψ_{el} and Ψ_{eu} denote the initial and final electronic-state functions. Equation (3-20) includes three items: ① the electronic component, the dipole interaction between the initial and final states, which is also called transition dipole moment; ② integral of the vibrational overlapping component between the two states; ③ spin component. The dipole moment operator $\boldsymbol{M}$ appears only in the electronic term, this is because that the vibrational process relates to the vibration of nuclei, which cannot respond rapidly enough to optical frequency, and hence has no dipole change; while the spin is insensitive to the electric field. An optical transition requires that any of the three components is not zero. If any of the components is zero, the transition probability will vanish, and this would be called a forbidden transition[20].

1. Electronic Factor

In electronic term, the integral $\langle \Psi_{el} | \boldsymbol{M} | \Psi_{eu} \rangle$ correspond to the dipole transition moment ($\boldsymbol{T}_{lu}$). For allowed transition, or dipole-allowed transition, $\boldsymbol{T}_{lu}$ is not zero:

$$\boldsymbol{T}_{lu} = \langle \Psi_{el} | \boldsymbol{M} | \Psi_{eu} \rangle \neq 0 \qquad (3\text{-}21)$$

When $\boldsymbol{T}_{lu} = 0$, the transition from el to eu is dipole-forbidden. First, the dipole of a molecule is closely related to

19. 由于基于光的电磁场的磁场影响很小,可以被忽略。因此,激发一个分子的概率(R_{lu}^2)主要取决于两个态之间的跃迁偶极矩:

20. 式(3-20)包括三项:①电子组分,即初始状态和最终状态波函数电子组分之间的偶极相互作用,也称为跃迁偶极矩;②两种状态之间的振动组分重叠积分;③自旋组分。跃迁偶极矩算符 $\boldsymbol{M}$ 只出现在电子项,这是因为原子核不能迅速响应光学频率,与其相关的振动过程就没有偶极矩变化;而电子自旋对电场是不敏感的。光学跃迁要求这三个组分中的任何一个都不为零。如果任何一个组分为零,那么跃迁概率将消失,这种就被称为禁阻的跃迁。

21. 首先，一个分子的偶极与分子的对称性密切相关。偶极允许的跃迁发生在具有不同对称性的两个电子态之间。

22. 第二，初态和终态两个波函数的重叠积分值要求足够大。发生在轨道中心位于相同分子位置的两个轨道之间的跃迁概率比较大（如 π-π^*）；而轨道中心位于不同空间位置的跃迁概率较小（如电荷转移跃迁、n-π^* 跃迁或者金属到配体的电荷转移跃迁）。

23. 在这些对称选择定则的应用中，要知道分子振动运动会降低分子的对称性，因此对称性禁阻的跃迁也会被观察到。

24. 决定跃迁大小的第二项是自旋波函数，即初态和终态之间自旋波函数的积分值 $\langle \Psi_{sl} \mid \Psi_{su} \rangle$。这个积分值只有两种，要么是 0，要么是 1，分别对应着跃迁前后电子的自旋改变和电子的自旋保持不变两种情况。因此具有相同多重态的轨道间的跃迁是自旋允许的跃迁，如 $S_0 \rightarrow S_1$，$T_1 \rightarrow T_n$ 是允许的；而在不同多重态间的跃迁是自旋禁阻的，如 $S_0 \rightarrow T_1$，$T_1 \rightarrow S_0$ 是自旋禁阻的。

its symmetry. Dipole-allowed transition occurs between two electronic states with different symmetry[21]. The ground state wavefunction of most molecules transforms in the same manner as the identity representation A_g, being an even parity. If the excited state wavefunction is also of an even parity, the integral (T_{lu}) would vanish. It will yield finite values only if the excited state wavefunction has an odd parity, that is, the symmetry of the initial and final states should be different for optical transition. Second, the integral value of the overlap between the initial and final state wavefunctions is required large enough. Transition rates between orbitals centered on the same parts of the molecules (e.g., π-π^* transitions) will be larger than these occupy different spaces (e.g., charge-transfer transitions, n-π^* transitions or metal-to-ligand charge-transfer transition)[22]. Similarly, the integral will scale with the degree of electron delocalization, that is, it will be large if the involved orbitals are not only well overlapping but also well extended (i.e., the transition dipole moment is along the molecular axis). For example, the absorption and fluorescence intensity from conjugated molecules is well known to increase with conjugation length. In applications of these symmetry selection rules it needs to remember that the symmetry of a molecule can be lowered by vibrational motions so that symmetry-forbidden transitions may nevertheless be observed[23].

2. Spin Factor

The second term that determines the transition probability is the spin wavefunction, that is, the value of the integral $\langle \Psi_{sl} \mid \Psi_{su} \rangle$. This integral takes only two values, which is 0 if the spins of initial and final states differ, and 1 if they are equal. Thus, transitions between two states with same multiplicity, such as $S_0 \rightarrow S_1$, $T_1 \rightarrow T_n$, are spin-allowed, yet transitions between two states with different multiplicity, such as $S_0 \rightarrow T_1$, $T_1 \rightarrow S_0$, are spin-forbidden[24]. In fact, transition between two states with different multiplicity exists, with rate $10^3 \sim 10^5$ times lower compared to transition between two states with the same multiplicity, due to the existing of spin-orbital coupling. Heavy metal can enhance spin-orbital coupling, faciliting transition between two states with different multiplicity. As discussed previously, if spin-orbital coupling is introduced, the small amount of singlet character in the essentially triplet excited state leads to intensity integral for the singlet-triplet transition nonzero.

3. Vibrational Factor: Franck-Condon Factor

Both the electronic factor and spin factor are the electronic part of the wavefunction. While the electronic part controls the overall intensity of the transition, there is the third parameter that influences the optical transition process, i.e., the integral of the vibrational wavefunction overlap, $\langle \chi_{el} | \chi_{eu} \rangle$, which can be said to control the spectral shape of the absorption and emission. The squared overlap integrals of the vibrational wave functions $|\langle \chi_{el} | \chi_{eu} \rangle|^2$ are referred as vibrational factor or Franck-Condon factor[25]. For allowed transition, it requires:

$$| \langle \chi_{lm} | \chi_{un} \rangle |^2 \neq 0 \qquad (3\text{-}22)$$

The above expression, namely Franck-Condon factor, indicates the overlap of the two vibrational wavefunctions (e.g. m and n). This value is related to the equilibrium position of the two vibrational states. The larger the value, the more the overlap, and the stronger the transition.

An electronic transition is so fast, with a typical frequency of 3×10^{15} s^{-1}, compared to nuclear motion with a typical frequency of 3×10^{13} s^{-1}, that the nuclei still have nearly the same position and momentum immediately after the transition. Hence when electron transition occurs, molecular conformation does not change, and such electron transition is called vertical or Franck-Condon transition.

As shown in Figure 3.10, the two curves are the potential surface versus molecular conformation characteristics. The higher curve possesses larger energy and represents excited state (S_1), while the lower curve is the ground state with lower energy (S_0). The points on the curve represent the conditions with minimum nucleus-momentum at certain vibrational level, and are called turning points. Due

25. 电子因子和自旋因子都属于波函数的电子项。虽然电子部分决定了跃迁的总体强度,但是还有影响光学跃迁的第三个因素,即初态和终态振动波函数的重叠积分 $\langle \chi_{el} | \chi_{eu} \rangle$,可以说振动项控制了吸收和发射的光谱形状。初态和终态振动波函数的重叠积分的平方 $|\langle \chi_{el} | \chi_{eu} \rangle|^2$ 称为振动因子,也称为 Franck-Condon 因子。

26. 如图 3.10 所示，这两条曲线是势能面与分子构象之间的关系曲线。较高的曲线能量较大，代表激发态（S_1），而较低的曲线是具有较低能量的基态（S_0）。曲线上的点表示在某一振动水平下具有最小核动量的条件，称为拐点。由于基态和激发态之间的能量和构象差异，两种状态的平衡位置是不同的。在拐点处电子跃迁的概率最大。比如，$a_0 c_0$ 振动能级的电子跃迁到拐点（如 a_0^*、a_1^* 和 a_2^*）的概率比跃迁到非拐点（如 a_0^* 和 c_0^* 之间的位置）的概率要大。

27. 在吸收过程的垂直跃迁之后，为了更加稳定，激子发生辐射衰变之前，存在基于振动的热损失（给出声子），这导致了激发态的降低和发射光谱的红移。如图 3.10 所示，吸收能量 E_a 大于发射能量 E_b。

28. 上述提到的吸收和发射之间的能量差异称为 Franck-Condon 移位。通常，光能的退降称为斯托克斯位移。因此，Franck-Condon 移位是由分子位移变化引起的一种斯托克斯位移。

to the energy and conformation differences between the ground state and the excited state, the equilibrium positions for the two states are different. At the turning point, the probability for electron transition is the maximum. For example, the probability for electrons at $a_0 c_0$ vibrational level jumping to turning point (e. g. a_0^*, a_1^*, and a_2^*) is larger than that jumping to non-turning point (e. g. position between a_0^* and c_0^*)[26].

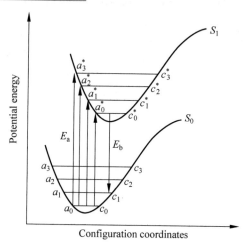

Figure 3.10　Illustration of Franck-Condon principle

After vertical transition in absorption process, for stabilization reason before the excited state subjects to radiative decay, there is a thermal loss via vibration (phonon is given out), leading to lowered excited state and hence red-shifted emission spectrum As shown in Figure 3.10, the absorption energy E_a is larger than emission energy E_b[27]. This process is also demonstrated by Kasha rule (Figure 3.11), which states that upon absorbing a photon, a molecule in its ground state may be excited to any of a set of higher electronic states; however, the following photon emission is expected in appreciable field only from the lowest excited state. The energy difference between absorption and emission mentioned above is called Franck-Condon shift. Generally, the reduction of optical energy is called Stokes shift. Thus Franck-Condon shift is a kind of Stokes shift caused by molecular confirmation shift[28].

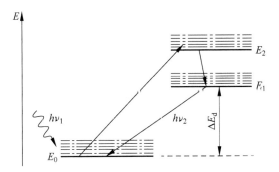

Figure 3.11　Kasha rule

The existence of vibrational energy levels leads to various electronic transitions between the ground state S_0 and the excited state S_1, which may form the fine structure in the absorption and emission spectra. On the other hand, according to Franck-Condon principle, if the environmental influence to the two states is small, the absorption and the emission spectra will be mirror symmetry (Figure 3.12)[29].

29. 振动能级的存在,导致基态 S_0 和激发态 S_1 之间的很多电子跃迁,这可能形成吸收和发射光谱的精细结构。另外,根据 Franck-Condon 原理,如果环境对这两个态的影响较小,吸收和发射光谱是镜像对称的(图 3.12)。

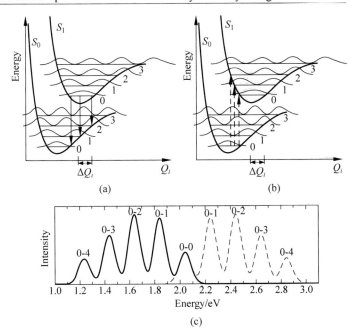

Figure 3.12　Schematic illustration of the fine structure and mirror symmetry between absorption and emission spectra: (a) Emission transition, (b) absorption transition and (c) spectra for emission (solid line) and absorption (dotted line) spectra

30. 镜像对称的主要原因是,它们涉及同一对能级之间相似的垂直跃迁过程,吸收是向上跃迁,发射是向下跃迁。同时,基态与激发态不同振动能级之间的跃迁概率也相似,不必考虑向上或者向下的方向。

The main reason for the mirror symmetry is that they involve similar vertical transition processes between the same pair of energy levels, with the absorption up and the emission down. Meanwhile, the transition probability between the vibrational levels of ground state and the excited state are also similar, disregarding its up- or down- direction[30]. Generally, for a stiff molecule with rigid bonds and correspondingly small displacement between the excited and ground state potential energy curve, the 0-0 transition is dominant, and both absorption and emission spectra exhibit fine structure and they are mirror symmetry. For a flexible molecule, a higher vibrational level such as 0-2 or 0-3 may form the peak of the distribution of transitions, and the symmetry and fine structure features in the absorption and emission spectra may be lost. The fine structure in absorption and emission bands is easily affected by the medium surrounding the molecule. Such as, in amorphous films or solutions, fine structure may be lost, just broad bands. Particularly, due the more active nature of the excited molecule, the fine structure in emission spectrum is generally less obvious than that in absorption spectrum[31].

31. 吸收和发射带中的精细结构很容易受到分子周围环境的影响。正如在无定形薄膜或者溶液中,精细结构可能丢失,仅仅是宽的谱带。尤其地,由于激发态比较活泼的本质,发射光谱中的精细结构通常没有吸收光谱中的明显。

3.4.3 Einstein Equations

In two energy levels, the light intermediated processes include absorption, spontaneous emission and stimulated emission, this step occurs when a photon of energy ($E_{un} - E_{lm}$) interacts with the electron in the level un. As shown in Figure 3.13, the subscripts lm and un represent the mth vibrational level of the lth electronic state and the nth vibrational level of the uth electronic state, respectively. Here, $B_{lm,un}$, $A_{un,lm}$, $B_{un,lm}$ are used to represent the corresponding

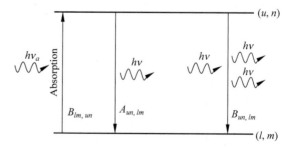

Figure 3.13　Optical transition processes between two energy levels

rate constant of absorption, spontaneous emission and stimulated emission, respectively.

Einstein discovered that under certain radiation field, the rate constants of the above three processes can be expressed as Einstein equation:

$$B_{un,lm} = B_{lm,un} \qquad (3\text{-}23)$$

$$\frac{A_{un,lm}}{B_{un,lm}} = \frac{8\pi h\nu^3 n_0^3}{c^3} \qquad (3\text{-}24)$$

where ν is transition frequency, h is Planck constant, n_0 is refraction index, c is light velocity. The Einstein coefficient B is directly related to experimental absorption parameters. If the irradiation density is constant over the vibronic bandwidth, for a transition $l0$-un, the Einstein coefficient B is also expressed as[32]

$$B_{l0,un} = \frac{c}{hn_0} \int \frac{\sigma(\nu)\,\mathrm{d}\nu}{\nu} \qquad (3\text{-}25)$$

32. 爱因斯坦系数 B 与实验的吸收参数直接相关。如果辐射密度在整个振动波长范围是常数，对于 $l0$ 态到 un 态的跃迁，爱因斯坦系数 B 可以表示为。

where $\sigma(\nu)$ is the cross section for capture of a photon of frequency ν by a molecule in the ground state. σ can also be expressed as

$$\sigma = \frac{2303\varepsilon}{N} = 3.81 \times 10^{-21}\varepsilon \qquad (3\text{-}26)$$

where N is the Avogadro's number, ε is the molar extinction coefficient. Then, the absorption rate constant

$$B_{l0,un} = \frac{2303c}{hn_0 N} \int \frac{E(\nu)\,\mathrm{d}\nu}{\nu} \qquad (3\text{-}27)$$

The molar extinction coefficient ε can be measured, and then B_{lu} can be calculated according to Equation (3-27), and the rate constant of spontaneous emission A_{ul} can be calculated according to Equation (3-24).

The lifetime for emission transition τ_R is defined as the reciprocal of the rate constant of radiative transition. So, it is possible to estimate the radiative lifetime of the excited state from the absorption spectrum[33].

33. 辐射跃迁的寿命 τ_R 定义为辐射跃迁速率常数的倒数。因此，可以根据吸收光谱来估算激发态的辐射寿命。

$$\tau_R^{-1} = \sum_n A_{u0,ln} = \frac{8\pi hc}{n_0}\langle \tilde{\nu}^{-3}\rangle^{-1} B_{lu} \qquad (3\text{-}28)$$

$$\langle \tilde{\nu}^{-3}\rangle^{-1} = \frac{\int I(\tilde{\nu})\,\mathrm{d}\tilde{\nu}}{\int \tilde{\nu}^{-3} I(\tilde{\nu})\,\mathrm{d}\tilde{\nu}} \qquad (3\text{-}29)$$

Here, $I(\tilde{\nu})$ is the fluorescence intensity in units of relative quanta per unit frequency interval at the wavenumber $\tilde{v}$. Lifetime for emission transition can be calculated according to Equation (3-28). Table 3.1 presents lifetime comparison of several materials between calculation and experiment. It is found that the calculated and experimental values are astonishingly similar.

Table 3.1　Lifetime for emission transition

Mater.	Cal/ ($\tau \times 10^9$ s)	Expt/ ($\tau \times 10^9$ s)
Perylene	4.29	4.79
Acridine	12.06	11.80
Fluorescein	4.37	4.02
Rhodamine B	6.01	6.15

3.4.4　Intensity Distribution of Absorption Spectrum

According to thermaldynamic statistics, most of the ground state molecules are in the zero vibrational state. Hence, absorption spectrum generally exhibits transitions of 0-0, 0-1, 0-3, etc. Due to the different transition probability, there will be great differences in the intensities of different vibrational bands. It is known from quantum theory that the transition probability for a given change of state is proportional to the square of the transition dipole moment and to an appropriate density-of-state factor. Transition probability partially determines the intensity of absorption band. In addition, the transition intensity is also proportional to the rate of transition[34]. The Einstein coefficient B for a given vibrational transition can be related to the transition moment by the expression:

$$B_{lm,un} = \frac{8\pi^3}{3h^2} \mid T_{lm,un} \mid^2 \tag{3-30}$$

In conventional terminology, the intensity of a given vibronic absorption transition is measured by its oscillator strength (f), which is defined as[35]

$$f = \frac{4.39 \times 10^{-9}}{n_0} \int \varepsilon(\tilde{v}) \, \mathrm{d}\tilde{v} \tag{3-31}$$

where n_0 is the refraction index of the media. Combining Equations (3-30) and (3-31), one has:

$$f = 4.70 \times 10^{29} \langle \tilde{v} \rangle \mid T_{lm,un} \mid^2 \tag{3-32}$$

34. 由于不同的跃迁概率,不同振动能带的强度将存在很大差异。从量子理论知道,给定两个态之间的跃迁概率与跃迁偶极矩的平方成正比,并且与其态密度因子成比例。跃迁概率部分地决定了吸收带的强度。另外,跃迁强度也和跃迁速率成正比。

35. 在经典术语中,特定振动吸收跃迁的强度可以通过其振子强度(f)来测量,定义为

where $\langle \tilde{v} \rangle$ is the mean wavenumber of the particular transition (in cm^{-1}). Adopting the B-O approximation, this equation can be written as a product of a pure electronic term and a vibronic overlap factor:

$$f_{lm,un} = f_{lu}\left(\frac{\langle \tilde{v}_{lm,un} \rangle}{\tilde{v}_{l0,u0}}\right) | \langle \chi_{lm} | \chi_{un} \rangle |^2 \qquad (3\text{-}33)$$

where f_{lu} is the oscillator strength for the 0-0 transition in the absence of vibrational states. For a displaced oscillator, in which the electronic transition causes a change ΔQ in the equilibrium position, the progressional intensity of the bands for the transitions $l0 \rightarrow un$ is

$$f_{l0,un} = f_{lu}\left(\frac{\langle \tilde{v}_{lm,un} \rangle}{\langle \tilde{v}_{l0,u0} \rangle}\right)\frac{Z^n}{n!}e^{-Z} \qquad (3\text{-}34)$$

$$Z = \frac{1}{2}k\frac{(\Delta Q)^2}{\eta\omega} \qquad (3\text{-}35)$$

where Z is the Huang-Rhys-parameter, k is the force constant, ω is angular frequency. When the two electronic states l and u have similar configurations near their respective equilibrium positions and the shift in the position of the potential energy $\Delta Q \approx 0$, only the 0-0 transition will be dominant. As the ΔQ increases, the position of the peak in the vibronic spectrum will be shifted to higher vibrational states (Figure 3.14)[36].

36. 当两个能级 l 与 u 的平衡位置附近的构象类似时,势能位置的移动 $\Delta Q \approx 0$,吸收光谱中将主要只有一个 0-0 跃迁。当 ΔQ 增大,振动光谱峰位将移至高振动态(图 3.14)。

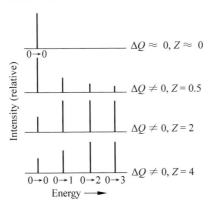

Figure 3.14   Intensity distribution of absorption peak as a function of ΔQ

3.4.5 Photoluminescence Quantum Yield

Generally, electrons possessing energy higher than the first excited state electron almost decay to the first excited

37. 通常地,具有比第一激发态电子能量高的电子,几乎都通过非辐射的内转换、系间窜越、等过程衰变到第一激发态。单线态可以通过系间窜跃以非辐射路径到达三线态,然后该激发态电子自旋翻转。第一激发单线态或者三线态,可以通过辐射或非辐射过程衰变到基态,如图 3.15 所示。因此,发光过程总是和非辐射过程竞争的,非辐射过程会影响发光效率。

state via radiationless decay processes of internal convention, inter system crossing, etc. A singlet excited state can nonradiatively pass to a triplet state by intersystem crossing, and then the spin of the excited electron is reversed. The first excited singlet or triplet state can decay to the ground state via radiation or radiationless process, as shown in Figure 3.15. Therefore, the emission process always competes with the nonradiation process; and the nonradiation process will influence the emission efficiency[37].

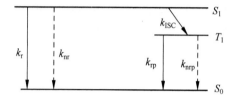

Figure 3.15   Decay processes of the first excited singlet or triplet state to the ground state

A measurement for emission efficiency is photoluminescence quantum yield (PLQY), which is defined as the ratio of the emission photons to the absorbed photons. Suppose that the population of the excited state is N, which is presumably originated from photon absorption. N_r and N_{nr} indicate the part of radiative and nonradiative process, respectively. According to Einstein equation, the decay rate (dN/dt) of the excited state is proportional to the population of that state (N):

$$\frac{dN}{dt} = -(k_r + k_{nr} + k_{ISC})N \qquad (3-36)$$

The equation for photoluminescence quantum yield can be expressed as

$$PLQY = \Phi_0 \frac{N_r}{N} = \Phi_0 \frac{K_r N}{(K_r + K_{nr} + K_{ISC})N}$$
$$= \Phi_0 \frac{K_r}{K_r + K_{nr} + K_{ISC}} \qquad (3-37)$$

where Φ_0 is the efficiency of forming that state initially. For optical excited singlet state, due to the singlet nature of the ground state and the requirement of spin conservation, the formation efficiency of excited singlet state can be up to 1. Meanwhile move the ISC process into the nonradiative processes (k_{nr}), Equation (3-37) becomes[38]

$$\text{PLQY}_F = \frac{k_r}{k_r + k_{nr}} \tag{3-38}$$

For photoexcited triplet emission, Φ_0 need to be reconsidered. Since at optical excitation, generally speaking, only singlet excited state is produced due to spin conservation rule, and the triplet excited state is obtained via ISC process with efficiency of (η_{ISC}), $\Phi_0 = \eta_{ISC}$. The photoluminescence quantum yield for phosphorescence (PLQY_{ph}) is[39]

$$\text{PLQY}_{ph} = \eta_{ISC} \frac{k_r}{k_r + k_{nr}} \tag{3-39}$$

An interesting phenomenon is that, as mentioned at the very beginning of this textbook, in general, heterocyclic aromatic compounds possess lone pair electrons, which occupy frontier orbitals (termed as n-orbital). In these materials, after optical excitation of these n electrons from n-orbital, emission occurs from the lowest excited state to n-orbital. Since n-orbital related optical transition is generally forbidden, both the absorption and emission efficiencies are expected low in heterocyclic aromatic compounds compared to pure aromatic hydrocarbons.

Excited state lifetime (τ) is the reciprocal of decay rate

$$\tau = \frac{1}{K_r + K_{nr}} \tag{3-40}$$

where K_r and K_{nr} are the radiative and nonradiative decay-rates, respectively. Thus the photoluminescence quantum yield can also be written as

$$\text{PLQY}_F = \frac{K_r}{K_r + K_{nr}} = \frac{\tau}{\tau_r} \tag{3-41}$$

$$\text{PLQY}_{ph} = \eta_{ISC} \frac{K_r}{K_r + K_{nr}} = \eta_{ISC} \frac{\tau}{\tau_r} \tag{3-42}$$

where, $\tau_r = 1/k_r$, is the natural radiative lifetime, which occurs without nonradiative decay process. Only when PLQY_F is equal to 1 or $\text{PLQY}_{ph} = \eta_{ISC}$, natural radiative lifetime and observed radiative lifetime are the same. Generally, photoluminescence quantum yield is smaller than 1[40].

38. 对于光激发的单线态激子,由于基态的单线态特性及自旋守恒的要求,其形成的效率可以达到 1。同时,将 ISC 过程并入非辐射过程 (k_{nr}) 中,式(3-37)变为

39. 对于光激发的三重态发射,Φ_0 需要重新考虑。由于在光激发下,一般而言,由于自旋守恒规则,仅产生单线态,三线态是通过 ISC 过程获得的,其效率为 η_{ISC},$\Phi_0 = \eta_{ISC}$。磷光的光致发光量子产率(PLQY_{ph})为

40. 其中,$\tau_r = 1/k_r$,是自然辐射寿命,它在没有非辐射衰减过程的情况下发生。只有荧光量子产率等于 1,或磷光量子产率等于系间蹿越效率时,自然辐射寿命等于表观辐射寿命。通常来说,光致发光量子产率小于 1。

3.5 Excitonic Interactions in Organic Aggregates

When isolated molecules aggregate to form solid, the optical spectra are very similar to those of the molecules, due to the weak van de Waals force between molecules. However, one observes the following most important differences[41]:

(1) A general shift in absorption and emission spectra, mostly to lower energies, i. e., solvent shift. This solvent shift is based on the interaction of the molecules within the crystal with their neighbouring molecules, i. e. on the van der Waals bonding. It is molecule specific and has a different magnitude for different excited states.

(2) A splitting of energy levels, which is called Davydov splitting and is due to the resonant interaction between nontranslationally equivalent molecules in excited states. The Davydov splitting is also different in magnitude for different excited states and transitions. The Davydov components of the spectra also differ in their polarisation[42].

(3) A broadening of absorption and emission spectra from discrete line to band due to the delocalization of excitation energy over all the molecules in an ideal crystal.

(4) A complete or partial lifting of degeneracies between excited molecular states and the breaking of selection rules which hold for transitions in the free molecules; because now not just the symmetry of the molecules, but that of the whole crystal is the determining factor. For example, the 0-0 transition of 48000 cm^{-1} is forbidden in the benzene molecule by reasons of symmetry; in benzene crystals, however, it is allowed[43].

(5) In transitions with very high oscillator strength, the shifts and Davydov splittings can become so large that the similarity to the spectrum of the free molecule vanishes.

In the followings, we shall discuss the main features and the decay processes of excited states in organic aggregates.

41. 当孤立分子聚集形成固体时,其光学光谱与分子的非常类似,这是由于分子与分子之间是非常弱的范德瓦耳斯力。但是,可以观测到下述最重要的差别:

42. 由于非平移等价的激发态分子间的共振作用而产生的称为达维多夫裂分的能级裂分。不同激发态和跃迁的达维多夫裂分在数值上也是不同的。光谱的达维多夫组分的极性也是不同的。

43. 激发态分子简并态的部分或者完全消除,以及打破原本在自由分子中成立的选择定制。因为现在不仅仅是分子的对称性,而是整个晶体的对称性是决定性因素。例如,在苯分子中,对称性禁阻的波数为 48000 cm^{-1} 的 0-0 跃迁,在苯晶体中是允许的。

3.5.1 Davydov Splitting

1. Phenomena

Figure 3.16 shows the absorption spectra of anthracene in an ethanolic solution and in an aggregate form of single crystal at 90 K. In solution, well discernable vibrational peaks can be seen, which can be assigned to 0-0, 0-1, 0-2, 0-3, 0-4 optical transitions of S_0 to S_1 level. Compared to solution, the absorption spectra of anthracene crystals are redshift due to polarization energy and solvent shift. For crystal, in the two absorption cases of irradiation parallel to the a-axis and b-axis, there are also differences, which are caused by Davydoy splitting[44].

44. 在溶液中,可以非常清晰地看到来自于 S_0 到 S_1 的 0-0、0-1、0-2、0-3、0-4 光学跃迁的振动峰。与溶液相比,由于极化能和溶剂位移的缘故,晶体蒽的吸收光谱发生了红移。在晶体中,平行于 a 轴或 b 轴入射蒽晶体所获得的吸收光谱也存在差异,这是由于达维多夫分裂造成的。

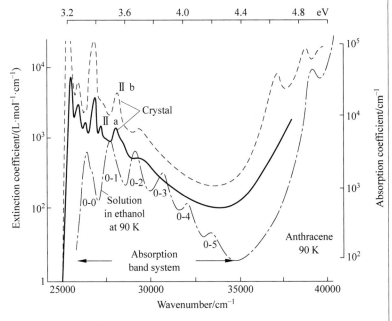

Figure 3.16 Absorption spectra of anthracene crystals and solution, where the absorption spectra of anthracene crystals are obtained by excitation of polarized light parallel to the a-axis or b-axis

Triplet excited state also subjects to Davydov splitting. However, as discussed in previous sections, due to the spin conservation rule for optical transition, triplet absorption of a pure organic material is very weak and difficult to be detected. One effective way to probe triplet levels is

45. 三线态激子也受达维多夫裂分支配。但是，如前面部分讨论的，由于光学跃迁自旋守恒的缘故，纯有机材料的三线态吸收很弱，难以被检测到。延迟荧光的激发光谱是一种探测三线态的有效方法，延迟荧光通常来自于三线态-三线态聚变，因此其激发光谱反映的是三线态电子。

excitation spectrum of delayed fluorescence, which generally originates from the fusion of triplet-triplet, and hence signifying the population of the triplet state[45]. Figure 3.17 shows the excitation spectra of the delayed fluorescence from anthracene crystal with polarized incident light. The Davydov splitting of anthracene first triplet is 21.5 cm^{-1}.

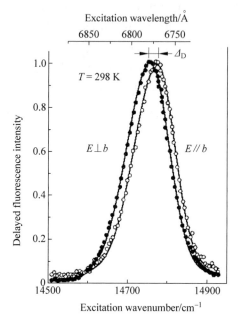

Figure 3.17　The excitation spectrum for the delayed fluorescence from anthracene crystal with polarized incident light

Why there exists Davydov splitting? We will explain it in three cases：①in physical dimer，②in one dimension crystal，and ③in crystal with two molecules in unit cells.

2. Mini Exciton Model

The mini exciton model is based on a physical dimer. When we say physical dimer, it describes the situation in which two identical molecules are closer to each other in a special spatial arrangement than to other like molecules but do not form a chemical bond between themselves. There is also concept of dimer, which means a chemical union between two identical molecules; and new bonds are created that did not exist in either of the original molecules. Figure 3.18 shows a physical dimer (a) and a dimer (b). Another related term is excimer. An excimer is a physical dimer composed of one electronically excited molecule paired with a ground state of the same molecule[46]. Excimer will be discussed later.

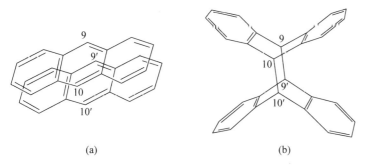

9

9'

10

10'

(a)

9

10

9'

10'

(b)

Figure 3.18 A physical dimer (a) and a dimer (b)

In the case of a physical dimer, there are a pair of equivalent molecules. We denote the ground-state wavefunction and energy of the two individual molecules by φ_1 and φ_2, and $E_1 = E_2 = E_0$. Similarly, those for the excited state are φ_1^* and φ_2^*, and $E_1^* = E_2^* = E^*$. Neglecting vibrational and spin wavefunctions for simplicity, and considering that intermolecular electron overlap is small, and therefore, the molecular units preserve their individuality, the dimer may be considered a two-particle system that is described by a Hamiltonian[47]

$$H = H_1 + H_2 + V_{12} \tag{3-43}$$

with H_1 and H_2 being the operators of the isolated molecule and with V_{12} being an intermolecular perturbation potential. This interaction potential is a Coulomb potential that may be approximated by the point-dipole terms of the multipole expansion, in which case it takes the form of Equation (2-2). The overall wavefunction of the ground state of the two-molecule system (Φ_G and E_G) can be approximated as the product of the wavefunctions of both molecules[48]:

$$\Phi_G = \varphi_1 \varphi_2 \tag{3-44}$$

With Equations (3-43) and (3-44), the ground state energy of the two-molecule system is then obtained by solving the Schrödinger equation as

$$E_G = \langle \varphi_1 \varphi_2 | H_1 + H_2 + V_{12} | \varphi_1 \varphi_2 \rangle = E_1 + E_2 + D_0 \tag{3-45}$$

The last term

46. 当我们谈及物理二聚体时，描述的是这样一种情形：两个靠近的相同分子之间的空间位置与其他类似分子之间的空间位置不同，二者之间却并没有形成化学键。还有一个二聚体的概念，指的是两个相同分子之间存在化学结合；形成了原来分子中不存在的、新的化学键。图3.18展示了一个物理二聚体(a)和一个二聚体(b)。另外一个相关的术语是激基二聚物。激基二聚物指的是一个激发态分子与一个相同分子的基态分子配对的物理二聚物。

47. 对于物理二聚体，存在两个相同的分子。我们用 φ_1 和 φ_2 以及 $E_1 = E_2 = E_0$ 分别表示两个分子的基态波函数和能量。类似地，激发态为 φ_1^*、φ_2^*，以及 $E_1^* = E_2^* = E^*$。为了简便，忽略振动波函数和自旋波函数，并考虑到分子间电子重叠很小，因此分子单元保留了各自的独立性，二聚体可以被认为是一个由哈密顿量描述的双粒子体系。

48. 双分子体系基态的总波函数(Φ_G 和 E_G)可以近似为两个分子的波函数的乘积：

$$D_0 = \langle \varphi_1 \varphi_2 | V_{12} | \varphi_1 \varphi_2 \rangle \qquad (3\text{-}46)$$

is negative and corresponds to the van der Waals interaction energy, that is, the polarization energy D_0 that lowers the ground state energy of the system of the two molecules compared to the ground state energy of the individual molecules. It ensures that an ensemble of molecules will condense to form a liquid or solid and it is caused by the zero-point oscillations of the molecules that induce dipoles in the environment. According to Equation (2-2), the magnitude of D_0 depends on intermolecular distances and orientations.

The excited state of a physical dimer means one excited state monomer interacts with the other ground state monomer. Actually, it is an excimer, and can be expressed as $\varphi_1^* \varphi_2$ and $\varphi_1 \varphi_2^*$. Assuming that the probability of a molecule being excited is equally great for either of the two molecules, and by the linear combination, the excited physical dimer can be finally expressed as[49]

$$\Phi_{\pm}^* = \frac{1}{\sqrt{2}} (\varphi_1 \varphi_2^* \pm \varphi_1^* \varphi_2) \qquad (3\text{-}47)$$

Inserting Equation (3-47) into the Schrödinger equation with the Hamiltonian of Equation (3-43) yields

$$E_+^* = \frac{1}{2} \langle \varphi_1^* \varphi_2 + \varphi_1 \varphi_2^* | H_1 + H_2 + V_{12} | \varphi_1^* \varphi_2 + \varphi_1 \varphi_2^* \rangle$$

$$= E_1^* + E_2 + \langle \varphi_1^* \varphi_2 | V_{12} | \varphi_1^* \varphi_2 \rangle + \langle \varphi_1^* \varphi_2 | V_{12} | \varphi_1 \varphi_2^* \rangle$$

$$= E_1^* + E_2 + D' + I \qquad (3\text{-}48)$$

and

$$E_-^* = \frac{1}{2} \langle \varphi_1^* \varphi_2 - \varphi_1 \varphi_2^* | H_1 + H_2 + V_{12} | \varphi_1^* \varphi_2 - \varphi_1 \varphi_2^* \rangle$$

$$= E_1^* + E_2 + \langle \varphi_1^* \varphi_2 | V_{12} | \varphi_1^* \varphi_2 \rangle - \langle \varphi_1^* \varphi_2 | V_{12} | \varphi_1 \varphi_2^* \rangle$$

$$= E_1^* + E_2 + D' - I \qquad (3\text{-}49)$$

E_1^* and E_2 are the energies of the (non-interacting) molecule **1** in the excited state and the (noninteracting) molecule **2** in the ground state. The third term that is given by

$$D' = \langle \varphi_1^* \varphi_2 | V_{12} | \varphi_1^* \varphi_2 \rangle \qquad (3\text{-}50)$$

is analogous to D_0 in Equation (3-46). It represents the van der Waals interaction (polarization energy) between molecule **1** in the excited state and molecule **2** in the ground state (or vice versa). One may think of it as the Coulombic energy of

49. 物理二聚体的激发态是指,一个激发态单体分子与另一个基态单体分子之间相互作用。实际上它是激基二聚物。可以表示为 $\varphi_1^* \varphi_2$ 和 $\varphi_1 \varphi_2^*$。假设两个分子被激发的概率是相同的,通过线性拟合,激发态物理二聚物最终可以表示为

interaction of the charge distribution presented in the excited state of molecule **1** with that of molecule **2** in the ground state. For non-polar molecules, because one of the monomer is in the excited state, which is unstable, generally, $|D'|$ is greater than $|D_0|$, and both are negative. As a result, the 0-0 transition of the dimer redshifts by an amount $D = D' - D_0$. This shift is termed the "solvent shift" in molecular physics considering one of the two molecules involved is treated as a solvent molecule[50]. The fourth term, which is given by

$$I = \langle \varphi_1^* \varphi_2 | V_{12} | \varphi_1 \varphi_2^* \rangle \tag{3-51}$$

represents the resonance interaction energy that determines the splitting between the two excited levels E_+^* and E_-^*. It can be understood as representing the interaction of the overlap charge density with either molecule, i. e., an exchange of the excitation energy between molecules **1** and **2**. For explanation, as a thought experiment, at time $t = 0$ only molecule **1** is excited, then the excitation energy oscillates coherently between molecule **1** and molecule **2**. The oscillation frequency is found from the energy difference: $\hbar \omega = \Delta E = 2I_{12}$[51]. Similar to D_0 and D', the value of I depends on the relative intermolecular distance and orientation. This is, for example, evident when the interaction potential V_{12} is expressed in a point-dipole approximation as in Equation (2-2). Note however, that I and D' differ in their sensitivity to these parameters. There may well be sizable coulomb interaction between the charge distribution present in the excited molecule **1** with the ground state of molecule **2** (and thus a finite D'), even though poor overlap between the two wavefunctions prevents significant resonance interaction (implying $I \approx 0$). Since the degree of the resonance energy I depends on wavefunction overlap, it is thus very sensitive to the intermolecular distance. The $2I_{12}$, i. e., energy difference between the steady-state solutions Φ_+^* and Φ_-^* obtained by comparing Equations (3-48) and (3-49), is called the Davydov splitting Δ_D. Its numerical value, Δ_D/hc, is generally between a few for triplet and a few thousand cm^{-1} for singlet in organic systems[52].

Figure 3.19 schematically summarizes the intermolecular

50. 它表示激发态分子 1 与基态分子 2 之间的范德瓦耳斯力相互作用（极化能）（反之亦然）。可以把它看作分子 1 激发态电荷分布与分子 2 基态电荷分布相互作用的库仑能。对于非极性分子，由于其中一个单体处于不稳定的激发态，一般$|D'|$大于$|D_0|$，且均为负值。因此导致物理二聚体的 0-0 跃迁红移了 $D = D' - D_0$。考虑到另外一个参与的分子是第一个分子的溶剂分子，这个位移在分子物理学中称为"溶剂位移"。

51. 它可以理解为代表每个分子重叠电荷密度的相互作用，即分子 1 和分子 2 激发能的交换。作为解释，想象一个实验，在 $t = 0$ 时，只有分子 1 被激发，然后这个激发能在分子 1 和分子 2 之间相干共振。该共振频率等于（2 个激发态的）能量差：$\hbar \omega = \Delta E = 2I_{12}$。

52. 由于共振能的大小 I 取决于波函数重叠，因此对分子间距离非常敏感。$2I_{12}$，即通过比较方程（3-48）和方程（3-49）得到的稳态解 Φ_+^* 和 Φ_-^* 之间的能量差，称为达维多夫分裂 Δ_D。对于有机体系的三线态和单线态，其数值 Δ_D/hc 通常分别在数个 cm^{-1} 和数千 cm^{-1}。

interactions that lead to shifts and splitting in the spectra. D_0 and D' are based to first order on the molecular polarizability in the ground and excited sates, and the resonance energy I_{12} in the singlet state is due to the resonance interaction between molecule **1** in an excited state and molecule **2** in its ground sate or vice versa. In the triplet state, I_{12} is determined in the main by the overlap of the orbitals of the two molecules, one of which is excited.

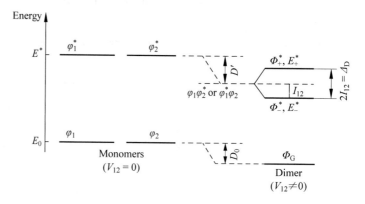

Figure 3. 19　Quantum chemistry explanation for Davydov splitting

In terms of polarized optical transitions from ground state to the two Davydov components, the determinant factor of the transition dipole moments $\boldsymbol{M}_\pm$ can be expressed as

$$\boldsymbol{M}_\pm = \langle \Phi_G \mid e\boldsymbol{r} \mid \Phi_\pm^* \rangle = \langle \varphi_1 \varphi_2 \mid e\boldsymbol{r} \mid \frac{1}{\sqrt{2}} (\varphi_1 \varphi_2^* \pm \varphi_1^* \varphi_2) \rangle$$

$$= \frac{1}{\sqrt{2}} (\langle \varphi_1 \varphi_2 \mid e\boldsymbol{r} \mid \varphi_1 \varphi_2^* \rangle \pm \langle \varphi_1 \varphi_2 \mid e\boldsymbol{r} \mid \varphi_1^* \varphi_2 \rangle)$$

$$= \frac{1}{\sqrt{2}} (\boldsymbol{M}_2 \pm \boldsymbol{M}_1) \tag{3-52}$$

Here $e\boldsymbol{r}$ is the electric dipole operator, $\boldsymbol{M}_1 = \langle \varphi_1 \mid e\boldsymbol{r} \mid \varphi_1^* \rangle$ and $\boldsymbol{M}_2 = \langle \varphi_2 \mid e\boldsymbol{r} \mid \varphi_2^* \rangle$ are transition dipole moment for molecule **1** and **2**, respectively. According to Equation (3-52), the transition dipole moment of the two states in a physical dimer results from the vector sum of individual molecules' transition dipole moments. In a consequence, it depends on the relative orientation of the two molecules[53]. As shown in Figure 3.20, few limiting cases are worth highlighting

53. 由式(3-52)可知,物理二聚体中两个态的跃迁偶极矩为单个分子跃迁偶极矩的矢量和。因此,它取决于两个分子的相对取向。

(1) H-Type Interaction: The two molecules are arranged in a coplanar stacked-manner (e. g., side by side, Figure 3.20(a)). In this case, the transition dipole moments for lower energy excited states E_-^* are arranged in an antiparallel manner and add to a total value of zero; while parallel arrangement prevails for E_+^* and the optical transition dipole moment is

$$M_+ = \frac{1}{\sqrt{2}}(M_2 + M_1) = \sqrt{2}M_1 = \sqrt{2}M_2 \qquad (3\text{-}53)$$

This means that light absorption can only occur from ground state into E_+^*, so the absorption spectrum appears hypochromically (blue) shifted relative to the absorption of the parent molecule. With regard to emission, one needs to keep in mind that excitation energy always relaxes to the lowest excited state, in this case E_-^*. As transitions from E_-^* to the ground state of the two coupled molecules carry no oscillator strength, the radiative decay rate k_r is zero and the energy can only be dissipated non-radiatively[54]. The lifetime of E_-^*, $\tau = 1/(k_r + k_{nr})$, is then determined solely by the nonradiative decay rate k_{nr}. In real systems, a slight mis-orientation and/or vibronic coupling in the excited state gives rise to a finite yet weak oscillator strength, so that usually a weak, long-lived emission is observed. Electronically coupled molecules of this category are called H-aggregates. As their cofacial arrangement is conducive to a large resonance interaction, they are prone to excimer formation.

(2) J-Type Interaction: The two molecules are arranged in a sequential collinear and parallel manner (head to tail or head to head, Figure 3.20(b)). The lower-energy state E_-^* is then realized for a parallel sequence of transition dipole moments (Equation (3-53)), while the higher-energy state E_+^* is associated with an antiparallel order and a net zero moment. Absorption to and emission from E_-^* are optically allowed, so that the absorption spectrum is bathochromically (red) shifted with respect to the monomer spectrum, and the fluorescence is fast and intense. Molecules coupled in such a manner are referred to as J-aggregates or Scheibe-aggregates[55]. The "J" derives from EE Jelley who characterized such coupled molecules that, by the way, were

54. 这意味着光吸收只能从基态到 E_+^*，因此吸收光谱相对于原来分子的吸收出现了蓝移。对于发射，必须牢记激发能总是弛豫到最低激发态，在这种情况下是 E_-^*。由于从 E_-^* 到基态两个耦合分子之间的跃迁振子强度为零，因此辐射衰减速率 k_r 为零，能量只能以非辐射方式耗散。

55. 这两个分子按照相继共线且平行的方式排列（头对尾或头对头，图 3.20(b)）。低能量态 E_-^* 为平行排列的跃迁偶极矩（式(3-53)），而高能量态 E_+^* 则与反平行排列相关，其净矢矩为零。到 E_-^* 的吸收和从 E_-^* 的发射，是光学允许的，因此吸收光谱相对于单体红移，并且荧光辐射快且较强。以这种方式偶联的分子被称为J聚集体或Scheibe聚集体。

56. 当两个组成分子的跃迁矩方向为任意时,低能跃迁和高能跃迁都获得一定程度的振子强度。

57. 当二聚体中两个非平移等价的分子以一定的角度排列,两个激发态都拥有非抵消的跃迁矩。它们的大小和方向可以通过式(3-52)计算,对偶极跃迁来说是将各自的矢量相加或者相减(图3.32(c))。

58. 因此 M_+ 和 M_- 的极化分别是平行和垂直于 b 轴的。因此这两个达维多夫组分在能量和极化程度上都不同。

discovered independently at almost the same time by G Scheibe.

（3） Arbitrary Orientation：Very often π-bonded chromophores are oriented in neither of the above two limiting ways，but they may be arranged with a variable angle between them (e.g. amourphos, Figure 3.20(c)). When the orientation of the two transition moments of the two constituent molecules is arbitrary，the lower and the higher energy transitions acquire a certain degree of oscillator strengths[56]. The ratio of the intensities of the components is set by the vectorial sum of the monomer moments，as indicated in Figure 3.20(c)，and the fluorescence would be emitted from the lower dimer state with lifetime one to two orders of magnitude longer than for the monomer.

（4） Ordered Orientation with Two Translational Inequivalent Molecules：In dimer with two translational inequivalent molecules that are arranged with an angle，both excited states have nonvanishing transition moments. Their magnitudes and directions can be computed with the aid of Equation (3-52)；for dipole transitions they are obtained by addition or subtraction of the individual vectors (Figure 3.20(c))[57]. There are thus two allowed transitions with the energy difference $2I_{12}$. This holds equally well for excitons in the crystal and for the mini-exciton. For example，in the monoclinic anthracene crystal lattice，the orientation **1** can be converted into orientation **2** by a mirror-glide operation (vice versa). The mirror-glide plane is the a-c plane. M_+ and M_- are therefore polarized parallel and perpendicular to the b axis，respectively. The two Davydov components thus differ in both energy and polarization[58]. This means that with incident polarized-light parallel and perpendicular to the b axis of the anthracene crystal，there are respectively one of the two Davydov components in the absorption spectra；and the polarized light with other incident direction will result in broadened absorption spectrum due to Davydov splitting. For further understanding，we can say that when irradiate a crystal with non-polarized light or irradiate amorphous material with polarized light，both transition moments of the two Davydov components are not zero. The absorption spectrum usually shows broadening by Davydov splitting. Some time, Davydov splitting can be observed in absorption spectra with high resolution[59].

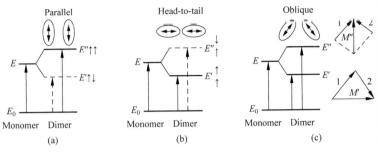

Figure 3. 20   A sketch showing the orientation and magnitude of the transition dipole moments M for various mini-exciton geometries（The solvent shift is left out here for simplicity）：（a）blue-shift，（b）red-shift，（c）Band-splitting

The magnitudes of the Davydov splitting $2I_{12}$ are found from Equation（3-51）. For singlet terms which belong to allowed optical transitions, one can calculate the interaction potential V_{12} and the resonance energy I_{12} to a good approximation as a dipole interaction of the transition dipole moment of the excited molecular state with the neighbouring molecule in the ground state[60]. We thus have

$$I_{12} \propto z \cdot \frac{M}{R^3} \approx \text{const.} \frac{f}{\Delta E_M} \qquad (3\text{-}54)$$

where, z is an orientation factor of the order of 1, M is the value of transition dipole moment of the optical transition in the molecule, f is the oscillator strength of the transition, R is the distance between the molecules in the crystal, ΔE_M is the molecular excitation energy. For oscillator strengths f between 0.001 and 0.1, numerical values for I_{12} between ca. 10 and ca. 1000 cm^{-1} are obtained in the singlet state. In the triplet state, the interaction energy between the excited and the non-excited state is smaller and cannot be computed as simple dipole interaction. The splitting is also smaller.

3. Crystals with One Molecule per Unit Cell

Extending the dimer model to encompass an "infinitely large" three dimensional crystal lattice is similar to the

59. 为了进一步理解，可以说当以非偏振光入射晶体或者以偏振光入射无定形材料时，两个达维多夫组分的跃迁矩都不为零。吸收光谱通常表现出源于达维多夫裂分的展宽。有时，达维多夫裂分可以在高分辨的吸收光谱中观察到。

60. 达维多夫裂分的大小 $2I_{12}$ 可通过式（3-51）得到。对于属于光学跃迁允许的单线态，相互作用势能 V_{12} 是可以计算的，并且共振能 I_{12} 可以通过激发态分子的跃迁偶极矩与周围基态分子的相互作用得到很好的近似结果。

61. 从二聚体模型扩展到含有"无穷大"三维晶体晶格的情形,类似于从两个共价原子扩展到晶体的离域能带结构。考虑一个线性晶体,其含有 N 个等同分子 i($i=1,2,\cdots,N$),单个分子的电子基态用电子波函数 φ_i^0、哈密顿算符 H_i 和基态能量 E_i^0 来描述。

delocalized band structure which is extended from two covalent atoms to crystal. Considering a linear crystal with N identical molecules i ($i=1,2,\cdots,N$), with their individual electronic ground states described by an electronic wavefunction φ_i^0, a Hamilton operator H_i, and a ground state energy E_i^{0} [61]. Analogous to the treatment of two coupled molecules described in above section, we neglect spin and vibrational wavefunction for simplicity. The coupling between the molecules arises from the same electrostatic interaction V_{ij} between two molecules i and j, and it may be expanded in dipole terms (Equation (2-2)) or, more appropriately, in multipole terms. The Hamilton operator is then the sum of the operators for the individual molecules H_i and the term V_{ij} that represents the interaction between them,

$$H = \sum_{i=1}^{N} H_i + \frac{1}{2}\sum_{\substack{i,j=1\\i\neq j}}^{N} V_{ij} \tag{3-55}$$

Notice that the ½ in the second term of Equation(3-55) is due to the pair of interaction, that is, the interactions of i to j and j to i represent one interaction not two.

The wavefunction for the crystal can be expressed as

$$\Psi_0 = A\prod_{i=1}^{N}\varphi_i^0 \tag{3-56}$$

$$\Psi_i^* = A\varphi_i^*\prod_{\substack{j=1\\j\neq i}}^{N}\varphi_j^0 \tag{3-57}$$

62. Ψ_0 是晶体的基态(所有分子都处于基态),Ψ_i^* 是激发态,指第 i 个分子处于激发态(晶体激发态是指第 i 个分子是激发态,其他分子是基态),φ_i^0 和 φ_i^* 分别为第 i 个分子的基态和激发态,A 是不对称因子。

Ψ_0 is the ground state of the crystal (all molecules are in the ground state). Ψ_i^* is the excited state, in which the i-th molecule is excited (the crystal excited state is referred to the situation that the i-th molecule is in the excited state, while the other molecules are in the ground state), φ_i^0 and φ_i^* are ground and excited states of the i-th molecule, A is asymmetrical factor[62]. Combining Equation (3-56) with Equation(3-55), the total energy of the ground state of the linear crystal is given by

$$E^0 = \langle\Psi_0\,|\,H\,|\,\Psi_0\rangle = \left\langle A\prod_{i=1}^{N}\varphi_i^0\,\middle|\,\sum_{i=1}^{N}H_i + \frac{1}{2}\sum_{\substack{i,j=1\\i\neq j}}^{N}V_{ij}\,\middle|\,A\prod_{i=1}^{N}\varphi_i^0\right\rangle$$

$$= \sum_{i=1}^{N}E_i^0 + D_0 \tag{3-58}$$

with

$$D_0 = \left\langle A \prod_{i=1}^{N} \varphi_i^0 \left| \frac{1}{2} \sum_{\substack{i,j=1 \\ i \neq j}}^{N} V_{ij} \right| A \prod_{i=1}^{N} \varphi_i^0 \right\rangle \qquad (3\text{-}59)$$

The overall energy shift D_0 in the ground state of the crystalline assembly of molecules is known as the gas-to-crystal shift for the ground state energy. It is a polarization energy that results from the van der Waals interaction between the assembled molecules. For a quantitative treatment，Ψ_0 would need to include further terms.

The excited state described by Equation（3-57）neglects the interaction between the i-th excited molecule and the other ground state molecules（$V = 0$）. There are N-fold degenerated states in the crystal excitation wavefunction（Ψ_i^*）, meaning the energy for the excitation of any ground state molecule is the same. States described by Ψ_i^* are localized and nonstationary（not eigenfunctions of the Hamiltonian operator）. The N-fold degeneracy of the crystal excited-states can be lifted and yield eigenfunctions by appropriate linear combinations（Equation（3-60））of the above nonstationary states and the introduction of variational principle（Equation（3-61）），and presuming that the linear array of molecules has periodic boundary conditions，as shown in Equation（3-62）[63].

$$\Phi^* = \sum_{i=1}^{N} \alpha_i \Psi_i^* \qquad (3\text{-}60)$$

$$\delta(\langle \Phi^* | H | \Phi^* \rangle / \langle \Phi^* | \Phi^* \rangle) = 0 \qquad (3\text{-}61)$$

$$\Phi^*(k) = \frac{1}{\sqrt{N}} \sum_{j=1}^{N} \Psi_j^* \exp(\mathrm{i}k\boldsymbol{R}_j) \qquad (3\text{-}62)$$

where $\boldsymbol{R}_j$ is the position vector of the j-th molecule，$\Phi^*(k)$ is the wavefunction of an excited state electron，that is，of a neutral，delocalized，excited electron with the wavelength λ（$k = 2\pi/\lambda$），k is the wavevector. The k values are：0，$\pm 2\pi/Nd$，$\pm 4\pi/Nd$，$\cdots$，π/d，d is lattice constant. The excited electron is a quasiparticle with the momentum $p = \hbar k$. For $k \rightarrow 0$，the wavelength is large compared to the lattice constant d. $k = 0$ corresponds to an infinitely long wavelength，i.e.，the excitation phase is the same within all N unit cells of the crystal. Equation（3-62）describes an excited state electron whose excitation energy is delocalized

63. 式（3-57）表述的激发态，忽略了第 i 个激发分子与其他基态分子之间的相互作用（$V=0$）。晶体激发波函数（Ψ_i^*）存在 N 重简并，意味着任何基态分子被激发的能量是相同的。Ψ_i^* 描述的状态是定域的非稳态（不是哈密顿算符的本征函数）。晶体 N 重简并的激发态可以通过上述非稳态的适当线性组合（式3-60），以及引入变分原理（式3-61），并且假设分子的线性序列具有周期性的边界条件，得到非简并的本征函数，求出的解如式（3-62）所示。

64. 其中 $\boldsymbol{R}_j$ 是第 j 个分子的位置矢量，$\Phi^*(k)$ 是一个激发态电子的波函数，即一个波长为 λ（$k = 2\pi/\lambda$）的中性、离域的激发态电子，k 是波矢。k 值为 $0, \pm 2\pi/Nd, \pm 4\pi/Nd, \cdots, \pi/d$，$d$ 为晶格常数。这个激发电子是一个动量为 $p = \hbar k$ 的准粒子。当 $k \to 0$，电子波长大于晶格常数 d。$k = 0$，对应于无限长的波长，即激发态在晶体的所有 N 个单元是相同的。式(3-62)描述了一个激发态电子，其激发能量在整个晶体上离域，但限于一组平移等价的分子。因此，这种"单一位点"波函数适用于线性晶体，或晶胞中仅有一个分子的晶体。例如，准一维激发态电子可以在二溴萘酚中观察到，六甲基苯的晶胞中只有一个分子。

over the whole crystal, but is limited to a set of translationally-equivalent molecules. This "one-site" wavefunction thus holds for linear crystals or for crystals with only one molecule in the unit cell. Quasi-one-dimentional excited state electrons can for example be observed in dibromonaphthalene. Hexamethlybenzene has only one molecule in its unit cell[64].

Equation (3-62) can also be expressed as cosinoidal formulas:

$$\Phi^*(k) = \frac{1}{\sqrt{N}} \sum_{j=1}^{N} \Psi_j^* \cos(k\boldsymbol{R}_j) \tag{3-63}$$

From the experimentally measurable energy difference between the excited state E^* and the ground state E^0, we obtain

$$\Delta E(k) = E^* - E^0 = E_{\text{mol}}^* + |D' - D_0| + 2I\cos(kd) \tag{3-64}$$

where E_{mol}^* is the excitation energy of a single molecule, and $D' - D_0$ is difference in the gas to crystal shifts of excited and ground state of the system. D' is given by

$$D' = \left\langle \Phi^*(j) \left| \frac{1}{2} \sum_{\substack{i,j=1 \\ i \neq j}}^{N} V_{ij} \right| \Phi^*(j) \right\rangle \tag{3-65}$$

The term of $2I\cos(kd)$ is the resonance energy between the N-fold excited states, which can be positive or negative values:

$$I = \left\langle \Phi^*(j) \left| \frac{1}{2} \sum_{\substack{i,j=1 \\ i \neq j}}^{N} V_{ij} \right| \Phi^*(i) \right\rangle \tag{3-66}$$

65. 根据式(3-64)，由于 $-1 \leqslant \cos(kd) \leqslant 1$，最大共振能量分裂为 $2I - (-2I) = 4I$。由于平移等价，即所有晶格分子是相同的，达维多夫分裂表现为具有 N 重状态、宽度为 $4I$ 的能带（图 3.21）。带的宽度由这些相同分子之间的相互作用程度决定。

According to Equation (3-64), as $-1 \leqslant \cos(kd) \leqslant 1$, the maximum resonance energy splitting is $2I - (-2I) = 4I$. Owning to the translational equivalence, i.e., all the lattice molecules are identical, the Davydov splitting presents as an expended band with N-fold states and width of $4I$ (Figure 3.21). The width of the band is determined by the degree of interaction between the identical molecules[65].

4. Crystals with Two Molecules per Unit Cell

The formalism considered so far is appropriate for describing the Davydov splitting in an ordered array of rigid single molecules, that is, a molecular crystal with only one molecule per unit cell. For most molecules, if there are two translationally inequivalent molecules in the unit cell, the crystal will comprise two inequivalent excited states. Similar to Equations （3-55）, （3-56） and （3-57） for Hamilton operator and wavefunction expressions, by simplifying and introducing boundary condition, the interaction energies, such as D_0, D' and Davydov splitting can be obtained[66].

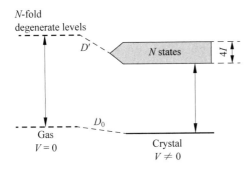

Figure 3.21　From isolated gaseous molecules to one dimensional crystals, the ground state and the first excited state energy levels changes, and the excited state is broadened by Davydov splitting of $4I$

Taking anthracene with N unit cell as an example, since there are two translationally inequivalent molecules in a unit cell, which are connected by the symmetry operation C_{2h}, by diagonalising the Hamilton matrix for each value of k, one obtains two excited states （Davydov components）. They can, if the interactions are predominantly of short range, be written in the simple form

$$\Phi_\pm^*(k) = \frac{1}{\sqrt{2}}\left[\Psi_\alpha^*(k) \pm \Psi_\beta^*(k)\right] \qquad (3\text{-}67)$$

$\Psi_\alpha^*(i)$ and $\Psi_\beta^*(i)$ represents the i-th non eigenfunctions of the crystal excited states, relating to the molecule α and β in the i-th unit cell. The width of the two Davydov bands can be obtained from Equation （3-66）, which is given by the interaction with the equivalent and the inequivalent molecules in the crystal.

$$E_\pm^*(k) = L_{\alpha\alpha}(k) \pm L_{\alpha\beta}(k) \qquad (3\text{-}68)$$

$L_{\alpha\alpha}$ and $L_{\alpha\beta}$ are the sum over translationally equivalent and translationally inequivalent interactions, respectively. The Davydov splitting Δ_D is defined as the energy difference of the $k=0$ levels in the two bands. Via transitions between the

66. 对于大多数分子,如果晶胞中含有 2 个非平移等价的分子,晶体将包含 2 个非等价的激发态。与式(3-55)、式(3-56)和式(3-57)所表示的哈密顿算符和波函数类似,通过简化并引入边界条件,可以获得分子的相互作用能,例如 D_0、D' 和达维多夫分裂。

67. 达维多夫裂分宽度被定义为 $k=0$ 时的两个能带之间的能量差异。通过零振动基态($k=0$)和激发能带之间的跃迁,$k=0$ 时的态是可以直接通过光学方式获得的。这遵从动量守恒。也就是说,与晶格矢量的倒数相比,入射光子的动量是很小的,可以忽略。该达维多夫裂分为

68. 这里加和是所选取的 i 分子针对所有非平移等价的邻近 j 分子进行的。对于一个像蒽一类的分子晶体,基于在(a-b)晶面内的、最近的 4 个邻近分子就可以得到很好的近似,因此得到 $8I_{12}$ 的数值,其中 I_{12} 是单位晶胞内非等价分子之间的共振相互作用。平移等价的位移,即相对于激发能带原来位置的达维多夫组分中心的位移,为 $L_{\alpha\alpha}(0)=\sum I_{\alpha\alpha}^{ij}$,它是通过所有平移等价邻近分子的加和得到的。近似到最近的 4 个分子,这个值表示为 $4I_{11}=4I_{22}$。分子能级到激发态能带的裂分在图 3.22 中示意地表示出来。

zero-vibrational ground state ($k=0$) and the excited band, the states with $k=0$ are directly accessible optically. This follows from momentum conservation. The momentum of the incident photon is namely very small compared to a reciprocal lattice vector and negligible. The Davydov splitting is[67]

$$\Delta_{\mathrm{D}}=2L_{\alpha\beta}(0)=2\sum_{i,j=1}^{N}I_{\alpha\beta}^{ij}\approx 8I_{12} \tag{3-69}$$

Here, the sum is carried out over all the translationally-inequivalent neighbours j of a chosen molecule i. In a molecular crystal of the anthracene type, these are to a good approximation the four nearest neighbours in the $(a-b)$ plane, thus the value being $8I_{12}$, where I_{12} is the resonance interaction between the inequivalent molecules within the unit cell. The translational shift, i.e. the shift of the center of the Davydov components relative to the origin of the excited band, is $L_{\alpha\alpha}(0)=\sum I_{\alpha\alpha}^{ij}$, with summation over all the translationally-equivalent neighbours. Approximating to four nearest neighbours, this value can be written as $4I_{11}=4I_{22}$. The splitting of the molecular energy levels into excited bands is shown schematically in Figure 3.22[68].

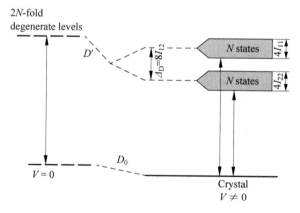

Figure 3.22 Davydov splitting from dimer to crystal with two molecules in the unit cell. The splitting between the two $k=0$ states is defined as the Davydov splitting. Its magnitude is determined by the resonance interaction between the inequivalent molecules within the unit cell, while the excited band width ($4I_{11}$ or $4I_{22}$) is determined by the resonance interaction between translationally equivalent molecules in different unit cells

For triplet，the width of the Davydov splitting Δ_D is within $10^0 \sim 10^1$ cm^{-1}，while for singlet，Δ_D is within $10^2 \sim 10^3$ cm^{-1}. Due to Davydov splitting as well as solvent shift，the lowest excitation energy of crystal is lower than that for isolated molecule. Table 3.2 shows values of Davydov splitting and other characteristic data in naphthalene and anthracene crystals[69].

69. 三线态达维多夫分裂的宽度，Δ_D 在 $10^0 \sim 10^1$ cm^{-1} 范围内，对于单线态，Δ_D 在 $10^2 \sim 10^3$ cm^{-1}。由于达维多夫分裂和溶剂位移，晶体最低激发能低于孤立分子的。表 3.6 给出了萘和蒽晶体达维多夫裂分值和其他特征数值。

Table 3.2 Some characteristic numerical values for S_1 and T_1 excitons in naphthalene and anthracene crystals

	0-0 region /cm^{-1}	Solvent shift /eV	Davydov splitting /cm^{-1}	Absorption constant of S_0/cm^{-1}	Exciton Lifetime (no trapping) /s
Anthracene S_1	25300	0.29	220	10^5	2×10^{-8}
Anthracene T_1	14750		21	3×10^{-4}	4×10^{-2}
Naphthalene S_1	31500	0.43	150	$10^{-4} \sim 10^5$	10^{-7}
Naphthalene T_1	21200		10	5×10^{-4}	< 0.5

3.5.2 Excimer and Exciplex

The term excimer denotes molecular configurations which absorb as monomers but in the excited state form (physical) dimers and fluoresce as such. One also speaks of resonance dimers. These excimers (excited dimers) have been extensively investigated in solution，as can be seen in the case of pyrene (Figure 3.23). The emission band with peaks smaller than 400 nm，where fine structures can be observed，belongs to monomer emission. While the structureless broad band with peak around 490 nm is assigned to excimer emission. With the increase of pyrene concentration，the excimer emission becomes stronger. Generally，compared to the monomer emission，the excimer emissions are weak and red-shifted，their vibronic structure is not or at best poorly resolved，and their lifetimes are longer[70].

Figure 3.24 schematically describes the formation of excimers，where lines of S_0，S_1 and Ex symbolize the potential curves of the molecules in their ground states and in the excited state，as well as the excimer state；B is the binding energy of the excimer. In the ground state，the binding

70. 激基二聚物这个词，表示这样的分子构象：吸收来自于单体分子，但其激发态形成(物理)二聚体，荧光也来自于此。有人也称之为共振二聚体。这些二聚体(激发态二聚体)在溶液中被广泛研究，就如芘这个例子所示(图 3.23)。峰位小于 400 nm 的发射带，可观测到精细结构，属于单体发射。而在 490 nm 左右，没有精细结构的发射带，属于激基二聚物的发射。随着芘浓度的增加，激基二聚物发射峰增强。通常地，与单体相比，激基二聚体发射较弱并且红移，其振动结构不可见或者往好里说分辨性较差，其寿命较长。

71. 图 3.24 示意地描述了激基二聚物的形成，其中线条 S_0、S_1 和 Ex 代表分子基态和激发态，以及激基二聚物态；B 是激基二聚物的结合能。在基态，这对参与者的相互束缚很弱，因此吸收来自于单体，发生在基态的平衡距离 r_M 位置，导致了 S_1 激发态。

between the two partners of a pair is weak, hence the absorption is from the monomers, occurring at the equilibrium spacing r_M in the ground state and leading to the excited state S_1[71]. In the excited state S_1, the binding of the pairs increases due to resonant exchange. In this process, the two partners change their mutual configuration somewhat and approach the state of maximum resonance stabilization, so that in particular, their planar spacing becomes smaller and reaching the cross point of line S_1 and Ex. This approach is accompanied by a radiationless energy relaxation to the equilibrium spacing of r_{Ex}, — in the case of pyrene, the spacing decreases from 3.53 Å to 3.37 Å. From the relaxed

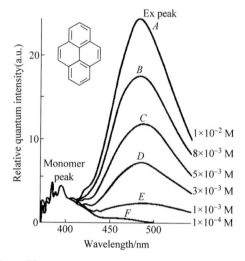

Figure 3.23　Fluorescence spectra of pyrene in cyclohexane at different concentrations

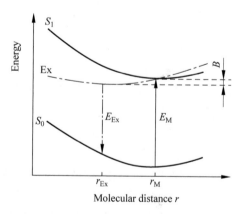

Figure 3.24　Potential curves and an absorption/emission transition for Ex in a crystal

state, the excimer emission leads back to the ground state S_0, whose potential at the smaller molecular spacing is repulsive. In the ground state, the energy relaxes by a further ca. 3000 cm^{-1} and the spacing increases by 0.16 Å back to that of the initial state. It is thus strongly Stokes-shifted relative to the absorption, in pyrene by ca. 0.5 eV[72].

Besides the observed excimer emission in concentrated solution, there are also crystals of aromatic hydrocarbons whose fluorescence originates from an excimer. For this to occur, the molecules in the crystal must be ordered pairwise in parallel planes, with a small distance between neighbouring planes. In the pyrene crystal, this distance is 3.53 Å. The molecules are thus already ordered as physical sandwich dimers within the crystal[73]. Figure 3.25 shows the emission spectra of a pyrene crystal at different temperatures. It is found that with the increase of temperature, the half width of the maximum peak is wider. This is because that thermal motion intensifies as the increase of temperature and disturbance to excimer increases.

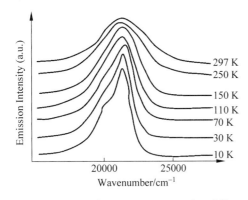

Figure 3.25 Excimer emission from a pyrene crystal at different temperatures

Excimer emission is observed from many crystals when they are subjected to pressure or when they are strongly deformed in some other manner. In the deformed crystals, molecular configurations that favour excimer emission are produced. Also in vapour-deposited films, which are initially amorphous with many defects at low temperatures, excimer structures can frequently be formed by a suitable annealing process possibly due to the removal of defects and formation

72. 在激发态 S_1，共振交换能使得这一对（分子）之间的结合增强。这个过程中，参与的双方在某种程度上相互改变构象，达到最大共振稳定性，这样一来，尤其是它们之间的平面间距变小，抵达 S_1 和 Ex 曲线的交叉点。这一过程伴随着非辐射能量衰减，达到平衡的间距 r_{Ex}-在芘的情况下，分子间距从 3.53 Å 降至 3.37 Å。从弛豫后的能态产生的激基二聚物发射过程，使得（电子）回到基态 S_0，此时这个较小的分子间距下，势能是相互排斥的。因此在基态，通过进一步释放大约 3000 cm^{-1} 的能量，并且分子间距增加 0.16 Å，回到最初态。因此，与吸收相比，斯托克斯位移很强，对于芘，大约是 0.5 eV。

73. 除了在浓溶液中观测到激基二聚物的发射，也存在芳香碳氢化合物的晶体，其荧光来自于激基二聚物。该过程的发生，晶体中的分子必须在平行的晶面成对地有序排列，相邻晶面之间的距离较小。在芘晶体中，这个距离是 3.53 Å。因此在晶体内部，分子已经作为物理夹层二聚物有序地排列了。

74. 许多晶体在受到压力或者当它们通过其他方式强烈形变时,可观测到激基二聚物发射。在形变的晶体中,产生了有利于激基二聚物发射的分子构象。气相沉积的薄膜,初始时是在低温下含有很多缺陷的无定形,适当的退火过程经常可以形成激基二聚物的结构,可能的原因是缺陷的移除和有序分子的形成。

75. 如果一个分子与一个具有相对高/低的 HOMO/LUMO 能级的不同分子相邻,前一个分子的激发,可能导致它将一个空穴/电子转移到后者的基态分子上,形成新了含有 2 个不同分子的激发态,称作激基复合物。激基复合物辐射衰变,就形成了激基复合物的发光。

of ordered molecules[74].

If a molecule is adjacent to a different molecule with relatively high/low HOMO/LUMO, the excitation of the former may lead to the transfer of a hole/electron to the ground state of the latter, forming a new excited state comprising two different molecules, which is termed as exciplex. The radiative decay from exciplex state gives exciplex emission[75]. As shown in Figure 3.26, the S_0 and S_1 represent the potential curves of one molecule in ground and excited states, and the S_0' and S_1' are those for a molecule of different type. E_X is the potential curve for an excited state forming between the two molecules. In the ground state, the interaction between the two molecules is negligible; hence the absorption is from one molecule, e.g., optical transition of $S_0 \rightarrow S_1$. In the excited state S_1, due to suitable energy levels, charge transfer occurs. For example, lower LUMO of S_1 leads to electron transfer from S_1' to S_1, higher HOMO of S_0' leads to hole transfer from S_0 to S_0', both of which lead to the formation of the exciplex. After relaxation to the equilibrium position, the exciplex may radiative decay to ground state, giving exciplex emission.

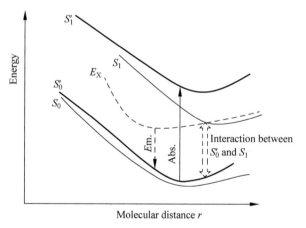

Figure 3.26　Formation and luminescence schemes of an exciplex

Features of excimer/exciplex include: ①new energy level appears attributed to the formation of excimer/exciplex, which is lower than the molecular excited state; ②there is no ground state for excimer/exciplex, that is, when the two molecules are all in the ground state, there is no interaction between them；③ the emission spectrum of excimer/exciplex is broad， structureless and redshift compared to the corresponding molecular emission；④ it has been found that under electric field or even without electric field， excimer/ exciplex tends to partially decompose[76].

3.5.3 Charge Transfer Exciton

In organic crystals， the term charge transfer exciton is applied to electronic excitations which are neutral but polar， and in which the electron-hole distance is one or two times greater than the distance of the molecular structural units. Charge transfer （CT） excitons thus lie in terms of their binding energies as well as in terms of their spatial extension between the Frenkel excitons and the Wannier excitons[77]. Previously， they were also referred to as ionic states. They are thus excitations in which an electron or a hole in one molecule is transferred to another molecule in the neighbourhood and thus gives rise to an ion pair.

Such CT excitons are found， in addition to the Frenkel excitons， in the aromatic molecular crystals such as anthracene which we have already treated， that consist of only one type of molecules. They are difficult to observe in the usual optical spectrum， because their absorption is very weak（typical oscillator strengths for the transition from the neutral ground state are $f = 10^{-4}$ to 10^{-2}）， and because they lie in the spectral range of the much stronger $S_0 \rightarrow S_n$ absorption， i.e. at excitation energies which are higher than those of the Frenkel excitons[78]. For anthracene， the absorption edge， that is the 0-0 threshold for the $S_0 \rightarrow S_1$ transition， lies at 3. 11 eV， and the CT excitation lies at 3.9 eV. These excited states become important and observable in the photoconductivity of anthracene and similar crystals. Their investigation is most expediently carried out in reflection or absorption in an applied electric field. The field is modulated and only the correspondingly modulated part of the absorption or reflection is measured. This spectroscopic method is called electroabsorption or electroreflection.

In crystals which are composed of two different partner molecules， CT excitations and with them CT excitons are frequently the predominant lowest excitation states and are

76. 基激二聚物/激基复合物的特点有：①由于激基二聚物/基激复合物的形成，比分子激发态能级低的新能级出现；②激基二聚物/基激复合物没有基态，也就是说当两个分子都处于基态时，它们之间没有相互作用；③激基二聚物/基激复合物发射峰较宽，没有精细结构，并且相对于单体红移；④在电场作用下，甚至没有电场的情况下，观测到一部分激基二聚物/基激复合物存在分解倾向。

77. 在有机晶体中，术语电荷转移激子指的是，虽然中性却存在极性的电子激发态，其电子和空穴的距离比分子结构单元内的距离大 1 到 2 倍。因此，在结合能以及空间伸展方面，电荷转移（CT）激子介于弗仑克尔激子和万尼尔激子之间。

78. 在如我们已经讨论过的蒽这样的、只含有一种分子的芳香分子晶体中，除了弗仑克尔激子，可观测到这样的 CT 激子。在常见的光学光谱中，它们是不易被观测到的，因为其吸收较弱（始于中性基态的典型跃迁谐振子强度 $f = 10^{-4} \sim 10^{-2}$），并且由于它们的光谱范围位于能量更大的 $S_0 \rightarrow S_n$ 吸收，即其激发能比弗仑克尔激子高。

79. 在含有两个不同参与分子的晶体中，CT 激发以及存在于其中的 CT 激子经常是主要的最低激发态，因此其单线态的能量最低跃迁来源于此。

80. 在混合堆叠的 DADADA 体系中（图 3.27（a）上部的中间部分），最低光学激发始于基态的给体 D，蒽，（从它的最高占据轨道，HOMO），并到达受体 A，PMDA 的最低空置轨道（LUMO）。

thus responsible for the lowest-energy transitions in the singlet system[79]. We will illustrate this using the example of the weak donor-acceptor complex anthracene/pyromellitic acid dianhydride，(A/PMDA) (Figure 3.27(a)). The ground state is neutral and nonpolar，with only a small charge-transfer fraction. The lowest optical excitation starts from the ground state of the donor **D**, anthracene，(from its highest occupied orbital or HOMO) and leads to the lowest unoccupied orbital (LUMO) of the acceptors **A**, PMDA，within the mixed stack DADADA (see the schematic view at the central of the top of Figure 3.27(a))[80]. A polar exciton is produced by such absorption，which is characterized by a CT absorption band，The optical transition is polarized along the stacking axis. Figure 3.27(b) shows the absorption and the fluorescence spectra，measured with the optical polarization parallel to the stacking axis. The absorption band，which is strongly broadened by phonon interactions，contains a sharp zero-phonon line as a 0-0 transition. The vibronic structure is barely visible，except for the sharp 0-0 zero-phonon line ZPL，because the strong electron-phonon interaction broadens the transitions severely.

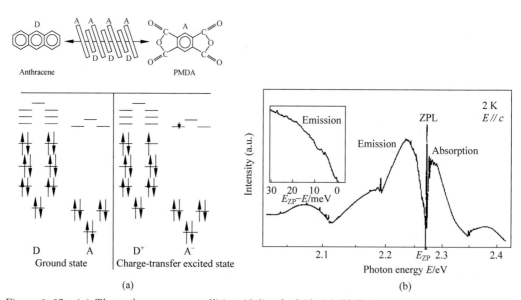

(a)

(b)

Figure 3.27 (a) The anthracene-pyromellitic acid diaanhydride (A-PMDA) mixed -crystal system as an example of a weak donor-acceptor crystal with mixed stacks. (b) The absorption and emission spectra of an anthracene-PMDA crystal at 2 K with the polarization of the light parallel to the stacking axis

The energy of a CT exciton, E_{CT}, i. e., the energy needed to excite this state, is given by

$$E_{CT} = IP_D - EA_A - P_{eh}(r) - C(r) \qquad (3\text{-}70)$$

with IP_D = ionization potential of the donor (at the site of the hole), EA_A = electron affinity of the acceptor (at the site of the electron), $P_{eh}(r)$ = polarization energy of the lattice due to the electron-hole pair at the distance r, $C(r)$ = Coulomb attractive energy between the electron and the hole at the distance r. The ionic complex partners produce a local lattice deformation and polarize their surroundings. This can also lead to self trapping of the excitonic energy of excitation. Note that the energy of the 0 - 0 transition in Figure 3.27(b) is smaller by ca. 8000 cm^{-1} than that of the 0 - 0 transition of Frenkel excitons in the anthracene crystal[81]. The lowest excited triplet state T_1 of this crystal belongs, in contrast, only to the anthracene molecule and corresponds to triplet Frenkel excitons with no polar character.

A local CT exciton can be represented as a linear combination of the states $\psi(A^- D^+ A^0)$ and $\psi(A^0 D^+ A^-)$. The optical transition from the electronic ground state $\psi(A^0 D^0 A^0)$ is allowed only into the state with odd parity, $\Psi_u = \frac{1}{\sqrt{2}}[\Psi(A^- D^+ A^0) - \Psi(A^0 D^+ A^-)]$. When an electric field is applied along the stacking axis, the transition into the state with even parity is also allowed. The energy difference between the two states is here equal to 12 cm^{-1}. It corresponds to the splitting of the exciton band at $k = 0$ (Figure 3.28)[82]. Transport of the CT exciton within the crystal requires the motion of a pair excitation states. Expressed in simple terms: the CT exciton is polar. The configuration with dipole moments along the direction of the applied field differs in its energy from that in the opposite direction.

Another D-A complex is strong D-A complexes or radical-ion salts, that is of crystals in which a charge separation already exists in the ground state. A typical such a material system is tetrathio-fulvalene (TTF, donor): tetracyano-quinodimethane (TCNQ, acceptor), who behaves as a conductor[83]. This textbook does not handle this content.

81. 请注意，图 3.27（b）中 0-0 跃迁能量比在蒽晶体中 0-0 跃迁的弗仑克尔激子大约低 8000 cm^{-1}。

82. 当沿着堆叠轴线方向施加电场时，偶宇称性的跃迁也是允许的。这里，两个跃迁之间的能量差异等于 12 cm^{-1}，相当于激子能带在 $k = 0$ 时的裂分（图 3.28）。

83. 另外一种 D-A 复合物是强 D-A 复合物，或者自由基离子盐，即属于在基态时电荷分离就已经存在的晶体。一个经典的材料体系是四硫富瓦烯（TTF，给体）：四氰基喹二甲基（TCNQ，受体），它是导体。

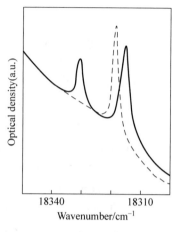

Figure. 3. 28　Detection of a CT exciton in anthracene-PMDA by means of the Stark effect. The sharp zero-phonon line at the absorption edge（dashed line，without an electric field）splits in the presence of an electric field (of ca. 4×10^4 V • cm^{-1}) into two components (solid line)

3.5.4　Surface Exciton

Molecules which are located at the surface of a crystal are not surrounded on all sides by other molecules as are those in the bulk. Therefore, surface excitons differ from bulk excitons. Due to the smaller number of neighbouring molecules, the shifts and splittings D and I etc. (described in the equations in Section 3.5.1) are smaller. The energies of the transitions lie at somewhat higher values than those of the bulk excitons[84]. This is demonstrated in Figure 3.29 using the example of anthracene. It shows the reflection spectrum from the（001）surface of an anthracene crystal. One can discern resonance and antiresonance lines which belong to the bulk excitons as well as to excitons in the first, second, and third surface layers. The identification of the lines is made possible by comparing the spectrum to the reflection from a crystal on whose surface a thin CH_4 layer had been deposited. The minima Ⅰ，Ⅱ and Ⅲ（the latter with an enlarged scale in the inset）in the broad reflection of the bulk exciton are identified as excitons from first（Ⅰ），the second（Ⅱ）and the third（Ⅲ）layers from the surface. The structures denoted by 1 and 6 correspond to the 0-0 transitions ∥ a and ∥ b of the bulk excitons. This yields a Davydov splitting of 225 cm^{-1}.

84. 晶体表面的分子不像块体中的分子那样被所有周边分子包围。因此，表面激子与块体中的不同。由于较少数量的相邻分子，其位移 D 和裂分 I 等（在 3.5.1 节的公式中被描述）都较小。某种程度上，其跃迁能比块体激子的大一些。

That of the surface exciton in surface I is equal to $218\ cm^{-1}$. The remaining structures 2 to 5 belong to phonon satellites.

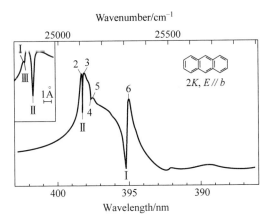

Figure 3. 29　Surface excitions in anthracene, observed in the reflection spectrum, E polarised with perpendicular incidence onto the (001) surface of the crystal (the cleavage surface of a Bridgman crystal)

In the usual crystals, surface excitons are hard to observe. Their absorption is naturally weak compared to that of the bulk and it lies in an energy range in which the bulk also absorbs. They are therefore best observed in reflection[85]. For naphthalene and anthracene, Davydov splittings and solvent shifts are seen in the 0-0 transitions $S_0 \rightarrow S_1$ which are $10\% \sim 20\%$ smaller than the known bulk transitions in these crystals. These surface excitons become important for very thin crystals, which consist, to a large extent, of surfaces.

3.5.5　Fusion/Annihilation and Fission/Splitting

Excited states can react with a ground state or other excitated state molecule. The most important processes are fusion/annihilation and fission/splitting. Fusion/annihilation between excitons has a lot of mechanisms, with or without emission, as described in the following[86]:

(1) T-T annihilation

$$T_1 + T_1 \rightarrow S_0 + T_n \rightarrow S_0 + T_1, \text{ triplet quenching}$$

$$(3\text{-}71)$$

(2) T-T delayed fluorescence

$$T_1 + T_1 \rightarrow S_0 + S_n \rightarrow S_0 + S_1, \text{ delayed fluorescence}$$

$$(3\text{-}72)$$

85. 在普通晶体中,表面激子是难以观测的。与块体相比,其吸收一定是弱的,且能量在块体的吸收范围,因此观测它们的最好手段是反射。

86. 激发态可以与一个基态或者其他激发态分子发生作用。最重要的过程是聚变和裂变。激子之间的聚变有许多机制,伴随或者不伴随发光,描述如下:

(3) S-T annihilation
$$S_1 + T_1 \rightarrow S_0 + T_n \rightarrow S_0 + T_1 + \text{heat} \qquad (3\text{-}73)$$
(4) S-S annihilation
$$S_1 + S_1 \rightarrow S_0 + S_n^* \rightarrow S_0 + S_1 \text{ or } S_0 + S_0^+ + e$$

$$(3\text{-}74)$$

In the case of Equation(3-71), starting from two triplet states, after annihilation, only a higher-energy triplet state (T_n) remains, which relaxes back to the lowest level T_1, and from it, light emission occurs, i. e. phosphorescence, or else a radiationless decay. To be sure, this process has combined two T_1 states into one with a correspondingly reduced emission intensity. In this case, we thus have a partial quenching[87]. In the case of Equation (3-72), in contrast, a higher singlet excitation state S_n is produced. This state relaxes towards S_1. From there, the fluorescence light is however emitted with a delay (Figure 3.30). Compared to fluorescence from directly formed S_1 by light absorption, the lifetime of fusion emission is longer and thus is called as delayed fluorescence, which exhibits the energy of singlet exciton and the lifetime of triplet state[88].

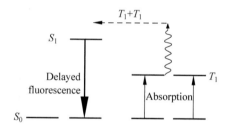

Figure 3.30　Delayed luminescence, a kind of fusion emission

Via Equation (3-72), one thus obtains, for example from anthracene, a blue delayed fluorescence if the T_1 excitons are excited directly with red light. Making use of delayed fluorescence, one can investigate in particular trapping states which are energetically close to the T_1 excitonic band, owing to the high probability that the trapping states will relax to the excitonic band T_1 following direct excitation. Pair spectroscopy takes advantage of this fact[89].

From the solution of the balance equations for mono- and bimolecular decay, one finds the dependence of the

87. 肯定地说,这个过程将两个三线态变为一个,其相关的发光强度也降低,因此这里存在部分猝灭。

88. 与直接吸收光形成的 S_1 态荧光相比,聚变产生的光发射具有长的寿命,因此称之为延迟荧光,其表现出单线态的能量和三线态的寿命。

89. 例如基于蒽,经由式(3-72)的过程,如果 T_1 激子被直接用红光激发,可以得到蓝色延迟荧光。利用延迟荧光,尤其可以研究在能量上与 T_1 激子带相近的陷阱态,这是由于在直接激发后,陷阱态弛豫到激发带 T_1 的概率非常高。配对光谱学利用了这一优势。

phosphorescence intensity I_p and the delayed fluorescence I_{DF} on the incident power I_E (or the absorption coefficient α):

(1) Under strong excitation: $I_p \sim \sqrt{\alpha I_E}$ and $I_{DF} \sim \alpha I_E$

(2) Under weak excitation: $I_p \sim \alpha I_E$ and $I_{DF} \sim (\alpha I_E)^2$

Studies of the delayed fluorescence are usually carried out in the range of strong excitation. The phosphorescence is in this case unimportant. The intensity of the phosphorescence and the delayed fluorescence produced by triplet-triplet annihilation as in Equation (3-72) can be influenced by an applied magnetic field. This takes the form of a modulation of the bimolecular rate constant γ_{TT} for the annihilation. The constant and thus the intensity of emission depend on the strength and the direction of the field relative to the crystal axes. For an example of anthracene, at a given orientation, the intensity of the delayed fluorescence from an anthracene crystal increases slightly with incrcasing magnetic field up to ca. 350 G, then decreases until it reaches a saturation value at ca. 2 kG, where it is about 80% of the intensity in zero magnetic field[90]. To explain the field dependence, we have to assume that the applied field has an influence on whether the spin selection rules can be obeyed in the fusion process. A kinetic explanation by Merrifield and Johnson assumes in the reaction $T_1 + T_1 \Leftrightarrow (T_1 \cdots T_1) \Leftrightarrow S_1 + S_0$ the existence of an intermediate pair state $(T_1 \cdots T_1)$, in which the two excitons repeatedly collide before they react. The possible spin correlations in this pair state have both triplet and singlet character. The triplet fraction in the pair state is also influenced by an applied magnetic field via the Zeeman interaction of the coupled individual spins with the field. The strength and direction of the field thus determine the relative fraction of triplet and singlet states in the pair. The singlet fraction leads to the states S_1 and S_0, and thus to delayed emission[91]. Therefore, the intensity and lifetime of the emission can be modulated by an applied magnetic field. This holds for all biexcitonic processes in which two triplet states participate. A similar magnetic-field dependence is found in photoconductivity which is stimulated by triplet excitons.

90. 如蒽的例子,在一定的方位,蒽晶体中延迟荧光的强度随着磁场强度的增加而略微增加,直到 350 G,之后随磁场强度降低,在大约 2 kG,达到饱和值,这时光发射强度大约是零磁场下的 80%。

91. 在这对能态中,可能的自旋关联既有三线态又有单线态的特点。这对能态中的三线态部分还通过相互耦合的个体自旋与磁场的塞曼作用,受到外加磁场的影响。因此磁场强度和方向决定了这一对能态的单线态和三线态的相对比例。单线态部分形成 S_1 和 S_0,进而产生延迟发射光。

92. 三线态激子通过与单线态激子发生反应，猝灭荧光。单线态激子给出能量，将较低三线态变为到较高三线态，这是允许的跃迁。

93. 因此，两个参与的 S_1 态中的至少一个，被无辐射退激，因而失去了荧光发射。随着激发强度的增加，即随着稳态 S_1 激子浓度的增加，例如在并四苯晶体中，可以观测到降低的荧光量子产率。在足够高的总能量下，这个过程甚至可以产生电子从晶体中发射出来；这就是激子湮灭导致的电离（方程（3-74）的第二个情形）。

There is also single-triplet annihilation according to the reaction scheme of Equation (3-73). Here, the triplet excitons act as fluorescence quenchers by reacting with the singlet excitons. The singlet excitons give up their energy to the triplet excitons in an allowed transition from a lower to a higher triplet state[92]. Due to their longer lifetimes, the steady-state triplet exciton concentration increases relative to the singlet concentration with increasing excitation intensity and the quenching process becomes stronger. This is important in the application of organic crystals as scintillation detectors for radiation. γ_{ST} is the rate constant for this quenching reaction.

In addition, one observes singlet-singlet annihilation or fusion according to the reaction scheme in Equation (3-74). S_n^* is here a vibronically-excited singlet state with $n > 1$, and γ_{SS} is the bimolecular rate constant. At least one of the two participating S_1 states is therefore deactivated without radiation and thus is lost for fluorescence emission. With increasing excitation intensity, i.e. with an increasing steady-state concentration of S_1 excitons, one observes for example a decrease in the fluorescence quantum yield of crystalline tetracene. With a sufficiently high total energy, this process can even cause the ejection of an electron from the crystal; that is, the exciton annihilation leads to ionization (the second case in Equation (3-74))[93]. In the case of anthracene, the ionization limit of 5.75 eV lies lower than twice the S_1 energy, 2×3.15 eV = 6.30 eV. Measurement of the kinetic energy of the emitted photoelectrons permits the verification of the fusion process. In general, the fusion of two excitons allows higher excited states to be reached with smaller energy quanta. Table 3.3 contains numerical values of the rate constants for exciton annihilation processes.

Table 3.3 Bimolecular exciton annihilation rate constants, in $cm^3 \cdot s^{-1}$, γ_{tot} is the total rate constant for triplet-triplet annihilation

	γ_{tot}	γ_{ST}	γ_{SS}
Anthrance	2×10^{-11}	5×10^{-9}	1×10^{-8}
Naphthalene	3.5×10^{-12}	5×10^{-11}	1×10^{-10}

Contrary to fusion, when an exciton with higher energy interacts with a ground state molecule, two identical excitons with lower energy can be produced. This is termed as fission or splitting (Figure 3.31). Triplet excitons originating from fission generally carry out nonradiative decay, however emission can also be observed some time. In this case the energy of the incident light is much higher than that of emission. In the meantime the emission is phosphorescence[94]. The process is described in Equation (3-75).

$$S_1 + S_0 \rightarrow T_1 + T_1 \qquad (3\text{-}75)$$

This process was discovered in tetracene. The energy of the first singlet excited state is 2.4 eV, which is nearly double to that of two triplet states (1.28 eV). With the assistant from thermal energy (ΔE), a singlet exciton can split into two triplet excitons[95]:

$$S_1 + S_0 + \Delta E \rightarrow T_1 + T_1 \qquad (3\text{-}76)$$

It is a competing process to $S_1 \rightarrow S_0$ emission and is partly responsible for the low fluorescence quantum yield from crystalline tetracene. Fission is also a biexcitonic process, which, like triplet fusion, can be modulated by an applied magnetic field with a similar dependence for fusion. The fluorescence intensity of tetracene is therefore magnetic field dependent[96].

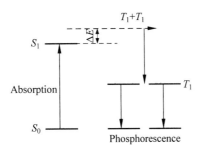

Figure 3.31   Fission emission

3.6   Energy Transfer Processes

Exciton transport refers to an exciton transferring from one molecule to anotherr. In other words, if an excited donor molecule $\mathbf{D}^*$ reverts to its ground state with the simultaneous transfer of its electronic energy to an acceptor molecule $\mathbf{A}$,

94. 与聚变相反,当具有较高能量的激子与基态分子相互作用时,可以产生具有较低能量的两个等同激子,称为裂变(图 3.31)。源自裂变的三线态激子通常进行非辐射衰变,但是有时候也可以观测到光发射。这种情况下,入射光能量远远高于发射光能量,同时发射的是磷光。

95. 这个过程发现于并四苯中。其第一单线态能量是 2.4 eV,几乎是其三线态能量(1.28 eV)的两倍。借助于热能(ΔE),单线激子可以裂变成两个三线态激子:

96. 和三线态聚变类似,裂变也是双激子过程,可以通过磁场进行调制,其磁场依赖性与聚变类似。因此,并四苯的荧光强度是磁场依赖的。

97. 激子的输运指的是激子从一个分子传递到另外一个。换句话说，如果一个处于激发态的给体分子 **D*** 回到基态，同时将能量传递给一个受体分子 **A**，使得 A 分子被激发，这个过程就被称为能量转移（ET）。

leading to the excitation of **A** molecule, the process is referred to as energy transfer（ET）[97]. When talking about energy transfer between two identical molecules, it is also referred as energy migration. Energy transfer can occur either radiatively through absorption of the emitted radiation（cascade energy transfer）or by nonradiative pathways including the Coulombic（Förster resonant energy transfer）or the exchange mechanism（Dexter energy transfer）.

3.6.1　Cascade Energy Transfer

Cascade energy transfer is a two-step process and does not involve the direct interaction of donor（**D**）and acceptor（**A**）:

$$D^* \rightarrow D + h\nu \tag{3-77}$$

$$h\nu + A \rightarrow A^* \tag{3-78}$$

The efficiency of cascade energy transfer depends on both enough optical path and a high quantum efficiency of emission from the donor in a region of the spectrum where the light-absorbing power of the acceptor is also high. Organic optoelectronic devices generally present as thin films, which are often too thin for the process of acceptor absorbing the donor emission to be considered.

3.6.2　Förster Energy Transfer

98. Förster 共振能量转移也称为荧光共振能量转移（FRET），它是由偶极-偶极之间的库仑作用引起的；因为 FRET 涉及电场相互作用而不是直接的电子交换，它是一种长程过程。共振能量转移的条件与级联能量转移类似。但共振能量转移过程没有真正的吸收和发射过程。如图 3.32 所示，分子 **A** 的跃迁偶极矩产生电场，诱导激发态分子 **D** 发生辐射衰变。在辐射发生之前，分子 **D** 的能量转移到距离为 50～100 Å 范围内的基态分子 **A** 上，从而使分子 **D** 衰变到基态，分子 **A** 被激发。

Förster resonant energy transfer also refers as fluorescence resonance energy transfer（FRET）, which is caused by dipole-dipole Coulombic interaction; and because it involves fields rather than direct electron exchange, it is a long-range process. The criteria for resonance energy transfer are similar to cascade energy transfer. But in resonance energy transfer, there are no real absorption and emission processes. As shown in Figure 3.32, the transition dipole moment of molecular **A** produces an electric field, which induces the radiative decay of the excited **D** molecule. Before radiation, the energy of **D** transfers to **A** molecule in ground state within the range of 50～100 Å, leading to the decay of **D** to ground state and the excitation of the **A** molecule[98]. FRET is typically relevant only for the singlet exciton transport, although, in some phosphorescent materials,

FRET for triplet states was also reported. Rate constant (k_{FET}) and Förster radius are the determinative factors for the efficiency of Förster energy transfer. The larger the above values, the more efficient the Förster energy transfer[99].

99. Förster 能量转的速率常数和 Förster 半径是 Förster 能量转移效率的决定因素。上述两个值越大，Förster 能量转移效率越高。

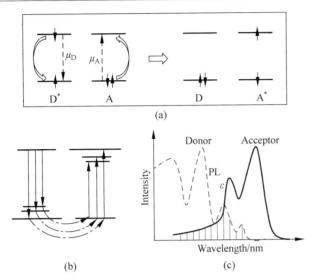

(a)

(b)   (c)

Figure 3. 32   Förster energy transfer mechanism: (a) The electronic structure of donor and acceptor before and after energy transfer. (b) The interaction between the donor and the acceptor in a dipole-dipole manner. (c) The overlap between the luminescence of the donor and the absorption spectrum of the acceptor

1. Rate Constant

Förster energy transfer is closely related to the dipole-dipole interaction between the **D** and **A** molecules, whose interaction energy is inversely proportional to (R_{DA})^3 (Equation (2-2)), where R_{DA} represents distance between the two molecules The rate constant for Förster energy transfer (k_{FET}) is proportional to the square of the dipole-dipole interaction energy, hence it is linearly dependent on $(1/R_{DA})^{6}$[100]. Combining with other factors, the k_{FET} can be expressed as

100. Förster 能量转移速率常数(k_{FET})与偶极-偶极相互作用能的平方成正比，因此它线性依赖于 $(1/R_{DA})^6$。

$$k_{FET} = k \times k_D^0 \times \frac{\kappa^2}{R_{DA}^6} J(\varepsilon_A) \qquad (3-79)$$

where, k is related to density and refraction index, κ is orientation factor, $J(\varepsilon_A)$ is spectra overlap, k_D^0 is the natural radiative decay rate of **D** (reciprocal to lifetime τ_0), which can be expressed by observed lifetime (τ) and photoluminescence quantum yield (Φ_{PL}).

$$k_D^0 = \frac{1}{\tau_0} = \frac{\Phi_{PL}}{\tau} \qquad (3\text{-}80)$$

In k_{FET} expression, the $J(\varepsilon_A)$ is the integrated spectra overlap between emission of **D** and absorption of **A** represented by extinction coefficient (ε):

$$J(\varepsilon_A) = \int_0^\infty F_D(\nu)\varepsilon_A(\nu)\frac{d\nu}{\nu^4} \qquad (3\text{-}81)$$

where, $F_D(\nu)$ is the fluorescent intensity of **D** at frequency of ν (spectrum is normalized), $\varepsilon_A(\nu)$ is the extinction coefficient of **A** at frequency of ν. Effective Förster energy transfer needs high $F_D(\nu)$ and $\varepsilon_A(\nu)$ in their overlap range. The third parameter in k_{FET}, κ^2, is the orientation factor taking into account the relative orientation of the donor emission and the acceptor absorption dipole, respectively, with value range of $0\sim4$. As shown in Figure 3.33, κ^2 attains its maximum of 4 for in-line and same orientation of the two dipoles. For randomly arranged **D** and **A** in anisotropic condition, κ^2 is taken as 2/3 [101].

101. k_{FET} 的第三个参数 κ^2，是考虑给体发射和受体吸收偶极子的相对取向的方位因子，其值的范围为 $0\sim4$。如图 3.33 所示，当两个偶极子在一条线上且取向相同时，κ^2 取最大值，为 4。对于在各向异性条件下随机排列的 **D** 和 **A**，κ^2 取值为 2/3。

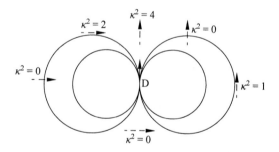

Figure 3.33　Process factor κ^2 value in Förster energy transfer. The dipole of donor **D** is indicated by a thick arrow in the middle, and the surrounding annular curve represents the spatial electric field generated by the donor **D** dipole. The dipole direction of receptor **A** is indicated by dashed arrows. **D** and **A** possess different relative position and direction, so that the κ^2 value is different, when the **D** dipole and **A** dipole in the same direction, κ^2 attains its maximum value of 4

Factors that influence the efficiency of Förster energy transfer are summarized as below： ① Good spectra overlap between emission of the donor and absorption of the acceptor. ② Strong absorption of the acceptor (The extinction coefficient of the acceptor is large enough). ③Strong to moderate emission of donor (transfer rate is proportional to donor's natural radiative rate). ④ Within distance of 100 Å. ⑤ Proportional to orientation factor (κ^2)[102]. Introducing material density and refraction index, k_{FET} can be further written as

$$k_{FET} = \frac{9000\ln(10)}{128\pi^5 n^4 N} \times \frac{1}{\tau_0} \times \frac{\kappa^2}{R_{DA}^6} \int_0^\infty F_D(\lambda)\varepsilon_A(\lambda)\lambda^4 d\lambda$$

(3-82)

where, n is refraction index, N is Avogadro constant.

2. Förster Radius

Förster radius (R_0) is the distance between **D** and **A** at which the energy transfer rate equals the rate of the radiative decay of the excited **D**[103]. This means:

$$K_{FET}(R_0) = \frac{9000\ln(10)}{128\pi^5 n^4 N} \times \frac{1}{\tau_0} \times \frac{\kappa^2}{R_0^6} \int_0^\infty F_D(\lambda)\varepsilon_A(\lambda)\lambda^4 d\lambda$$

$$= \frac{1}{\tau_D} = \frac{1}{\Phi_{PL}\tau_0}$$

(3-83)

where, R_0 is Förster radius. From Equation (3-83), one has

$$R_0^6 = \frac{9000\ln(10)}{128\pi^5 n^4 N}\Phi_{PL}\kappa^2 \int_0^\infty F_D(\lambda)\varepsilon_A(\lambda)\lambda^4 d\lambda$$

$$= \frac{0.5291}{n^4 N}\Phi_{PL}\kappa^2 \int_0^\infty F_D(\lambda)\varepsilon_A(\lambda)\lambda^4 d\lambda$$

$$= 8.78 \times 10^{-25}\Phi_{PL}\kappa^2 \int_0^\infty F_D(\lambda)\varepsilon_A(\lambda)\lambda^4 d\lambda \quad (3-84)$$

Introducing R_0 into k_{FET}'s expression gets:

$$k_{FET} = \frac{1}{\tau_D}\left(\frac{R_0}{R_{DA}}\right)^6$$

(3-85)

It can be found that the larger the Förster radius (R_0), the more efficient the Förster energy transfer. When $R_{DA} < R_0$, Förster energy transfer can compete with D emission. Small Förster radius means poor energy transfer, small distance is needed for efficient energy transfer[104].

3.6.3 Dexter Energy Transfer

Dexter energy transfer is a short-range energy transfer

102. 影响 Förster 能量转移效率的因素总结如下：①给体发射与受体吸收之间的良好光谱重叠。②受体的强吸收(受体的消光系数足够大)。③给体具有中强发射(转移速率与给体的自然辐射率速率成比例)。①给体与受体之间的距离范围在 100 Å 以内。⑤与方位因子(κ^2)成比例。

103. Förster 半径(R_0)是指能量转移速率与激发态 **D** 辐射衰变速率相等时，**D** 和 **A** 之间的距离

104. 可以发现，Förster 半径(R_0)越大，Förster 能量转移效率越高。当 $R_{DA} < R_0$：Förster 能量转移可与 D 的发射竞争。较小 Förster 半径意味着能量转移较差，需要较小的距离才能进行有效的能量转移。

105. Dexter 能量转移是一种短程的能量转移过程,它依赖于分子轨道的重叠,是通过直接电子交换发生的。它是指电偶极禁阻的跃迁,也就是说,那些发生在多极矩相互作用中,或者是电子交换产生的。

process, which relies on the overlap of molecular orbitals and proceeds via direct exchange of electrons. It refers to the electric dipole-forbidden transitions, namely, those occur by multipole interactions or by electron exchange[105].

Figure 3.34 shows Dexter energy transfer between donor (**D**) and acceptor (**A**). Due to the interaction between LUMOs of donor and acceptor, and the interaction between HOMOs of donor and acceptor, electron in LUMO of donor transfers to LUMO of acceptor, meanwhile, electron in HOMO of acceptor transfers to HOMO of donor, resulting in excited acceptor and ground state donor. The Dexter mechanism applies both to singlet and triplet excitons, however, the degree is quite different. In pure molecular crystals, the coupling energy between a singlet exciton and a ground state molecule is usually in the wavenumber range of $100 \sim 10000$ cm^{-1}. The time required for the excitation energy transferring from one molecule to another is $10^{-14} \sim 10^{-12}$ s, which is shorter than the period of lattice vibration. In this case, energy may be rapidly transported along a chain of molecules to such a position as to complete Dexter energy transfer; if the coupling between adjacent molecules is very low, for example, the coupling energy of a triplet exciton and an adjacent ground state molecule is about 1 cm^{-1}, energy transfer efficiency of Dexter is very low. There is a strong competition between exciton energy transfer process and the decay process through phosphorescence radiation.

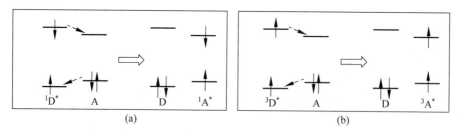

(a)   (b)

Figure 3.34　Dexter energy transfer from singlet excited state to singlet ground state (a) and from triplet excited state to singlet ground state (b)

Requirements for Dexter energy transfer: ① better spectra overlap between the **D** emission and the **A** absorption to provide orbital overlap of the two pairs of concerned energy levels; ② spin conservation; ③ within 10 Å separation between **D** and **A** to render spatial vicinity[106]. The rate for Dexter energy transfer (k_{DET}) is given by D. L. Dexter:

$$k_{DET}(\text{exchange}) = K \cdot J_{DA} \cdot \exp\left(-\frac{2R_{DA}}{L}\right) \quad (3\text{-}86)$$

where, K indicates the interaction between two orbitals; J_{DA} is the integration of the spectra overlap (normalized); R_{DA} is distance between **D** and **A**; L is their average van de Waals radius.

Dexter energy transfer can be described by Marcus theory, which is proposed for electron transfer. The difference is that Dexter energy transfer involves four frontier orbitals, while classical electron transfer process only involves two orbitals. With Marcus theory, k_{DET} can be expressed as

$$k_{DET} = \frac{2\pi}{\hbar} J_{DA}{}^2 \sqrt{\frac{1}{4\pi k_B T \lambda}} \exp\left(-\frac{E_a}{k_B T}\right) \quad (3\text{-}87)$$

where, J_{DA} represents electron coupling, k_B is Boltzmann constant, E_a is activation energy (Figure 3.35) and can be expressed as

$$E_a = \frac{\lambda}{4}\left(1 + \frac{\Delta G^0}{\lambda}\right)^2 \quad (3\text{-}88)$$

where, ΔG^0 is Gibbs free energy (i.e. formation energy), λ is reorganization energy.

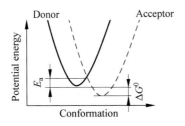

Figure 3.35　Activation energy E_a and Gibbs free energy ΔG^0 for Dexter double-electron transfer

The electron reorganization energy relates to the vibrational relaxation when an electron transfers from one orbital to another. If **D** and **A** are identical, the involved two

106. Dexter 能量转移要求：①**D** 发射和 **A** 吸收之间具有好的光谱重叠，来达到两对相关能级的轨道重叠；②自旋守恒；③**D** 和 **A** 的距离在 10 Å 内，以便它们空间上靠近。

107. 电子重组能与电子从一个轨道转移到另一个轨道时的振动弛豫相关。如果 **D** 和 **A** 是相同的，电子转移涉及的两个轨道能级也是相同的，如 LUMO 对 LUMO 和 HOMO 对 HOMO，但是它们的构象是不同的，因为一个分子在基态，另一个在激发态。因此，这种情况下的重组能是指与激发态分子构象相关的垂直能量差异（图 3.36（a））另外，此时 $\Delta G^0 = 0$，根据式(3-88)，活化能是 $E_a = \dfrac{\lambda}{4}$。

energy levels for electron transfer is also identical，e. g. LUMO versus LUMO and HOMO versus HOMO，but their configuration is different because one molecule is in ground state and the other is in excited state. Therefore，the reorganization energy in this case is the vertical energy difference relative to the conformation of the excited molecule（Figure 3. 36（a））. In addition，now，$\Delta G^0 = 0$，according to Equation（3-88），the activation energy is

$$E_a = \frac{\lambda}{4} \quad {}^{107}.$$

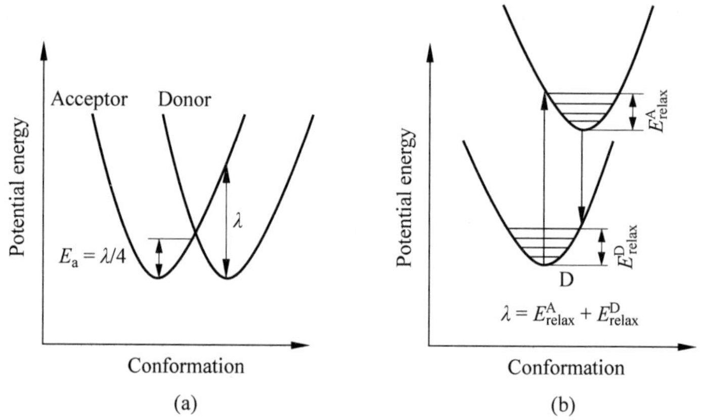

Figure 3. 36　Reorganization energy，λ，in Dexter energy transfer，with **D** and **A** are identical（a）and different（b）

108. 当 **D** 和 **A** 不同时，根据 Franck-Condon 原理的垂直跃迁特点，在 Dexter 能量转移后，**D** 和 **A** 都有振动松弛，因此总的重组能是 **D** 和 **A** 振动松弛能量的总和，如图 3.36(b)所示。

When **D** and **A** are different，according the vertical transition feature by Franck-Condon principle，after Dexter energy transfer，**D** and **A** all subject to vibrational relaxation，thus the total reorganization energy is the sum of that from **D** and **A**，as shown in Figure 3.36(b)[108].

In Marcus electron transfer theory，as shown in Figure 3.37，Dexter energy transfer has three situations：①Normal region，$|\Delta G^0| < \lambda$. In this region，Dexter energy transfer rate increases as the ΔG^0 increases. ②Zero activation energy region，$|\Delta G^0| = \lambda$. In this region，activation energy is zero，Dexter energy transfer rate reaches its maximum. ③Inverted Marcus region，$|\Delta G^0| > \lambda$. In this region，Dexter energy transfer rate decreases as the ΔG^0 increases.

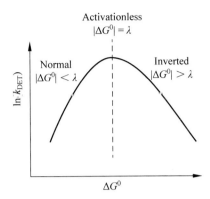

Figure 3.37 Dependence of Dexter energy transfer rate on ΔG^0

3.7 Dynamic Processes in Excitons

3.7.1 Exciton Diffusion

Exciton transport in organic semiconductors is important for many optoelectronic applications, including organic solar cells and organic light emitting diodes. Exciton can be transported by diffusion, that is, exciton diffuse from high concentration to low concentration[109]. Normal diffusion can be described by the following equation:

$$\frac{\partial n}{\partial t} = D \nabla^2 n - \frac{n}{\tau} + G \qquad (3\text{-}89)$$

where, n is the exciton density; D is a diffusion coefficient, ∇^2 is Laplace operator, τ is the exciton lifetime, G is the exciton generation rate. Exciton diffusion length (L_D) is a measure of how far on the average an exciton moves from its initial point during its lifetime[110]. As shown in Figure 3.38, diffusion length (L_D) is the distance between X_0 and X, and for isotropic material it can be expressed as

$$L_D = \sqrt{ZD\tau} \qquad (3\text{-}90)$$

where, τ is the exciton lifetime, D is a diffusion coefficient ($cm^2 \cdot s^{-1}$), Z is equal to 2, 4 or 6 in case of one-, two- or three-dimensional diffusion, respectively. For estimation, $Z = 1$.

When hopping of the exciton is mediated by Förster energy transfer, the diffusion coefficient can be expressed as

109. 有机半导体中激子输运在包括有机太阳能电池和有机电致发光的许多光电应用中都很重要。激子可以通过扩散输运,即激子从高浓度向低浓度扩散。

110. 激子扩散长度(L_D)是一个激子在其寿命期间,距离其初始位置平均远近的度量。

$$D = \frac{R^2}{6\tau_{\text{hop}}} = A \frac{1}{\tau_0} \frac{R_0^6}{6d^4} \qquad (3\text{-}91)$$

where, τ_{hop} is the hopping time between the chromophores, τ_0 is the intrinsic exciton lifetime that is not limited by diffusion limited quenching at defects, d is the distance between the chromophores, R_0 is Förster radius that can be obtained from Equation (3-84), A is a constant that accounts for the distribution of molecular separations d. Then, the exciton diffusion length in a solid is

$$L_D = \sqrt{D\tau} = \frac{R_0^3}{d^2} \sqrt{A \frac{\tau}{6\tau_0}} \qquad (3\text{-}92)$$

where, τ is photoluminescence lifetime in solid film. Equations (3-91) and (3-92) can be used to estimate exciton diffusion parameters in organic semiconductors.

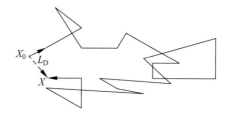

Figure 3.38　Random motion and diffusion length of excitons

111. 在有机材料中,单线态和三线态激子的扩散系数 D 通常分别为 $10^{-4} \sim 10^{-2}$ cm^2 · s^{-1} 和 $10^{-6} \sim 10^{-3}$ cm^2 · s^{-1}。扩散长度 L_D 一般是单线态在 $5 \sim 20$ nm;三线态在 $30 \sim 300$ nm。当然也有报道提到过单线态激子扩散长度大于 100 nm,三线态激子扩散长度大于 1 μm(在个别单晶中甚至大于 10 μm)。在传输过程中,激子可以与基态分子或者其他激发态分子反应,它也可以与陷阱、载流子、声子、光子或者表面反应。这些过程影响激子寿命。

In organic materials, typical values of diffusion coefficient D are 10^{-4} to 10^{-2} cm^2 · s^{-1} for singlet excitons and 10^{-6} to 10^{-3} cm^2 · s^{-1} for triplet excitons. Typical diffusion lengths L_D are ~ 5 nm to 20 nm for singlets and ~ 30 to 300 nm for triplets, although >100 nm diffusion lengths for singlets and >1 μm (and even >10 μm in several single crystals) for triplets have also been reported. In the transport process, the exciton can react with a ground state molecule and other excited state molecule, and it also can react with a trap, a carrier, a phonon, a photon or the surface. These processes affect the lifetime of the exciton[111].

3.7.2　Interaction of Exciton with Various Species

1. Interaction with Exciton

The interaction of exciton with another exciton is also referred as annihilation, which has been partially discussed in Section 3.5.5. Here is the more comprehensive summarization for exciton annihilation.

1) Singlet-Singlet Interaction

The singlet-singlet annihilative reactions produce a highly excited singlet state (S_n^*) and a ground state molecule. The vibronic state S_n^* can either convert internally to S_1, and decay back to the ground state giving off light, or it can ionize[112]:

$$S_1 + S_1 \rightarrow S_n^* + S_0 \rightarrow S_1 + S_0 + \text{phonon} \qquad (3\text{-}93)$$

$$S_1 + S_1 \rightarrow S_n^* + S_0 \rightarrow e + S_0^+ + S_0 \qquad (3\text{-}94)$$

where S_n^* is a vibronically excited singlet state ($n \geqslant 1$).

Generally, only one singlet exciton is lost in the fusion process because the decay of S_n^* produces one S_1, and because the ionization process is less efficient than the electronic de-excitation process; in addition, subsequent to the ionization process, geminate carrier recombination takes place, and this recombination tends to produce singlet states.

2) Triplet-Triplet Interaction

The triplet-triplet annihilative (TTA) reaction produces a higher energy singlet (or triplet) state and a ground state molecule.

$$T_1 + T_1 \underset{k_{-1}}{\overset{k_1}{\rightleftharpoons}} (T_1 T_1) \overset{k_2}{\longrightarrow} S_1^* , \quad T_2 \text{ or } T_1^* + S_0 + \text{phonon}$$

$$(3\text{-}95)$$

TTA process is often a most severe one due to the long lifetime of triplet state. The total quenching rate for triplet-triplet interaction is

$$r_{\text{tot}} = \frac{k_1}{1 + \dfrac{k_{-1}}{k_2}} \qquad (3\text{-}96)$$

3) Singlet-Triplet Interaction

As a result of the forbidden nature of the singlet-triplet transition, the lifetime of the triplet exciton is quite long, making possible the generation of large concentrations of these excitons in condensed systems. Triplet excitons act as efficient quenchers of singlet excitons; thus producing a higher energy triplet state and a ground state molecule[113]:

$$S_1 + T_1 \rightarrow T_1^* \quad \text{or} \quad T_2 + S_0 \qquad (3\text{-}97)$$

In this case, the singlet exciton disappears, passing its energy to the triplet exciton in an allowed transition from a lower to an upper triplet state.

112. 单线态-单线态湮灭反应产生一个更高的激发单线态(S_n^*）和一个基态分子。这个电子振动态 S_n^*，可以通过内转换回到 S_1 态，然后衰变到基态并发光；它也可以发生电离。

113. 由于单线态-三线态跃迁的禁阻特性，三线态激子的寿命比较长，在聚集体系中可能产生较高浓度的三线态激子。三线态激子是单线态激子的有效猝灭剂，这样就产生一个更高能量的三线态和一个基态分子。

The singlet-triplet annihilative reaction can also produce photon generated carrier and a ground state molecule.

$$S_1 + T_1 \rightarrow e + S_0^+ + S_0 \tag{3-98}$$

2. Interaction with Ground State Molecule

An exciton interacts with a ground state molecule can produce two triplet excited state, this is the fission of singlet exciton (Section 3.5.5).

$$S_1 + S_0 \rightarrow T_1 + T_1 \tag{3-99}$$

An exciton interacts with a ground state molecule also can form excimer or exciplex (Section 3.5.2).

$$S_1 + S_0 \rightarrow \text{Excimer} \rightarrow S_0 + S_0 + h\nu \tag{3-100}$$

$$S_1 + S_0' \rightarrow \text{Exciplex} \rightarrow S_0 + S_0' + h\nu \tag{3-101}$$

3. Interaction with Trap

A singlet exciton interacts with a trap can produce a trapped singlet exciton (S_{1G}), a phonon and a ground state molecule[114].

$$S_1 + \text{trap} \rightarrow S_{1G} + S_0 + \text{phonon} \tag{3-102}$$

The trapped singlet exciton (S_{1G}) can change to triplet via fission process:

$$S_{1G} + S_0 \rightarrow T_1 + T_{1G} \tag{3-103}$$

These resulted triplet and trapped triplet can react reversely, back to the initial state of $S_{1G}^* + S_0$, or leading to $T_{1G}^* + S_0$:

$$T_1 + T_{1G} \rightleftharpoons (T_1 T_{1G}) \rightarrow S_{1G}^* \text{ or } T_{1G}^* + S_0 \tag{3-104}$$

4. Interaction with Carrier

A singlet exciton can be quenched readily by a trapped (e_t/h_t) or free (e/h) charge by transferring its energy to the ion[115].

Exciton quenching:

$$S_1 + e/h \rightarrow S_0^* + e/h + \Delta \tag{3-105}$$

$$T_1 + e/h \rightarrow S_0^* + e/h + \Delta \tag{3-106}$$

Detrap carrier:

$$S_1 + e_t/h_t \rightarrow S_0^* + e/h + \Delta \tag{3-107}$$

$$T_1 + e_t/h_t \rightarrow S_0^* + e/h + \Delta \tag{3-108}$$

5. Interaction with Surface

When an exciton reaches the surface of a crystal after

114. 一个单线态激子与陷阱作用,可以产生一个被捕获的单线态激子、一个声子和一个基态分子。

115. 一个单线态激子能很容易地被一个捕获(e_t/h_t)或自由(e/h)的电荷猝灭,将其能量传递给离子。

diffusing through the bulk material, it can either be reflected back into the crystal or it can be quenched. A variety of quenching processes are available at the surface. One process involves a reaction with an impurity molecule whose energy levels lie below those of the donor. Thus, if an acceptor is placed on the surface of a donor crystal, a singlet exciton of donor can transfer its energy to the acceptor molecule, producing an electronically excited state of acceptor. If a metal film covers the surface, the exciton can transfer its energy to the free electrons inside the metal. If an aqueous salt solution makes contact with the surface, the exciton can react with the solution and dissociate into a positive and negative charge.

Charge transfer

$$S_1(T_1) + \text{surface trap} \rightarrow S_0^+ + \text{trap}^- \qquad (3\text{-}109)$$

$$S_1(T_1) + \text{surfac trap} \rightarrow S_0^- + \text{trap}^+ \qquad (3\text{-}110)$$

Energy transfer

$$S_1(T_1) + \text{surface trap} \rightarrow S_0 + \text{trap}^* \qquad (3\text{-}111)$$

If an excited organic molecule locates at metal surface, generally within 20 Å, the energy level of the organic molecule can be changed, due to dielectric constant discontinuity at the interface; and the lifetime of the exciton will be shorten, originating from the nonradiative decay caused by metal induction[116].

6. Interaction with Photon

An exciton is of course possible to become more highly excited by the absorption of additional energy, such as a photon. In the particular case of the singlet exciton in anthracene, the absorption of a photon of suitable energy creates an excited state that can ionize and produce a photo-current. Indeed, it is by measuring these currents that one can study the photon-singlet exciton interaction.

The interaction between exciton and photon can lead to photoionization.

$$S_1 + h\nu \rightarrow S_0^+ + e \qquad (3\text{-}112)$$

$$T_1 + h\nu \rightarrow S_0^+ + e \qquad (3\text{-}113)$$

116. 如果激发态有机分子位于金属表面，通常在 20 Å 以内，有机分子的能级可能被改变，这是由于界面处的介电常数不连续；激子寿命缩短，这是由于金属诱导引起的非辐射衰变。

141

7. Interaction with Phonon

When Frenkel exciton interact with phonon, its band width can be reduced; and the effective mass of exciton will increase with the increment of exciton-phonon coupling. These influences are more serve to exciton of higher energy; and the wider the exciton band, the smaller the influence. Exciton-phonon coupling also influences the intensity distribution of the corresponding electronic level.

本章小结

1. 内容概要：

本章聚焦于有机材料的能量相关过程，它们是有机材料光电应用需要考虑的主要因素。在简单介绍之后，首先讨论了能量相关的基本概念，然后学习了分子内的电子跃迁过程，包括光吸收产生激发态以及激发态的衰变过程。为了更好地理解光学跃迁，我们全面展示了光学跃迁的选择定则。之后，内容移向有机聚集态上，讨论了 Davydov 裂分、激基二聚物/激基复合物、电荷转移/表面激子以及聚变/裂变。能量转移机是下一个内容，包括级联能量转移、Förster 能量转移和 Dexter 能量转移。本章的最后，总结了激子的各种动力学过程。

2. 基本概念：基态、激发态、激子、单重态激发态、三重态激发态、电子跃迁、光学吸收、内转换、系间窜越、发射、荧光、磷光、三线态-三线态湮灭延迟荧光、热激活延迟荧光、自旋-轨道耦合、自旋守恒、光学选择定则、爱因斯坦方程、光致发光量子产率、激发态寿命、达维多夫分裂、激基二聚物、溶剂位移、激基复合物、电荷转移激子、表面激子、激子裂变（分裂）、激子聚变（湮灭）、级联能量转移、非辐射能量转移、Förster 能量转移、Förster 半径、Dexter 能量转移、激子扩散。

3. 主要公式：

朗博-比尔定律表达式：$A = \lg\left(\dfrac{I_0}{I}\right) = \lg\left(\dfrac{1}{T}\right) = -\lg T = OD = \varepsilon \cdot c \cdot l$ (solution) $= ad$ (film)

分子的总波函数：$\Psi_t = \Psi_e' \chi_v \Psi_r$

分子中电子的能量表达式：$E_t = E_e + E_v + E_r$

给定电子能级的振动能量表达式：$E_v^* = \dfrac{1}{2}(2m + 1)E_v$

跃迁偶极矩：$R_{lu}^2 \propto |\langle \Psi_{el} | \boldsymbol{M} | \Psi_{eu} \rangle|^2 |\langle \chi_{el} | \chi_{eu} \rangle|^2 |\langle \Psi_{sl} | \Psi_{su} \rangle|^2$

爱因斯坦方程：$B_{un,lm} = B_{lm,un}$，$\dfrac{A_{un,lm}}{B_{un,lm}} = \dfrac{8\pi h\nu^3 n_0^3}{c^3}$

爱因斯坦系数：$B_{l0,un} = \dfrac{c}{hn_0}\displaystyle\int \dfrac{\sigma(\nu)\,\mathrm{d}\nu}{\nu}$

吸收截面：$\sigma = \dfrac{2303\varepsilon}{N} = 3.81 \times 10^{-21}\varepsilon$

吸收速率常数：$B_{10,\text{un}} = \dfrac{2303c}{hn_0 N} \displaystyle\int \dfrac{\varepsilon(\nu)\,\mathrm{d}\nu}{\nu}$

理论发光寿命：$\tau_R^{-1} = \dfrac{8\pi hc}{n_0}\langle\tilde{\nu}^{-3}\rangle^{-1} B_{lu}$，$\langle\tilde{\nu}^{-3}\rangle^{-1} = \dfrac{\displaystyle\int I(\tilde{\nu})\,\mathrm{d}\tilde{\nu}}{\displaystyle\int \tilde{\nu}^{-3} I(\tilde{\nu})\,\mathrm{d}\tilde{\nu}}$

光致发光量子产率：$\text{PLQY}_F = \dfrac{K_r}{K_r + K_{nr}} = \dfrac{\tau}{\tau_r}$，$\text{PLQY}_{ph} = \eta_{ISC}\dfrac{K_r}{K_r + K_{nr}} = \eta_{ISC}\dfrac{\tau}{\tau_r}$

达维多夫分裂的共振能：$I_{12} \propto z \cdot \dfrac{M}{R^3} \approx \text{const.}\dfrac{f}{\Delta E_M}$

电荷转移激子的激发能：$E_{CT} = IP_D - EA_A - P_{eh}(r) - C(r)$

Förster 能量转移的速率常数：

$$K_{\text{FET}} = k \times k_D^0 \times \frac{\kappa^2}{R_{DA}}J(\varepsilon_A)$$

$$= \frac{9000\ln(10)}{128\pi^5 n^4 N} \times \frac{1}{\tau_0} \times \frac{\kappa^2}{R_{DA}^6}\int_0^\infty F_D(\lambda)\varepsilon_A(\lambda)\lambda^4\,\mathrm{d}\lambda = \frac{1}{\tau_D}\left(\frac{R_0}{R_{DA}}\right)^6$$

Dexter 能量转移的速率常数：

$$k_{\text{DET}} - \frac{2\pi}{\hbar}J_{DA}^2\sqrt{\frac{1}{4\pi k_B T\lambda}}\exp\left[-\frac{E_a}{k_B T}\right],\quad E_a = \frac{\lambda}{4}\left[1 + \frac{\Delta G^0}{\lambda}\right]^2$$

激子扩散长度：$L_D = \sqrt{ZD\tau}$

Glossary

Absorbance	吸光度
Absorbance constant	吸收常数
Absorption	吸收
Absorption cross section	吸收截面
Absorption extinction coefficient	吸收消光系数
Acceptor	受体
Activation energy	活化能
Allowed	允许的
Annihilation	湮灭
Anthracene	蒽
Approximation	近似
Configuration	轮廓,结构,组成
Conformation	构象,结构,形态
Davydov splitting	达维多夫分裂
Decay process	衰变过程

Delayed fluorescence	延迟荧光
Detrap	解除被捕获
Dexter energy transfer	Dexter 能量传递
Diamagnetic	反磁性的
Diffusion	扩散
Diffusion coefficient	扩散系数
Diffusion length	扩散长度
Dimer	二聚体
Donor	给体
Dopant	掺杂物
Dynamic process	动力学过程
Eigenenergy	本征能量
Eigenvalue	本征值
Electromagnetic field	电磁场
Electron-hole pair	电子-空穴对
Emission	发射
Emission quantum yield	发光量子产率
Exchange	交换
Excimer	基激二聚体
Exciplex	激基复合物
Excitation	激发
Excited state	激发态
Exciton	激子
Fine structure	精细结构
Fission/splitting	裂变
Fluorescence	荧光
Forbidden	禁止
Förster energy transfer	Förster 能量传递
Franck-Condon principle	夫兰克-康登原理
Free energy	自由能
Frenkel exciton	弗仑克尔激子
Frontier orbital	前沿轨道
Fusion/annihilation	聚变/湮灭
Ground state	基态
Hamilton operator	哈密顿算符
Heavy metal	重金属
Identical	同一的
Incident light	入射光
Infrared	红外

Integrated spectra	积分光谱
Inter system crossing	隙间穿越
Internal conversion	内转换
Intrinsic photoconducting	本征光导
Kinetic energy	动能
Lattice vibration	点阵振动
Lifetime	寿命
Light excitation	光激发
Mirror symmetry	镜像对称
Monomer	单体
Multiplicity	多重性
Nonradiative decay	非辐射衰变
Normalized	规范化的,归一化的
Optical transition	光学跃迁
Orientation factor	取向因子
Oscillator strength	谐振子强度
Paramagnetism	顺磁性
Pauli exclusive theory	泡利不相容原理
Phonon	声子
Phosphorescence	磷光
Photoionization	光电离
Photoluminescence quantum yield	光致发光量子产率
Physical dimer	物理二聚物
Polariton	极化声子/激元,电磁耦合波子
Polarized light	偏振光
Polaron	极化子
Probability	概率
Pyrene	芘
Quantum chemistry	量子化学
Quasiparticle	准粒子
Quenching	淬灭
Radiationless	非辐射的
Radiative	辐射的
Radiative decay	辐射衰变
Radius	半径
Random	随机的
Rate constant	速率常数
Redshift	红移
Refractive index	折射率

Reorganization energy	重组能
Resonance	共振
Resonance energy transfer	共振能量转移
Rotational	转动的
Schrödinger equation	薛定谔方程
Selection rule	选择定则
Selfionization	自电离
Singlet	单线态
Splitting	分裂
Spontaneous emission	自发发射
Statistically	统计上地
Stimulated emission	受激发射
Stokes shift	斯托克斯位移
Surface	表面
Symmetry	对称性
Tetracene	并四苯
Thermal activated	热活化的
Thermally activated	热激活的
Transition dipole	跃迁偶极
Transport	传输
Trap	俘获，陷阱
Triplet	三线态
Ultraviolet	紫外的
Variational principle	变分原理
Vertical transition	垂直跃迁
Vibrational	振动的
Visible light	可见光
Wanneir exciton	万尼瓦激子
Wavefuction	波函数
Wavenumber	波数
Wavevector	波矢
Zero energy	零点能

Problems

1. True or false questions

（1）In triplet state，the spin directions of the two unpaired electrons are antiparallel.

（2）In singlet state，the spin directions of the two electrons are the same.

（3）In the ground state of organic molecules, electrons are generally paired, and the total spin number is zero.

（4）In an excited organic molecule, nonradiative decay processes always exist.

（5）Radiative decay processes refer to the electron transition from higher energy level to lower energy level without light emission.

（6）An internal conversion process can not occur between two different electronic levels.

（7）The rate of internal conversion is faster than the radiative decay process.

（8）The internal conversion is a nonradiative decay process between electronic energy level with different multiplicity.

（9）The inter system crossing is a nonradiative decay process between electronic energy level with different multiplicity.

（10）Compared to pure organic materials, metal complexes exhibit more strong spin-orbital coupling, as well as more strong phosphorescence.

（11）Spin-orbital coupling renders triplet state lost its feature and exhibiting some singlet characteristic, which can promote inter system crossing and phosphorescent emission.

（12）Generally, inter system crossing is forbidden, spin-orbital coupling can release the forbidden feature.

（13）For pure organic material, under optical excitation almost all excitons are singlet.

（14）Frenkel exciton has the smallest radius.

（15）Wannier exciton is generally found in inorganic semiconductors.

（16）An electron-hole pair bonded in adjacent molecules is Frenkel exciton.

（17）Franck-Condon shift is a Stokes shift caused by molecular conformation shift.

（18）According to Franck-Condon principle, and if the environmental influence to the two states is small, the absorption and the emission spectra will be mirror symmetry.

（19）In spectra of optical absorption or emission, the rotational energy some time is observable as fine structure.

（20）In spectra of optical absorption or emission, the vibrational energy some time is observable as fine structure.

（21）The fine structure of the absorption spectrum is related to the vibrational levels of the excited state.

（22）Dipole allowed optical transition occurs between two electronic states with same symmetry.

（23）Transition between orbitals that are centered on the different part of the molecules will be efficient.

（24）The electronic part controls the overall intensity of the electronic transition.

（25）Electronic transition between initial and final states with same multiplicity is spin-forbidden transition.

(26) For the same vibrational energy level, at turning point, the probability for electron transition is maximum.

(27) When two monomer forms physical dimer, both the ground state and excited state of the dimer are lowered compared to those of the monomer.

(28) According to mini exciton model for the explanation of Davydov splitting, solvent shift leads to the blue shift of absorption spectrum.

(29) In a physical dimer comprising two monomers, there exists chemical bond between the two monomers.

(30) Under the explanation of Davydov splitting, compared to monomer, the blue-shifted absorption from the dimer is caused by H-aggregation.

(31) Generally speaking the Davydov splitting from singlet state is wider compared to that from triplet state.

(32) For a one dimensional crystal with N molecules, the wavefunction of this crystal in excited state can be expressed by a product of the N molecular wavefunctions of the ground state.

(33) According to Davydov splitting, the excited state of a one-dimensional crystal with N molecules can have N degenerated states.

(34) According to Davydov splitting, if there are two translationally inequivalent molecules in one unit cell, the excited state will have two Davydov bands.

(35) Compared to monomer emission, the excimer emission is redshift, structureless, but short lifetime.

(36) There is no ground state for excimer, but exists ground state for exciplex.

(37) Electric field generally assists the dissociation of exciplex.

(38) Excimer is formed by two different molecules.

(39) The interaction between two lower energy excitons produces an exciton with higher energy and a ground state molecule. This is exciton fission process.

(40) An exciton with higher energy interacts with a ground state molecule, producing two identical excitons with lower energy. This is exciton annihilation process.

(41) Radiative energy transfer is a one-step process and involves the direct interaction of donor and acceptor.

(42) Förster energy transfer is a resonance energy transfer by dipole interaction between the donor molecule and the acceptor molecule.

(43) In Förster energy transfer, there are real absorption and emission processes.

(44) Förster energy transfer is a short range energy transfer process.

(45) FRET is typically relevant only to the singlet exciton.

(46) Dexter energy transfer is a resonance energy transfer by dipole interaction between the donor molecule and the acceptor molecule.

(47) Dexter energy transfer is a short range energy transfer process.

(48) In organic semiconductors, excitons can diffuse from high concentration to low concentration.

2. What is exciton? According to exciton radius, please classify it, and point out the radius (r) and energy range (E) of each exciton.

3. Please compare the magnitude of energy between the excited triplet state and the corresponding single state, and give brief explanation.

4. Please summarize the differences between singlet and triplet states.

5. Please give definition for transmittance (T) and absorbance (A), and use equation to express them; please give the expression of Lambert-Beer law both in solution and film.

6. Please describe optical absorption features of organic materials

7. Please explain: (1)internal conversion (IC), (2)inter system crossing (ISC) and point out their similarity and difference.

8. Please explain prompt fluorescence, phosphorescence, and delayed fluorescence.

9. Please describe the levels within one electronic level in an organic molecule, and point out the energy range (eV) for transition within each type.

10. What is hot electron transition? Is it common in organic materials? Why?

11. According to quantum chemistry, neglecting rotation and under Born-Oppenheimer approximation, please give the expressions of the probability for a molecule to be optically excited. And explain the factors that determine optical transition, and give detail requirements for optical transition between two states.

12. What is Franck-Condon factor? Please explain Franck-Condon principle and Franck-Condon shift.

13. Based on Figure 3.13, please point out the three light intermediated processes, and briefly state the content of Einstein equation.

14. What is photoluminescence quantum yield? Please use the radiative decay rate and the nonradioactive decay rate to express the photoluminescence quantum yield of fluorescence and phosphorescence.

15. Why the emission efficiency of heterocyclic aromatic compounds is generally lower than that of pure aromatic hydrocarbons?

16. Please explain the fine structure that overrides on a pure electronic level. For the fine structure of the absorption spectrum and emission spectrum of a molecule, what is each related to? Why the fine structure of emission spectra in some molecules could be lost?

17. According to Figure 3.16, please point out the spectrum difference in the case of (1) solution and crystal, (2) two perpendicularly irradiated incident light to crystal. Please also point out the reasons for the well discernible fine structure in the solution absorption spectrum.

18. Using quantum chemistry manner, please: (1)explain the absorption redshift of

the physical dimer compared to monomer; (2)deduce the transition dipole moment of a physical dimer.

19. Based on mini excition model, please explain the H-aggregation, J-aggregation, and the spectrum broadening phenomena.

20. Please explain the Davydov band in one dimensional crystal comprising N molecules, and with only one molecule in a unit cell.

21. Please explain the two Davydov band in one dimensional crystal comprising N molecules, and with two molecules in a unit cell.

22. What are excimer and exciplex? Please point out the features of excimer and exciplex emission compared to the monomer emission.

23. What are fusion and fission? Please present these two processes with figures.

24. Please state the similarity and difference between fission process and Davydov splitting

25. Please explain Förster radius and state its relationship with energy transfer efficiency.

26. According to Förster energy transfer rate shown in Equation (3-82), please describe the main influencing factors for this process.

27. Please explain orientation factor in Equation (3-82).

28. What is Dexter energy transfer? Which exciton transport mainly through DET mechanism, why?

29. Please state the main requirement for Dexter energy transfer.

30. Using equation, please briefly summarize main dynamic processes suffered by exciton in organic materials, and give a simple explanation for each.

CHAPTER 4 ELECTRICAL PROPERTIES AND RELATED PROCESSES IN ORGANIC MATERIALS

In this chapter, firstly, concepts for electricity, such as conductivity and mobility, are provided, along with the description of energy diagram and carrier type/transport in semiconductors, as well as the introduction for the electrical property of organic materials. In the second part, carrier mobility especially that for organic materials is discussed comprehensively; In terms of temperature and electrical field dependent mobility, models both for crystalline and amorphous solids are addressed. The third part gives discussion about the three valid electrical contacts between metal electrode and organic semiconductor, along with the relationship between current and voltage under each condition. Then, this chapter talks about injection limited current and space charge limited current in organic semiconductors after carrier injection; the recombination of carriers in organic semiconductors are briefly discussed as last.

4.1 Conductivity and Carriers in Solids

4.1.1 Basic Concepts for Electricity

1. Conductivity

Collectively, conductivity (σ) is the primary parameter to evaluate the electrical conducting property of materials, which is defined as conductance per unit length ($S \cdot m^{-1}$), and can be expressed as[1]

$$\sigma = \frac{1}{RL} = \frac{j}{F} \qquad (4\text{-}1)$$

where R is resistance, the reciprocal of conductance, L is the length of the sample, j is current density, F is electric field intensity. According to conductivity, materials can be divided into conductor, semiconductor, insulator and superconductor, as shown in Table 4.1.

1. 总地来说,电导率(σ)是评价材料导电性的基本参数,定义为单位长度下的电导($S \cdot m^{-1}$),可以表达为

Table 4.1　Electrical properties of different materials

	Insulator	Semiconductor	Conductor	Superconductor
Band gap /eV	> 5	$0.1 \sim 3$	very small	—
Conductivity /(S·cm^{-1})	$10^{-22} \sim 10^{-15}$	$10^{-8} \sim 10^{-2}$	$10 \sim 10^4$	∞

2. Mobility

In the simplest case of a conducting solid, the current is due to only a simple type of charge carrier with a charge q, e.g., electrons ($q = -e$), whose density is n. If the charge carriers move with a drift velocity v_D, then an electric current density flowing through the conducting solid is

$$j = qnv_D \tag{4-2}$$

Generally, drift velocity of charge-carrier is proportional to electric field F:

$$v_D = \mu F \tag{4-3}$$

The scale factor is termed as mobility, μ, meaning carrier velocity under unit electric field. Combining Equations (4-1) to (4-3), the specific conductivity is given by

$$\sigma = \frac{j}{F} = \frac{qnv_D}{v_D/\mu} = qn\mu \tag{4-4}$$

Conductivity is therefore the product of the two independent physical quantities n and μ. Equation (4-4) simply indicates that we must increase n and μ, if we need larger conductivity. To obtain sufficient current in organic films, due to the intrinsic low carrier mobility, we need to inject external charge carriers effectively from electrodes to increase carrier density.

Carrier density n and mobility μ are, on the one hand, material specific properties; on the other hand, depend on the applied external field and the temperature[2]. In particular, in various semiconductors, the mobility varies greatly and can either increase or decrease strongly on the variation of semiconductor temperature. Besides temperature dependence, carrier mobility is also largely dependent on the energy band. Generally, the higher electron delocalization and broader the band, the higher the mobility. In inorganic

2. 一方面,载流子密度 n 和迁移率 μ 是材料的特定属性。另一方面,这两个材料参数受施加的外部电场和温度影响。

semiconductors, for example Si or Ga, carrier delocalization expands in the solid, so mobility is high. In organic semiconductors, carrier delocalization is usually limited and their mobility is low. It needs to point out that the internal field can be determined by $\Gamma = V/d$ only when the charge carriers are distributed homogeneously within the solid. Further, Ohm's law holds only when n and μ are both constant and not dependent on the field strength.

4.1.2　Energy Diagram of Semiconductors

One pillar of semiconductor physics is band theory. In the condensed version used to describe the behavior of electronic devices, the entire energy diagram of crystalline material can be considered to effectively consist of N_V states at a discrete energy E_V, named the valence band (VB) and N_C states at E_C, i.e., the conduction band (CB), even while knowing that the real band diagram is much more complicated[3]. Figure 4.1 exhibits a general scheme of occupied (filled) VB and unoccupied (non-filled) CB electron states or bands versus the density of states (DOS) for several classic materials including insulators, p-type semiconductors (p-SC), intrinsic semiconductors (i-SC), n-type semiconductors (n-SC), semimetals, and metals at both zero absolute temperature and nonzero absolute temperature. The Fermi level E_F of these materials is also indicated. At nonzero temperature, due to certain energetic radiations (thermal activation), some electrons in occupied states can be excited to unoccupied states, which would shift the E_F level up.

The basic underlying idea is that only the states close to the band edges are significant for the electrical behavior. For most chemical reactions, and for most electronic and optoelectronic processes, it is predominated by the transferring of valence electrons between the VB/HOMO and the CB/LUMO. These bands/orbitals are also therefore called frontier bands/orbitals, and the information such as energy levels, shapes, orientations, and spatial distances of the frontier bands/orbitals is very crucial for the electronic and

3. 半导体物理学的一个支柱是能带理论。在用于描述电子器件行为的浓缩版本中,晶体材料的整个能量图可以被认为是有效地包括在离散能量 E_V 下的 N_V 个状态组成的价带(VB),和在能量为 E_C 的 N_C 个状态组成的导带,尽管实际的能带图要复杂得多。

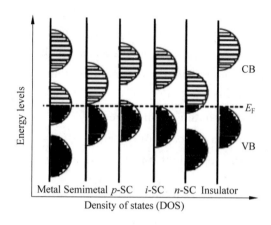

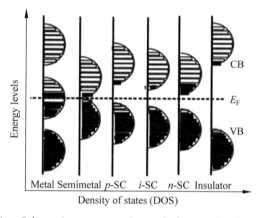

Figure 4. 1　Schematic representations of electron bands and density of states（DOS）of several classic materials including metals，semimetals，semiconductors（p-type，n-type，and intrinsic），and insulators in（a）ground state or at absolute zero temperature and（b）excited state or nonzero temperature. E_F presents Fermi energy level

4. 基本的深层含义是，只有靠近能带边缘的状态才对电子行为有意义。在大多数化学反应中，以及在大多数的电子和光电子过程中，它主要是价电子在 VB/HOMO 和 CB/LUMO 之间的转移。因此，这些能带/轨道也被称为前沿能带/轨道，诸如前沿能带/轨道的能级、形状、方向和空间距离等信息对于材料的电子和光电特性非常重要。

optoelectronic properties of the materials[4]. The difference between this effective CB and VB is then called the band gap，$E_g = E_C - E_V$. At low temperatures，the VB is completely full，occupied with N_V electrons，whereas the CB is completely empty. Although the VB has many electrons，none of them can contribute to current in this situation. This is because for every electron with velocity $+ v$ there always exists an electron occupying a state with an exact opposite velocity $- v$，even when an electrical field is present. For low temperatures，the average velocity and thus conductivity of the material is therefore zero and this distinguishes semiconductors from metals. Conductivity in semiconductors can be achieved by promoting electrons from the VB to the CB or by directly injecting electrons into the CB or removing them from the VB. In either case, this allows for a mismatch of electrons with opposite speeds and conduction can occur. Moreover, it is customary to name missing electrons as "holes" in the VB. In this way, either electrons in the CB or holes in the VB can contribute to current[5].

In band theory, the bands are solutions of the Bloch equation. In order for materials to form or exhibit electronic bands, it is crucial that the materials possess a periodic potential structure to satisfy the Bloch's theorem requirements. In a classic single crystal semiconductor, since the Bloch function can be applicable to the whole crystal (except the boundary or edge regions of the crystal), the band size (BS) is therefore roughly the same as the single crystal size. On the contrary, in the amorphous domains where the bands are poor or do not exist, charge transport follows incoherent or hopping mechanisms, as will be discussed later in this chapter. The mean free path (MFP, l) or mean free distance (MFD) of electron, is defined in semiconductor physics as an average non-scattering (ballistic) electron transport length between two consecutive scattering centers as expressed by

$$l = v\tau \tag{4-5}$$

where v is the electron velocity, τ is the average electron transport relaxation time, also called mean free time (MFT), defined as the average time an electron travels between the two scattering centers.

In certain polycrystalline materials where both amorphous (grain boundaries) and crystalline domains coexist, bands may exist in the crystalline domains. Therefore, the band size may be defined as the average size of the actual periodic domain or path (or an effective conjugation size corresponding to the size of a particle box) where Bloch function is applicable and where electron transport is of coherent or tunneling type. While the effects of amorphous domains can approximate the discrete levels at the edge of the conduction and valence bands[6].

5. 半导体的导电性可以通过将电子从 VB 激发到 CB 或直接将电子注入 CB 或从 VB 中移除来实现。任何一种情况都会引起具有相反速度电子的(数量)失配,产生导电。此外,习惯上将 VB 中缺失的电子命名为"空穴"。通过这种方式,CB 中的电子或 VB 中的空穴都可以产生电流。

6. 在某些多晶材料中,无定形(晶界)和结晶区域共存,能带可存在于结晶区域中。因此,能带大小可以定义为实际周期性区域或路径(或者相当于粒子盒尺寸的有效共轭尺寸)的平均尺寸;在那样的区域内,布洛赫函数是适用的,并且电子在其中的传输是相干输运或隧穿过程。而无定形区域的影响可以近似为在导带和价带边缘形成的离散能级。

7. 因此，相比于作为导带和价带边缘离散能级的近似，更充分的描述是：存在（电）传导和能带态的分布。正如离散能级是一个很好的对晶体材料中非晶态的近似一样，无定形材料中载流子的分布可以通过两个指数衰减的函数很好地进行近似处理，一个用于传导态（"价带"和"导带"），另一个用于陷阱态，其中导电态通常也称为"尾态"。

Besides lacking of closely packed periodic potential structures，which make the bands difficult to form or ever exist，amorphous inorganic semiconductors are full of unterminated "dangling" bonds. These bonds create deep electronic levels. Moreover，the surroundings of these dangling bonds are not constant throughout the surface or the bulk of the material. The same applies to the conduction levels of conduction and valence bands. As a result，instead of the approximation of discrete levels at the edge of the conduction and valence bands，a more adequate description is one in which a distribution for conduction and band states exists. Just as the discrete levels for amorphous states in crystalline materials are a good approximation，the distributions of carriers in amorphous semiconductors can be well approximated by two exponentially decaying functions，one for the conduction states（"valence" and "conduction" bands）and one for the trap states，where the conductive states are also often called "tail states"[7]. Between the tail states of the conduction and valence bands，a large density of trap states exists. In many cases，these trap states can outnumber the conduction states and in the extreme case all charges will eventually wind up trapped. However，any charge must starts in a conductive state and we can expect strong transient behavior. Moreover，the sheer abundance of these trap states can cause a symmetry in trapping and de-trapping times，compared with the situation in crystalline materials，where trapping is normally fast and de-trapping slow. Long trapping times cause transient behavior in a long time domain.

The effects of traps on the electrical characteristics of devices can be summarized as：①Low conductivity，which is also possible to analyze as low mobility. ②Thermal activation of current. ③Long-lived transient effects. For exponentially distributed density of states，power-law transients are expected in constant bias. ④In admittance，a nonconstant Mott-Schottky plot is used to profile the acceptor concentration. ⑤In thin film transistor（TFT）devices，output and transfer curves are nonlinear and the devices have a thermal activation energy depending on bias. ⑥Anomalous time of flight transients.

4.1.3 Carrier Types

In semiconductor materials, according to the nature of carriers and material purity, there are n-type and p-type carriers, and also intrinsic and extrinsic carriers (Figure 4.2); according to carrier generation route, there are photo-generated carriers, thermal-generated carriers, chemical doped carriers, electrode injected carriers, field assisted carriers and so on[8].

1. Intrinsic Carriers

Intrinsic carriers refer to those in a pure semiconductor without impurity and defect, and hence electrons and holes are generated in pairs, the density of hole (p) in VB and density of electron (n) in CB are the same (Figure 4.2(a)). Generally, there are two ways to generate intrinsic carriers, as discussed in the following.

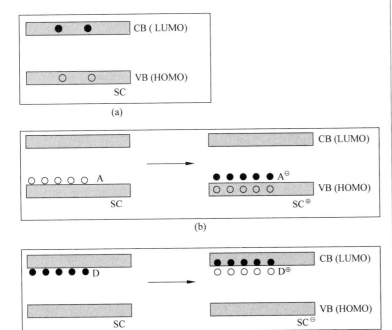

Figure 4.2 Schematic diagram for intrinsic carriers (a) and extrinsic carriers by chemical doping (b) and (c)

1) Thermally Activated Carrier

At high enough temperature, the lattice vibrations can

8. 在半导体材料中,根据半导体中载流子本质和材料纯度,有 n-型 和 p-型 载流子,也有本征和非本征载流子(图4.2);根据载流子生成方式,有光生载流子、热致载流子、化学掺杂载流子、电极注入载流子、场助载流子等。

9. 在足够高的温度下，晶格振动可以从半导体的 VB 热激发电子到 CB，同时产生相同数量的电子和空穴。在半导体中这样产生的电子和空穴可以通过复合消失。

thermally excite electrons from the VB to the CB of semiconductors, simultaneously producing electrons and holes in the same quantity. The electrons and holes thus created in the semiconductor can then disappear by recombination[9]. This process allows an electron in the CB to take the place of a hole in the VB by losing an amount of energy theoretically equal to the energy of the semiconductor bandgap. This energy can be released either thermally (phonon emission) or optically (photon emission). When the semiconductor is in thermal equilibrium, the concentrations of electrons in CB and holes in VB are dynamically stable because the rates of generation and recombination of electron and hole are constant. This concentration is the function of temperature T:

$$p = n = AT^{3/2}\exp\left(-\frac{E_g}{2k_B T}\right) = \sqrt{N_C N_V}\exp\left(-\frac{E_g}{2k_B T}\right) \quad (4\text{-}6)$$

where A is a constant, T is temperature, E_g is bandgap of semiconductor, k_B is Boltzmann constant, N_C and N_V are the volume density of states within $k_B T$ of the edge of CB and VB, respectively.

2) Optically Pumped Carrier

10. 这种类型的载流子也称为光生载流子，它是指 VB 中的电子通过光激发跳到 CB 上，从而产生电子和空穴对；一部分电子-空穴对可以解离，产生等量的自由电子和空穴。

This type of carrier is also termed as photo-generated carriers, referring to that electrons in the VB jump to the CB via photon excitation to produce electron and hole pairs, and part of which are dissociated in equal quantity of free electrons and holes[10]. This process is similar to the thermal generation of carrier generation and is an intrinsic process.

The carrier generation process in molecules by photon excitation is shown in Figure 4.3. The first step is the absorption of photon with the results of singlet-exciton formation: S_1, S_2, S_3, ⋯ The second step is autoionization process with a quantum yield of Φ_0, and positively-charged molecule-ions and quasi-free electrons with kinetic energy (hot electrons) are produced. In the third step, the hot electrons are thermalized by scattering processes and form metastable charge pairs (CP states) within the sphere of action of the Coulomb potential of the positive ions, i.e., radical-anion-radical-cation pairs with distance of r_{th}.

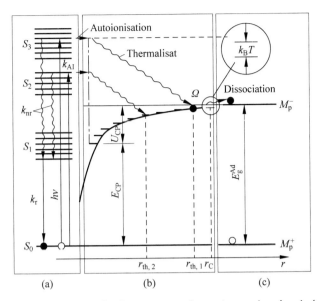

Figure 4.3   Process of photo-generated carrier and related levels/ processes. (a) Absorption and corresponding intramolecular recombination (IS, ISC, k_r and k_{nr}); (b) bound charge-carrier pairs (charge-carrier pair, CP); (c) ionized states: M_p^+ and M_p^-, E_{CP}: energy of electron in electron-hole pair, U_{CP}: bonding energy of charge-carrier pair, E_g^{Ad}: adiabatic bandgap

Since the electrons and holes are thermalized but not yet "free", the CP states are charge-transfer (CT) excitons whose energy is E_{cp}. Thus, a charge transfer process from one molecule to a second molecule at a distance ($r_c > r_{th}$) has occurred[11]. r_c is the distance at which the magnitude of the Coulomb energy is just equal to $k_B T$ (thermal energy). In the final step, these bound charge-carrier pairs dissociate by thermal activation. The thermal activation energy (E_d^{ph}) of dissociation of bound electron-hole pairs can be expressed as

$$E_d^{ph} = E_{CP} - E_g^{Ad} = -U_{CP} = \frac{q^2}{4\pi\varepsilon\varepsilon_0 r_{th}} \qquad (4-7)$$

where, U_{CP} and r_{th} are the bound energy and the distance between electron and hole, respectively; E_{CP} is the energy of electron in electron-hole pair. It indicates that E_{CP} increases when the r_{th} increases, and the E_d^{ph} decreases. Total quantum yield for photogeneration of charge carrier can be expressed by modified Onsager model:

11. 由于电子和空穴被热化但尚未"自由"，CP 态是能量为 E_{cp} 的电荷转移（CT）激子。因此在一定距离（$r_{th} <$ r_c）时发生了从一个分子到另一个分子的电荷转移过程。

$$\eta_{\mathrm{p}} = \eta_{\mathrm{n}} = \eta_0 \left(1 + \frac{qF}{2k_{\mathrm{B}}T}\right) \exp\left(-\frac{U_{\mathrm{CP}}}{k_{\mathrm{B}}T}\right) \qquad (4\text{-}8)$$

where F is the strength of electrical field. $F = 0$, represents the condition without electric filed; $F \neq 0$, represents the condition of electrical field assisted photogeneration of carrier.

2. Extrinsic Carriers

Extrinsic carriers refer to those which are produced by the injection, chemical doping, impurity/defect and other external factors except thermal and optical routes. At this time, electrons and holes don't occur in pairs, densities of holes in VB and electrons in CB are not equal[12]. The common ways to produce extrinsic carriers are discussed as following:

1) Injected Carriers

Injected carriers are generated via electrode injection, i.e., electrons are injected from cathode and holes injected from anode when the voltage is applied. In general, the density of injected electrons or holes (n or p) are determined by both electric field (F) and the energy barrier (ϕ_{b}) between the work function of cathode and CB of semiconductor for electrons, and the barrier between the work function of anode and VB of semiconductor for holes:

$$n \text{ or } p \propto \frac{1}{F} \exp\left(-\frac{\phi_{\mathrm{b}}}{k_{\mathrm{B}}T}\right) \qquad (4\text{-}9)$$

where T is temperature and k_{B} is Boltzmann constant.

2) Doped Carriers

The idea of doping is to deliberately contaminate the material with chemically foreign dopants which either easily donate electrons to or accept electrons from the host material (in chemistry jargon these agents are called reducers and oxidizers whereas in solid-state physics terms they are called donors and acceptors, respectively)[13]. Introducing a dopant energy level within the bandgap, can facilitate the valence electron in VB jumping to the dopant level to produce a p-type carrier, or facilitate the electron in dopant level jumping to CB to produce a n-type carrier. Figure 4.4 schematically shows the process for donor- and acceptor-type doping in a semiconductor. As can be seen, the electrons in the CB

12. 非本征载流子是指由注入、化学掺杂、杂质/缺陷和其他外部因素（热和光学方法除外）产生的载流子。此时，电子和空穴不是成对出现的，VB 中的空穴密度和 CB 中的电子密度不相等。

13. 掺杂的想法是将外来的化学掺杂剂故意加入材料中，这些掺杂剂很容易向主体材料提供或从主体材料接受电子（在化学术语中，这些试剂被称为还原剂和氧化剂，而在固态物理学术语中，它们分别被称为给体和受体）。

originate from the donor levels (E_D), and the holes in VB come from the acceptor level (E_A).

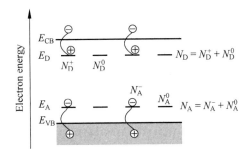

Figure 4. 4　Schematic diagram of electron process of doped carriers. E_{CB}: energy level of CB, E_{VB}: energy level of VB, E_D: energy level of donor, E_A: energy lever of acceptor, N_D: total density of donors, N_D^+: density of donors that attributes electron to E_{CB}, N_D^0: density of unexcited donors, N_A: total density of acceptors, N_A^-: density of acceptors that accept electrons from E_{VB}, N_A^0: density of acceptors that without electron accepting

After chemical doping, the charge neutrality condition of the semiconductor can be expressed as follow:

$$N_A^- + n = N_D^+ + p \qquad (4\text{-}10)$$

where n and p are the carrier density of electrons in CB and holes in VB. Assuming a completely ionization of impurity,

$$N_A^- \approx N_A \quad \text{and} \quad N_D^+ \approx N_D \qquad (4\text{-}11)$$

At thermal equilibrium, the product of the concentration of electrons and holes conforms with the mass action law ($n \times p = n_i^2$, n_i represents density of intrinsic carriers at a fixed temperature). Since the concentration of carriers created by ionization is dominant, the concentration of n and p are no longer equal[14]. For an n-type semiconductor, because $N_D \gg n_i$ and $N_D \gg N_A$, we have

$$n \approx N_D \quad \text{and} \quad p \approx \frac{n_i^2}{N_D} \qquad (4\text{-}12)$$

For a p-type semiconductor, we obtain the similar relationship:

$$p \approx N_A \quad \text{and} \quad n \approx \frac{n_i^2}{N_A} \qquad (4\text{-}13)$$

Figure 4.5 shows the electron concentration of $n(T)$ and

14. 在热平衡时，电子和空穴浓度的乘积满足质量作用定律（$n \times p = n_i^2$，n_i 代表一定温度下本征载流子浓度）。由于离化产生的载流子的浓度占主导地位，因此 n 和 p 不再相等。

the Fermi energy in an n-type semiconductor as function of temperature. At low temperature, it is called "Freeze out" region, the energy E_F basically remains unchanged and locates at level between bottom of CB and donor level. In this region, electrons can be excited from donor level to CB with proper thermal energy. Charge carrier concentration increases with temperature approximating as $\ln(n) \sim T^{-1}$ with the slope of $-(E_{CB} - E_D)/2k_B$. In the "exhaustion" temperature range, the donors are already completely oxidized and donor electrons are totally thermal excited to CB, thus carrier density dose not change with temperature. At even higher temperatures, intrinsic charge carriers can be released; the carrier concentration increases sharply with the increase of temperature[15]. The slope of $\ln(n) \sim T^{-1}$ is $-(E_g)/2k_B$ in this region. In organic semiconductors, it is difficult to generate intrinsic carriers through thermal activation owing to the narrow width of the bandgap E_g, this region generally can't be observed. Statistically, the concentration of doped carriers follows certain rules. For the case of a pure n-type semiconductor, that is a semiconductor which contains only donors and no acceptors, the charge-carrier concentration $n(T)$ can be calculated to a good approximation:

$$n(T) \approx \frac{2N_D}{1 + \sqrt{1 + 4\dfrac{N_D}{N_C}\exp\left(\dfrac{E_d}{k_B T}\right)}} \tag{4-14}$$

here, N_D is density of donors, N_C is the density of states in the CB and $E_d = E_{CB} - E_D$ is the energy spacing between the donor levels (E_D) and the lower edge of the CB, k_B is Boltzmann constant, T is temperature.

3) Impurity/Defect Trapped Carriers

Impurity or defect in materials can produce trap center, which can capture carrier. On the other hand, these trapped carriers can be released by thermal or photo excitation, leading to carrier generation. Carrier can also be produced via collision between exciton and impurity/surface[16].

15. 在"耗尽"温度范围内,给体已经完全被氧化,给体电子完全转移到半导体的CB,因此载流子密度不随温度变化。在更高的温度区域,本征载流子可以被释放;载流子浓度随着温度的升高而急剧增加。

16. 材料中的杂质或缺陷会产生捕获中心,可以捕获载流子。另外,这些被捕获的载流子可以通过热或光激发被释放,产生载流子。载流子也可以通过激子和杂质/表面之间的碰撞产生。

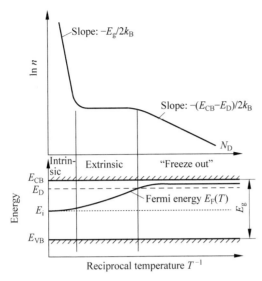

Figure 4.5　The dependence of the electron carrier concentration $n(T)$ and the Fermi energy $E_F(T)$ in an n-type semiconductor as functions of the temperature T. E_i is the Fermi energy of the intrinsic

4) Carriers Generated via Field Assisting

The influence of electric field on carrier generation includes the following cases: ①Electric field can separate a pair of Coulombic bound carriers before their recombination (Onsager Effect); ②Electric field can change the ship of the trap state, facilitating trapped carrier to release; and ③Assisted with thermal and electric conduction, the thermally activated carriers generally increase rapidly.

4.1.4　Carrier Transport in Solids

Generally, there are three extreme types for carrier transport, i.e., band like transport, hopping transport and charge density wave transport.

1. Band Like Transport

If the interaction energy of carrier with the nearest neighbor is large compared to any other energy present due to dynamic or static disorder, charge transport takes place through a band. The charge carrier delocalizes to form a propagating Bloch wave that may be scattered by lattice vibrations[17]. The charge carrier mobility is then given by:

17. 如果载流子与其最近邻的相互作用能比任何其他由于动态或静态无序而产生的能量大,则电荷是通过能带传输。离域的载流子形成传播的布洛赫波,该波可被晶格振动散射。

$$\mu = e\tau / M_{\text{eff}} \tag{4-15}$$

where τ is the mean scattering time, e is the elementary charge and M_{eff} is the effective mass of the charge carrier.

Band like transport can occur only if the bands are wider than the uncertainty of the charge carrier's site energy. Thus, band like transport refers to that, in solids with a wide band width, carriers move in allowed energy bands as plane wave with high mean free path and delocalization. The main requirements for band like transport are: ① energy band where carriers are located is smooth, ② carriers are delocalized in energy band and the interaction between carriers and lattices is small, and ③ the width of the band (W) should be larger enough:

$$W > \frac{h/2\pi}{\tau} \tag{4-16}$$

here τ is relaxation time between two scatterings. By zero order reasoning, the charge carrier mobility for band like transport must satisfy

$$\mu_{\text{band-like}} > ea^2 W / \hbar k_B T \tag{4-17}$$

where e is the elementary charge, a is the lattice constant.

As shown in Figure 4.6, in a perfect crystal solid, a free carrier is delocalized and depicted as the straight line. The free carrier moves in crystal as a plane wave without scattering. In a real crystal, there are always lattice vibrations (i.e., phonons) that disrupt the crystal symmetry. These phonons scatter the electron and thereby reduce its mobility. Lowering the temperature will therefore increase the mobility[18]. Equation (4-18) indicates the relationship between mobility μ and temperature T for bandlike transport.

$$\mu \propto \frac{1}{T^n}, \ 0 < n < 3 \text{ for inorganics}, \ 0 < n < 2 \text{ for organics} \tag{4-18}$$

where n is a constant. This is theoretically accounted from the increasing of scattering with temperature by acoustic phonons, by impurities or by electron-interactions. The reduction of mobility with temperature as expressed in Equation (4-18) is commonly taken as indication that band-type transport prevails.

18. 在真正的晶体中,总会有晶格振动(即声子)破坏晶体的对称性。这些声子散射电子,从而降低其迁移率。因此,降低温度将增加迁移率。

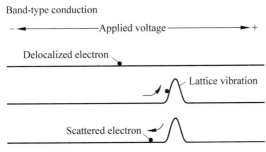

Figure 4.6　Band-type conduction

Theoretically，the temperature-dependent mobility of band like transport can be described as Figure 4.7. There are mainly two parameters that lead to temperature-dependent mobility in band like transport，i. e.，the phonons and the shallow traps. At high temperature，the mobility of electron in CB decreases with increasing of T due to the increase of lattice scattering (phonons). At low temperature region, phonons are frozen; the mobility of electron in CB will be independent of phonons. Therefore，the influence of temperature to shallow trapped electrons will be dominant. The mobility will increase with the increase of temperature due to the overcoming of shallow traps by charge carriers at relatively high temperatures[19].

19. 在低温区,声子被冻结,电子在导带中的迁移率将与声子无关。因此,温度对被捕获在浅陷阱的电子的影响将起主要作用。由于在相对较高的温度下电子可以克服浅陷阱的束缚,迁移率将随着温度的升高而增加。

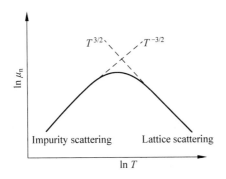

Figure 4.7　The theoretical temperature dependence of mobility due to both lattice and impurity scatterings

2. Hopping Transport

If any of the other energies due to the effects of dynamic or static disorder become significant compared to the nearest neighbor interaction energy，the transport band made up of

20. 然后,载流子被定域在单个位点,并通过一系列的非相干过程进行传输。这称为跃进运动。对于有机材料的情形,当载流子到达分子时,由于太多的结构弛豫、太多振动或者太多位点能量或位点间距离的初始变化,传输的能带可能会被破坏。

delocalized wavefunctions is destroyed. In this case, transport can no longer be described in terms of band motion but rather becomes incoherent. Then, a charge carrier is localized at individual site and proceeds by a sequence of non-coherent transfer events. This is referred to hopping motion. In the case of organic materials, a transport band can be destroyed because of too much geometric relaxation when the carrier gets onto the molecule, too much vibration, or too much initial variation in site energy or inter-site distance[20]. Thus, if the lattice of solid is irregular, in this case, the carrier can become localized on a defect site (or in a potential well originating from the polarization of the lattice due to the carrier). For this type of carrier, the lattice vibrations are essential when the carrier moves from one site to another as shown in Figure 4.8.

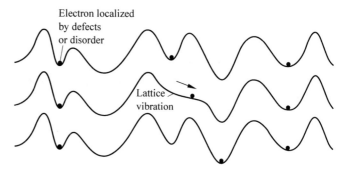

Figure 4.8   Hopping transport of carriers

21. 声子的释放有助于载流子的输运,可为载流子提供能量以克服势垒并从一个分子上跳到另一个分子上。这是一个热激活过程,迁移率随着温度的升高而增加。

In hopping transport, traveling charges couple strongly with phonons, and the mobility of the hopping charge is dominated by the so-called electron-phonon coupling. Here, phonons involve both delocalized lattice (inter-molecular) phonons and localized (intra-molecular) phonons at each molecule in a molecular solid. The release of phonons helps the transport of carrier, providing energy for carriers to surmount the barrier and hop from one molecule to another molecule. This is an activated process and the mobility increases with increasing temperature[21]. Equation (4-19) indicates the relationship between mobility μ and temperature T (E_a is the activation energy). Mobility is much lower for hopping transport than that of band like transport.

$$\mu \propto \exp\left(-\frac{E_a}{k_B T}\right) \qquad (4\text{-}19)$$

The hopping mobility μ of an electron or hole must, in principle, depend on the applied electric field because a field lowers the activation energy for jumps in the field direction. Field dependent mobility will be discussed later.

3. Charge Density Wave Transport

This is an unusual type of carrier transport. As shown in Figure 4.9, it involves a cooperative motion of a group of carriers, and has been proposed to explain high conductivity in quasi-one-dimensional solids. For these solids with charge density wave transport, at low temperatures, spontaneous periodic distortions appear in the lattice upon the introduction of free carriers. When the conduction electrons concentrate in a sequence of periodic clusters, it becomes possible for them to move as a unit. This is called a charge density wave[22]. The atoms in the lattice site move periodically around their average position, but the electrons can move along the periodically-moved atoms under the influence of the external field.

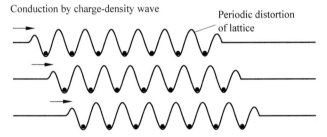

Conduction by charge-density wave

Periodic distortion of lattice

Figure 4.9   Charge density wave transport of carrier

4.1.5   Electrical Properties of Organic Materials

1. Preliminary Historical Remarks

The initial investigation on electrical property of organic materials was photoconduction study with anthracene. In 1906, photoconduction in anthracene crystals was discovered by the Italian scientist Pochettino; its dark conductivity was studied (simultaneously with that of silicon) in 1910 by

22. 对于具有电荷密度波传输的固体,在低温下,在引入自由载流子后,晶格中会自发地周期性畸变。当传导电子集中在一系列周期性畸变的团簇中时,它们就有可能作为一个整体移动。这称为电荷密度波。

23. 此后，蒽晶体被作为芳香分子晶体光导及许多其他电学、光学和光电特性的原型，得到最广泛的研究。1913年 W.E. Pauli 研究了荧光有机固体的荧光激发光谱和光电效应。发现，光电子发射所需要的最小能量大于光致荧光的激发能量。

German scientists Konigsberger and Schilling，and a prototype of an organic photovoltaic cell was presented in 1913 by Vollmer in his habilitation thesis at the University of Leipzig. After that，anthracene crystal was mostly investigated as the prototype for photoconductivity and many other electrical，optical and optoelectronic properties of aromatic molecular crystals. The excitation spectra of fluorescence and the photoelectric effect in fluorescing organic solids were investigated in 1913 by W.E. Pauli and it was found that the minimum photon energy required for the initiation of photoemission of electrons is greater than that for the photoexcitation of fluorescence. [23]

The main period of research on the study of electrical properties of molecular crystals such as naphthalene and anthracene，however，took place in the years from 1950 to 1980. In 1953，H. Mette and H. Pick purified anthracene using a melting technique and grew their crystals from the melt of purest starting material. They measured the dark conductivity of anthracene crystals at temperatures between 80 ℃ and 210 ℃，and it is the first time to demonstrate the anisotropy of the specific electrical conductivity. In 1959/1960，O. H. LeBlanc and R. G. Kepler investigated the transient photoconductivity of high-purity anthracene crystals. The charge carriers were generated by a UV pulse and the mobility of charge carriers was measured. They found that the mobility values were between $0.3 \sim 3$ cm^2 • V^{-1} • s^{-1} both for holes and electrons at room temperature. The values of mobility were dependent on the crystal orientation relative to the electric field and increased with cooling. In 1961，P. Mark and W. Helfrich studied the dark conductivity of about 50 mm thick crystals of naphthalene，anthracene，and other aromatic hydrocarbons. In their work，the charge carriers （holes） were injected from an anode into the crystals. They were the first to detect the space-charge-limited currents in organic molecular crystals. In 1966，N. Riehl and H. Bässler measured the temperature dependence of the dark current of holes in anthracene crystal. It was found that the stationary current depended only on the temperature of the anode. The activation energy was 0.7 eV. It is considerably smaller than the excitation energy of singlet excitons in the anthracene crystal (3.1 eV), and particularly, it is smaller than the gap between the valence and the conduction bands[24]. In 1977, A. J. Heeger, A. G. Macdiarmid and H. Shirakawa found that doping polyacetylene with electron acceptor, the conductivity could increase 9 orders of magnitude, up to 103 S · cm^{-1}. This is the first time to discover the conductivity in polymer. In 1980, D. Jérome et al., discovered the superconducting property of organic materials at low temperature based on derivative of tetra methyl tetra selena fulvalene (TMTSF).

The first commercial application of organic materials serving for carrier transport appeared in 1980s: those are the use of organic photoconduction materials in xerography. As shown in Figure 4.10, the active components in a xerography machine are charge generation layer and hole transport layer.

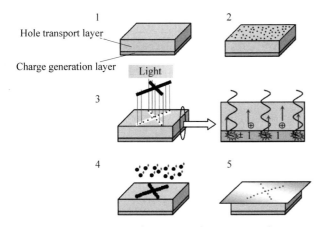

Figure 4.10　Working principle in xerography

When the device works, negative charges are filled in the surface layer; then light irradiates the surface layer through the cover information pattern, allowing the region without cover (here is the X-shape) to transmit light; at the bottom of the carrier generation layer, only the region with light irradiation can be excited to produce charge (holes); the produced holes can transport to the surface via hole transport layer and be neutralized by the surface negative charge; therefore, the left negative charges in the surface represent the information in the cover, which can then attract toner charcoal powders with positive charges; following the transfer of

24. 1966 年，N. Riehl 和 H. Bässler 测量了蒽晶体中空穴的暗电流的温度依赖性。研究发现，稳态电流仅取决于阳极的温度，活化能为 0.7 eV。它比蒽晶体中单线态激子的激发能（3.1 eV）小得多，特别是它小于价带和导带之间的能隙。

charcoal powders on to a white paper, photocopy completes[25]. Originally, the photoconductive material used in xerography is anthracene, which is then replaced by α-selenium. Now, most of the functional materials in xerography are organic hole-transport materials.

By now, organic materials cover all the conductivity range as shown in Figure 4.11. An organic material can either be an insulator, a semiconductor or a conductor. For example, highly-pure anthracene crystals are insulators, whose conductivity values lie at the lower edge of the scale here; polyacetylenes $(CH)_x$ are semiconductor, through doping, the conductivity of polyacetylenes can be increased by many orders of magnitude; TTF/TCNQ is a conductor.

25. 产生的空穴可通过空穴传输层输运到表面,并与表面负电荷中和;因此,表面留下的负电荷代表覆盖物的信息,然后(这些负电荷)可以吸引带正电的碳粉;将碳粉转移到白纸上,完成复印。

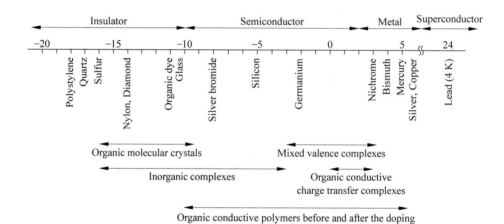

Figure 4.11　Electric conductivity σ (S·cm^{-1}) of miscellaneous materials in the form of lgσ at room temperature

2. Energy Diagrams for Organic Semiconductors

To investigate the electrical property of organic materials, their energy diagram needs to be addressed first. While the basic ingredient of organic materials is the carbon atom, the basic ingredient of organic semiconductors is conjugation, which is referred to a chain of carbon atoms with alternating single and double bonds[26]. This has two important results, namely the opening of a band gap (i.e., a splitting of the energy levels in the range of semiconductors), and the delocalization of charges in these levels.

In a conjugated organic material, the four electrons on

26. 有机材料的基本成分是碳原子,而有机半导体的基本组成是共轭。共轭是具有交替的单键和双键的碳原子链。

each carbon atom in the conjugation chain can be considered to reside in three sp^2 hybridized orbitals and one in p_z orbital. The three sp^2 electrons are used to form covalent bonds via σ molecular orbitals to neighbor carbon atoms in the chain on either side and to the side-group (for instance a simple hydrogen atom). The remaining electron in the p_z orbital is then used in a covalent bond via a π molecular orbital with a neighboring carbon atom in the chain on one side only. The result is a chain of alternating single (σ only) and double (σ and π) bonds. Figure 4. 12(a) shows the energy diagram of an interaction between two carbon atoms forming a double bond. After filling the levels from low to high (using Hund's Rules for the spins) it can be recognized that four electrons (two from each carbon atom) are used in bonding, two in σ molecular orbitals, and two in π molecular orbitals[27]. The remaining four electrons are in nonbonding (nb) orbitals and are still available for bonding to the rest of

27. 三个 sp^2 电子用于与链上每一侧的邻近碳原子或者支链(例如一个简单的氢原子)形成 σ 型分子轨道的共价键。在 p_z 轨道上剩余的电子,与一侧的链上邻近碳原子形成 π 型分子轨道的共价键。通过从低到高的填充(对于自旋使用洪特规则)之后,可以看到四个电子(每个碳原子两个)用于成键,两个在 σ 分子轨道,两个在 π 分子轨道。

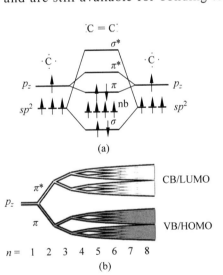

(a)

(b)

Figure 4. 12　(a) Energy diagram of two interacting carbon atoms. sp^2 and p_z atomic orbitals of the two individual carbon atoms combine to form π, σ and nonbonding (nb) molecular orbitals. (b) A band structure starts emerging with a narrowing band gap when the conjugation length of alternating single and double bonds is increased. A HOMO and LUMO can be recognized as the VB and CB in inorganic semiconductors, respectively

28. 因此，单键和双键链的形成导致了具有半导体范围内的 HOMO 和 LUMO 能级裂分（能隙）的能量结构。

the chain and the ligands. The basic feature is the splitting between the π and π^* molecular orbitals caused by the interaction between the p_z atomic orbitals. Interactions between p_z orbitals further away in the chain cause additional, smaller splitting of the levels, as schematically indicated in Figure 4.12(b). A highest occupied molecular orbit (HOMO, π) and lowest unoccupied molecular orbital (LUMO, π^*) can be recognized. The formation of a chain of single and double bonds thus causes an energy structure with a HOMO and LUMO splitting ("band gap") in the range of semiconductors[28]. From Figure 4.12 it is also clear that a material with only single σ bonds, like polyethylene, will have a much wider band gap and will not easily fall into the category of semiconductors. In contrast, polyacetylenes with double bonds will have narrower band gap following into the category of semiconductor.

In terms of charge carriers (free holes and electrons) for transporting, the energy diagrams for organic materials are different to those for inorganic semiconductors. Figure 4.13 shows the distribution of carrier and trap levels in organic materials. Figure 4.13(a) indicates in an isolated molecule, the energy levels of S_0, S_1 and T_1 represent electronic ground state, first singlet state and first triplet state, respectively; and IP_g and EA_g are the ionization potential and electron affinity of isolated molecule, respectively; Figure 4.13(b) shows the energy level distribution of electrons and holes in ideal organic crystal, IP_c and EA_c are ionization potential and electron affinity of ideal crystal, respectively, E_h and E_e are energy level of holes and electrons, respectively. Numerically (shown in Equations (4-20) and (4-21)), in organic crystals, the energy level of electrons (E_e), LUMO and the electron affinity are equal, the energy level of holes (E_h), HOMO and the ionization potential are equal. Due to the polarization during the formation of crystal, compared to isolated molecule, energy level of electrons markedly decreases and energy level of holes obviously increases. Electron polarization energy (P_e) and hole polarization energy (P_h) have the same absolute valve, and can be obtained from Equations (4-22) and (4-23), respectively. When electrons or holes occupy LUMO or HOMO, molecules become polarons. Since molecules in organic materials are bound weakly, polarons can only polarize adjacent molecules, and polarons are so called as small polarons[29]. As shown in Figure 4.13(c), energy level distribution of carriers in ionized organic crystals can be simply described as Gaussian distribution $G_e(E)$ and $G_h(E)$ deviated from average energy of $\bar{P}_e$ for electrons, and $\bar{P}_h$ for holes. ΔP_h and ΔP_e represent deviation of polarization energy for hole and electron compared to their average polarization energy, respectively. Similarly, $G_t(E)$ in Figure 4.13(d) represents trap density distribution.

29. 当电子或空穴占据 LUMO 或 HOMO 轨道时,分子成为极化子。由于有机材料中的分子相互束缚很弱,极化子只能使相邻分子极化,因而极化子被称为小极化子。

$$E_e = E_{LUMO} = EA \tag{4-20}$$

$$E_h = E_{HOMO} = IP \tag{4-21}$$

$$P_h = IP_c - IP_g \tag{4-22}$$

$$P_e = EA_g - EA_c \tag{4-23}$$

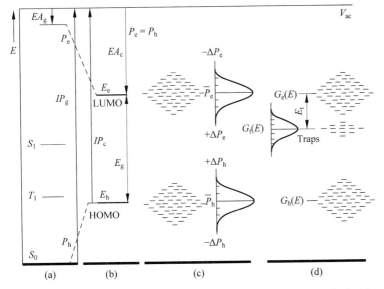

Figure 4.13   Energy diagram of an organic material. (a) Isolated molecules, (b) ideal crystal, (c) the ionized crystal with a statistical distribution of polarization energies, (d) trap states

3. Carriers in Organic Materials

In organic materials, absorption of light can produce photon-generated carrier. This is generally attributed to

30. 在有机材料中,光的吸收可以产生光生载流子,这通常归因于 $\pi \to \pi^*$ 或 $n \to \pi^*$ 的电子跃迁。

electron transition of either $\pi \to \pi^*$ or $n \to \pi^*$ [30]. In terms of doped carriers, doping in organic materials is not an important issue. This is because for the typical organic devices such as OLEDs and OFETs, doping is not an important parameter. For both these types of devices, purer materials work better. This is in comparison with some devices of classical inorganic semiconductors, such as pn-junction diodes and bipolar transistors, where doping is essential and device-dominating. Doping is not well studied in organic semiconductors. Often iodine is used when doping is needed. Also, oxygen (and water or air) seems to introduce electrically active levels either in the form of traps or doping. These effects of doping seem to be reversible.

4. Carrier Transport in Organic Materials

31. 有机半导体中的电荷载流子输运,受控于给体位点 LUMO 中电子到电子受体位点的空置 LUMO 的转移。等价地,空穴可以在 HOMO 能级之间转移。位点可以是分子或聚合物的共轭段。位点之间的电子耦合是发生这种电荷转移过程的必要条件,但不是充分条件。

Charge carrier transport in an organic semiconductor is controlled by the transfer of an electron in the LUMO of a donor site to the empty LUMO of an electron accepting site. Equivalently, a hole can be transferred among the HOMO levels. A site may be a molecule or a conjugated segment of a polymer. Electronic coupling among the sites is a necessary yet not sufficient condition for this charge transfer process to occur[31].

In a perfectly ordered crystal at $T = 0$ K, an electron (or, equivalently a hole) would move coherently within a band of states constructed from the LUMO (or HOMO) orbitals of the constituent molecules. In such a material, the band width W $\approx 10 k_B T$ and the lattice constant a ≈ 0.6 nm, according to Equation (4-17), if μ exceeds about 5 $cm^2 \cdot V^{-1} \cdot s^{-1}$, band transport can be considered to prevail. In many molecular crystals the mobilities measured at or near room temperature are of the order of 1 $cm^2 \cdot V^{-1} \cdot s^{-1}$, indicating that there is a cross-over between band and hopping motion.

For bandlike transport in organic materials, the requirement of the electronic coupling being large compared to dynamic or static disorder can be fulfilled in molecular crystals at low temperature. Compared to inorganic crystals where covalent interactions prevail, electronic coupling is weak in molecular crystals and the resulting bands are rather narrow, typically in the range of 50~500 meV. As a result, bandlike transport in organic materials is not as effective as that in inorganic semiconductors[32].

When the material is not crystalline but microscopically disordered as realized in a molecular or polymeric glass, the carrier transport is controlled by: ① the electronic coupling among the constituent molecular units, ② the coupling to intra-molecular as well as inter-molecular vibrations, and ③ the static intra- as well as inter-molecular disorder. In non-crystalline organic semiconductors, the variations in the site energy and in the distance between sites, and concomitantly, in the intramolecular coupling are large compared to the value of intermolecular coupling energy. This has a considerable impact on the mobility of charge carriers[33]. It implies that charge transport occurs as a random walk by incoherent hopping between neighboring transport sites. Thus, in disordered organic materials, carriers mainly exist as radical-ions. Holes are radical-cations, electrons correspond to radical-anions. Generally, Carrier transport in disordered organic materials follows hopping transport model, i.e., the carrier transport performs as oxidation-reduction of molecules, which is the most accepted theory.

As to the basic conduction mechanism in organic semiconductors, there is a debate between 'free' carriers (electrons in the CB and holes in the VB) on one side and polarons (or bipolarons) on the other side. A polaron is a charge together with its local lattice deformation and polarization, as shown in Figure 4.14. It seems most obvious that once the theory from inorganic semiconductors has been borrowed, free carriers are started to assume; and only when it is not possible to explain the data otherwise, polarons are introduced[34].

Another on-going debate is the distinction between hopping conduction and Poole-Frenkel conduction (Figure 4.15). In hopping conduction models, only localized states exist or play a role, for instance, when the delocalized bands are too far away or the temperature is too low to allow for thermal excitation of trapped carriers. The charges spend almost

32. 对于有机材料中载流子的能带传输,电子耦合能量大于动态或静态无序能量的要求,在低温下的分子晶体中可以实现。与共价键相互作用为主的无机晶体相比,分子晶体中的电子耦合作用较弱,产生的能带宽度相当窄,通常在 50~500 meV。因此,有机材料中的能带传输不如无机半导体有效。

33. 在非晶有机半导体中,与分子间相互作用的能量相比,位点能量和位点之间的距离以及伴随的分子内相互作用的变化都很大。这对电荷载流子的迁移率产生了相当大的影响。

34. 最明显的是,一旦半导体的理论被借用,就开始假设自由载流子,只有在无法以其他方式解释数据时才引入极化子。

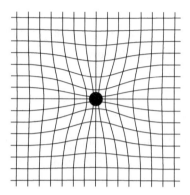

Figure 4.14　A polaron: a charge with its associated lattice distortion

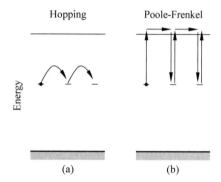

Figure 4.15　Distinction between hopping conduction (a) and Poole-Frenkel conduction (b). In the former charges occasionally jump ("hop") from trap to trap. In the latter carriers spend most of their time trapped but occasionally are excited to the delocalized (conduction or valence) band from where they can contribute to current

35. 电荷的传输是通过局部状态之间的瞬时跃进发生的。这也可以导致电场和温度依赖的有效迁移率。相反,在普尔-弗仑克尔传导理论中,所有传导都是通过附加有效陷阱即深层局部状态的导(和价)带进行的;为了对电流有所贡献,被捕获在深层能级中的电荷,首先必须被激发到导带,即不可以从陷阱到陷阱直接跃进。

all of their time on these localized states. Transport of charge occurs by instantaneous hops between localized states. This can also cause a field and temperature dependence of the effective mobility. In contrast, in Poole-Frenkel conduction, all conduction is through the conduction (and valence) bands with additional efficient traps, i.e., deep localized states; to contribute to current, a charge trapped on a deep level has first to be promoted to a conductive band, i.e., no direct hopping from trap to trap is possible[35]. The interesting thing is that the defect band can, in principle, also originate from the polarons. In other words, the traps can be self-traps. A charge can create a lattice distortion that makes it so immobile as to effectively be locked in place, thus behaving like a trap.

Comments by Rakhmanova and Conwell, however, indicate that Poole-Frenkel conduction seems more likely for organic materials. Also, Waragai et al. rejected hopping conduction for their transistors because of the unrealistic values found for the parameters. Similarly, Nelson et al. rejected the idea of hopping conduction for their pentacene transistors. Poole-Frenkel conduction is more adequate for conditions when the devices start having substantial currents and conduction is no longer a perturbation[36].

Other conduction mechanisms contain a two-band formalism; one conduction band and one defect band (where the designation "band" for the defect states is somewhat misplaced; they are comparable with the hopping states described above). When the defect band is dense enough, hopping can occur from defect to defect and the associated current can be substantial. This can be nearest-neighbor hopping, or variable-range hopping[37]. No bold statements are made about the exact conduction mechanism in organic solids. Maintaining a two-band formalism with only one band contributing to current, and without stating what the bands represent exactly, will not take anything away from the descriptive power. The discussion about the conduction mechanism is deferred. Many existing theories are consistent with measured data, yet, the reverse logic cannot be applied, namely that the data are not proof of the theory.

4.2 Mobility in Organic Materials

4.2.1 Organic Crystals

As discussed in Section 4.1.5, organic crystals often exhibit nondispersive transport, behaving as those between bandlike and hopping transport. The range of mobility in organic crystals often falls in the values of 10^{-2} cm$^2 \cdot$ V$^{-1} \cdot$ s^{-1} to several tens cm$^2 \cdot$ V$^{-1} \cdot$ s^{-1}, though extremely high value of several hundreds have been observed in very pure organic crystals at low temperature. Mobility in organic crystals is less dependent on temperature (Equation (4-18)).

36. 普尔-弗仑克尔传导更适合于当器件开始具有较大电流并且传导不再是扰动的情形。

37. 其他传导机制包含双能带的描述;包括一个导带和一个缺陷带(其中缺陷状态的称谓"带"有些不适宜,它们与上述跃进状态相当)。当缺陷带足够密集时,可能会从一个缺陷跃进到另一个缺陷,并且相关的传导电流可能很大。这可以是最近的邻位跃进或可变范围的跃进。

38. 到目前为止，我们只考虑了电荷载流子和低能振动之间的所谓非局域声子耦合，即长波长的分子间振动（声子）。然而，在较高温度下，分子内振动也变得很重要。原因是，一旦在分子中加入或移除电子，π 电子的分布就会发生变化。

39. 然而，如果电荷载流子的平均自由程约为一个晶格常数的数量级，则传导将通过跃进过程发生。

This can be understood by the role of temperature to carrier motion in organic crystals. At finite temperature, the motion of carriers suffers scattering from the vibrations of the lattice. However, as long as this scattering process is only a weak perturbation to the overall coupling between the adjacent molecules, charge transport can still be described in terms of a band model. This yields a mobility that decreases with increasing temperature because the number of scattering events increases with temprature. So far, we only considered the so-called non-local phonon coupling between a charge carrier and low energy vibrations, that is, long wavelength, inter-molecular vibrations (phonons). However, at higher temperatures intra-molecular vibrations also become important. The reason is that once adding or removing an electron to or from a molecule the distribution of the π-electrons changes[38]. This can be expressed in terms of so-called local phonon coupling, considering, though, that in this case the "phonon" is a molecular vibration and the coupling is of the vibronic type. If the strength of this vibronic coupling becomes comparable to the electronic inter-site coupling, a band model becomes inappropriate. In the extreme case, a charge carrier is scattered at each site, that is, there is a transition from a band transport to a hopping-type motion. This is accompanied by a change of the temperature dependence of the charge carrier mobility. The boundary between band and hopping conductivity is naturally not well-defined. However, if the mean free path of the charge carriers is of the order of magnitude around one lattice constant, then the conduction will take place via a hopping process[39].

A prototypical example of band-like transport in a molecular crystal is contained in the work of Karl and coworkers who measured electron and hole mobilities in an ultrapure naphthalene crystal in the temperature range of 4～300 K. Figure 4.16(a) shows that both electron (along the crystallographic b direction) and hole (along a direction) mobilities increase with decreasing temperature featuring a $\mu \propto T^{-n}$ law with exponents of 1.42 (electrons) and 2.9 (holes), respectively. Below 30 K the hole mobilities level off and saturate at values of the order of 200 cm^2 · V^{-1} · s^{-1} depending on the electric field. It is worth remembering that it took Karl and his coworkers more than a decade to improve the techniques for material purification and crystal growth in order to eliminate trapping effects that would otherwise obscure intrinsic transport at temperatures as low as 4 K. The results unambiguously prove that at temperatures below 300 K charge transport is band-like. These holds generally for high-purity single crystals. Another example is shown in Figure 4.16(b), that is the temperature dependence of the electron mobility μ in a 370 μm thick perylene crystal. At $T < 30$ K, because of the purification and crystal growth, the mobility is limited by shallow traps and increases with increasing temperature. At the temperature range of 30 K to 300 K, μ decreases with increasing temperature. In this particular case, μ increases from about 1 cm^2 · V^{-1} · s^{-1} at room temperature to around 100 cm^2 · V^{-1} · s^{-1} at 30 K. The carrier mobility μ differs fundamentally in ultrapure aromatic molecular crystals from those in less-perfect organic crystals or disordered organic solids or polymers[40].

40. 超纯的芳香族分子晶体中的载流子迁移率 μ 与不太完美的有机晶体或无序有机固体或聚合物中的载流子迁移率有根本的不同。

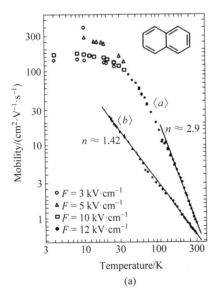

(a)

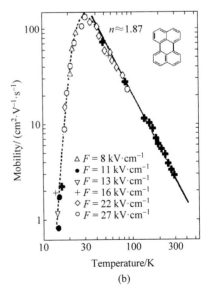

(b)

Figure 4.16   The temperature dependence of the carrier mobilities for (a) an ultrapure crystal of naphthalene and (b) a crystal of perylene at different field strengths F

4.2.2 Disordered Organic Materials

For disordered organic materials, carrier transport generally exhibits the following features: ① dispersive transport, ②mobility is low in the range of $10^{-7} \sim 10^{-5}$ cm$^2 \cdot$ V$^{-1} \cdot$ s^{-1} (some abnormal cases were also observed such as nondispersive transport with high mobility of $10^{-4} \sim 10^{-2}$ cm$^2 \cdot$ V$^{-1} \cdot$ s^{-1}), ③carrier transport shows the feature of thermal activation, hence mobility is strongly dependent on temperature (exponential dependence), and ④ mobility is dependent on applied electrical field.

A representative experimental result is shown in Figure 4.17. The temperature dependence of the hole mobility μ of a disordered MPMP film (thickness is 8.7 μm) at different field strength F, illustrates that the mobility data fit to exp T^{-2} dependence. The mobilities are, as in all non-crystalline solids, orders of magnitude smaller than in solid crystals, and the value decreases with decreasing temperature. Furthermore, the mobility depends on the electric field strength F^{41}.

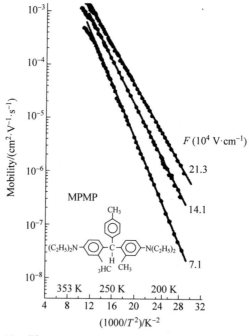

Figure 4.17　The temperature dependence of the carrier mobilities for in a vapor deposited MPMP organic film at different field strengths F. MPMP: bis (4-N, N-diethylamino-2-ethylphenyl)-4-methylphenylmethane

41. 与所有非晶固体一样,这些迁移率比固体晶体小几个数量级,并且该值随着温度的降低而减小。此外,迁移率依赖于电场强度 F。

1. Relationship between Mobility and Electric Field

The hopping mobility μ of an electron or hole in disordered organic semiconductors depends on the applied electric field because a field will reduce the activation energy for jumps in the field direction[42]. Experimentally, one usually finds that μ obeys a so-called Poole-Frenkel-type（PF model）field dependence.

$$\mu(F) = \mu_0 \exp(\beta_{PF} \sqrt{F}) \qquad (4\text{-}24)$$

μ_0 and β_{PF} are temperature dependence parameters. PF model characterizes the increase of the mobility with field and is adjusted through the fit parameters μ_0 and β_{PF}. The functional form of PF model resembles the field dependence of the Richardson-Schottky type of charge injection that describes the escape of a charge carrier from the Coulomb potential of a counter charge. However, in the case of charge transport, one has to postulate that the test system contains traps that are charged when they are empty in order to account for this field dependence. This is an unrealistic assumption because the phenomenon is ubiquitously observed with chemically very different classes of materials[43]. On the other hand, the $\ln(\mu) \propto \sqrt{F}$ dependence has been recognized as a genuine feature and is treated as another expression of PF model for amorphous organic semiconductors. This has been verified by Monte Carlo simulations, albeit only at high electric fields, while experimental results bear out this phenomenon already at lower fields of $10^4 \sim 10^5$ V $\cdot$ cm^{-1}.

2. Relationship between Mobility and Temperature

In contrast to ultrapure crystals, the charge carriers in disordered molecular solids are localized on the molecules and for transport, they must be thermally activated in order to hop from molecule to molecule. Therefore, in the disordered molecular solids, the mobility becomes greater with increasing temperature. The process of electrical conductivity is then termed hopping conductivity. If considering a carrier in organic semiconductor as a polaron, an Arrhenius-type temperature dependence for μ can be used to predict mobility of organic semiconductors：

42. 无序有机半导体中电子或空穴的跃进迁移率 μ 依赖于所施加的电场,因为电场会减少载流子沿着电场方向跃进时的激活能。

43. 然而,对于电荷传输,为考虑这种电场的依赖性,必须假设待测材料体系包含带有荷电的空陷阱。这是一个不切实际的假设,因为这种现象(载流子的电场依赖关系)无处不在地被具有不同化学性质的材料所观察到。

$$\mu = \mu_0 \exp\left(-\frac{\Delta}{k_B T}\right) \qquad (4\text{-}25)$$

μ_0 is pre-factor, Δ is activation energy. In a higher disordered material, Δ is larger and the mobility is smaller. For example, in films of polymer P3HT (poly-3-hexylthiophene), highly disordered packing ($\Delta = 0.13$ eV) leads to small μ of $10^{-5} \sim 10^{-4}$ cm$^2 \cdot$ V$^{-1} \cdot$ s^{-1}, while highly order film ($\Delta = 0.02 \sim 0.04$ eV) allows high mobility of $\mu = 0.1$ cm$^2 \cdot$ V$^{-1} \cdot$ s^{-1}.

When $\dfrac{\Delta}{k_B T} \ll 1$, temperature dependence for μ will be taken over by $T^{-3/2}$ dependence. For a realistic value of 100 meV for Δ, the temperature at which $\mu(T)$ merges into a $T^{-3/2}$ dependence is above 600 K. This implies that one should be cautious with the custom of approximating the mobility by Equation (4-25) and determining the pre-factor mobility μ_0 by extrapolating a $\ln(\mu(T))$ versus $1/T$ plot to infinite temperature[44].

In cases where the electron-phonon coupling is weak and the temperature is low ($T < 300$ K), the mobility is controlled by band-type transport (Figure 4.18(a)). In the case of strong electron-phonon coupling (Figure 4.18(b)), as the temperature increases, there is a superposition of the (decreasing) contribution of band motion and increasing contribution of thermally activated hopping. Eventually, the latter takes over and, concomitantly, (T) passes through a minimum. Finally, $\mu(T)$ will approach a $T^{-3/2}$ law.

44. 对于 Δ 的实际值为 100 meV 的情况,迁移率 $\mu(T)$ 的温度依赖关系满足 $T^{-3/2}$ 时的温度高于 600 K。这意味着人们应该谨慎对待惯用的通过式(4-25)来近似得出似迁移率,以及谨慎对待通过将 $\ln(\mu(T))$ 与 $1/T$ 外推到无限温度来确定迁移率的前置因子 μ_0。

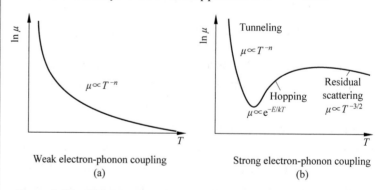

Weak electron-phonon coupling
(a)

Strong electron-phonon coupling
(b)

Figure 4.18　The type of temperature dependence of the charge carrier mobility predicted by the polaron model in the case of weak (a) and strong (b) electron-phonon coupling

In disorder-controlled carrier transport，Gaussian model (detail is discussed in the following section) can also describe the carrier transport by thermal activation：

$$\mu = \mu_0 \exp\left[-\left(\frac{T_0}{T}\right)^2\right] \quad (4\text{-}26)$$

here T_0 describes energy disorder.

Both polaron Arrhenius like equation and Gaussian model can well simulate the relationship between mobility and temperature for organic semiconductors in amorphous form.

3. Activation Model by WD Gill

In 1974，WD Gill combined the factors of temperature and electric field to mobility and gave the relational formula of the mobility with temperature T and electric filed F：

$$\mu = \mu_0 \exp\left[-\frac{\Delta E_a}{k_B T}\right] \exp\left[\beta \sqrt{F}\left(\frac{1}{k_B T} - \frac{1}{k_B T_0}\right)\right] \quad (4\text{-}27)$$

where μ_0 is pre-parameter of μ, ΔE_a is activation energy，β is a factor based on electric field，F is strength of the electric field，T is temperature，k_B is Boltzmann constant，and T_0 is the temperature when field strength dependence disappears. Temperature activation and electric field activation are all taken into account in activation model，and μ_0 obtained here is reasonable.

L. Shein got another activation mobility expression via the investigation of hole mobility：

$$\mu(E,T) = \mu_0 \exp\left[-\left(\frac{T_0}{T}\right)^2\right] \exp\left[-F^{1/2}\left(\frac{\beta}{T} - \gamma\right)\right] \quad (4\text{-}28)$$

where γ is an empirical parameter.

4. Gaussian Disorder Model（GDM）

A simple concept to explain charge carrier mobility in a disordered organic solid on a microscopic level is the Gaussian disorder model（（GDM），also known as Bässler model）and its subsequent extensions. It supposes that the degenerated energy levels involving carrier transport spit into localized states, charge carriers hop within these local states that feature a Gaussian energy distribution and a Gaussian distribution of inter-site spacing[45]：

45. 它假设参与载流子传输的简并能级裂分为定域状态,电荷载流子在这些定域状态内跃进,这些定域状态具有高斯型的能量分布和位点间距分布:

$$g(E) = \frac{1}{\sqrt{2\pi\sigma^2}} \exp\left(-\frac{E^2}{2\sigma^2}\right) \tag{4-29}$$

where E is the energy of an individual molecular orbital, and the standard deviation σ (the "energy disorder parameter") characterizes the width of the Gaussian distribution. σ is caused by the fluctuation of lattice polarization energy, indicating energy fluctuation of localized state. This fluctuation results mainly from the van der Waals coupling with the neighboring sites. For conjugated polymers, there is an additional contribution due to the variation in the lengths and thus the energies of the conjugated segments of the chain. It is important to recognize that the broadening of density of state (DOS) expressed through the value of σ is a result not so much of the magnitude of intermolecular interactions but rather of their randomness[46]. For instance, when studying the hole mobility of hole transporting molecules, such as TAPC molecules, embedded as dopants in a matrix, one finds that the hole mobility decreases by two orders of magnitude when the non-polar polystyrene matrix is replaced by a polar polycarbonate matrix. The reason is that the random orientation of the dipole moments on the carbonyl groups roughens the energetic landscape. The effect vanishes if the dipole moments are topologically aligned, for example, in the case in a molecular crystal composed by polar molecules.

The fact that the DOS distribution is of Gaussian shape is straightforward because the interaction energy of a charged chromophore imbedded in an amorphous polarizable environment depends on a large number of coordinates and interactions, each varying statistically. Based upon the central limit theorem this leads to a Gaussian envelope function[47].

Figure 4.19 considers charge carriers generated at an arbitrary site within a DOS distribution, where E_m and E_n represent different discrete energy levels. Each of the carriers hops from one site to the next, thereby relaxing toward the tails of the distribution. While initially the energetically downhill hops will dominate the carrier path, eventually a balanced equilibrium between downhill and

46. 重要的是要认识到，以 σ 表达的态密度(DOS)的展宽与其说是分子间相互作用的大小，不如说是其随机性的结果。

47. 态密度 DOS 分布是高斯形状的事实，非常直接，因为嵌入在可极化的非晶态环境中的带电发色团的相互作用能，取决于大量的坐标系和相互作用，每个坐标系和相互作用的变化服从统计规律。基于中心极限定理，这可导致一个高斯包络函数。

thermally activated uphill jumps will be established and a quasi-equilibrium is obtained.

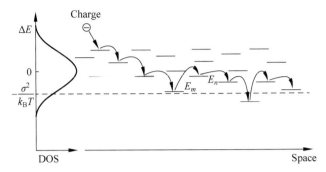

Figure 4.19　Illustration of a charge carrier, generated at an arbitrary energy, that hops within a Gaussian DOS. The dotted horizontal line is the energy at which the charge carriers tend to equilibrate in the long-time limit

In the case of polaronic transport between identical molecules, there is no difference between a forward jump from site i to site j and the backward jump from site j to site i. Both sites are isoenergetic and have the same polaronic binding energy. In consequence, the associated hopping rate is the same for the forward and backward jump. When introducing static energetic disorder, this symmetry is broken. If the forward jump is downhill in energy, the backward jump is evidently uphill, and vice versa. This requires an asymmetric jump rate. A simple approach is to consider that for the downhill direction, excess energy is simply dissipated while for the uphill transfer, an activation energy is required in the form of a Boltzmann factor[48]. This introduces an energy dependence that needs to be multiplied with the transfer rate due to the electronic coupling between the two sites. From m to n state in Figure 4.19, the hopping rate can be expressed as:

$$v_{mn} = v_0 \exp(-2\gamma \Delta R_{mn}) \exp\left[-\left(\frac{E_n - E_m}{k_B T}\right)\right], \quad E_n > E_m$$

$$\tag{4-30}$$

$$v_{mn} = v_0 \exp(-2\gamma \Delta R_{mn}), \quad E_n \leqslant E_m \tag{4-31}$$

where v_{mn} is hopping rate from m state to n state; ΔR_{mn} is the jump distance between sites m and n; γ is the inverse localization radius of the electron wavefunction. $2\gamma \Delta R_{mn}$ is

48. 一种简单的方法是考虑对于下坡方向移动,多余的能量只是简单的消散,而对于上坡方向的转移,需要玻尔兹曼因子形式的激活能。

related to the electronic coupling matrix element between adjacent sites. ν_0 is a frequency factor (attempt-to-hop frequency). Equations (4-30) and (4-31) imply that the hopping sites are point sites, where the dielectric coupling can depend on the mutual orientation of the transport molecules. The model assumes that the Boltzmann factor is equal to 1 for a downward jump regardless of electric field and that there is always an energy accepting phonon mode available to dissipate the excess energy released when a charge carrier hops to a lower energy site. In contrast, thermal activation is required for a jump to a site at higher energy. When the Gaussian DOS with width of Σ (σ) (positional disorder parameter) is used together with the above rate in order to account for charge transport, one obtains a model that is purely disorder-based without any contributions from polaronic effects.

When measuring the field-dependent charge carrier mobilities at different temperatures, one observes that the slopes of the $\ln \mu \propto \sqrt{F}$ plots can decrease significantly and can become negative. The reason is related to positional static disorder in addition to the energetic static disorder. Position static disorder varies the strength of electronic coupling among the hopping sites[49]. It is incorporated in the hopping rate of Equations (4-30) and (4-31) by changing the parameter $2\gamma \Delta R_{mn}$ to $2(\gamma \Delta R)_{mn}$ and allowing $2(\gamma \Delta R)_{mn}$ to vary statistically with a Gaussian distribution function

$$g(\gamma \Delta R) = (1/\sqrt{2\pi\Sigma}) \exp\left\{-\frac{[\gamma \Delta R - (\gamma \Delta R)_0]^2}{2\Sigma^2}\right\}$$

The positional disorder parameter Σ can, of course, be different from the energetic disorder parameter σ. It turns out that when $\Sigma > \sigma/(k_B T)^2$, the field dependence of μ can become negative.

The qualitative explanation of field dependence of mobility is that, in systems with large positional disorders, a charge carrier can find a more favorable detour in order to avoid a jump over a large energy barrier (see Figure 4.20 and the explanation in the figure caption). Since this detour can involve jumps against the field direction, it will be blocked at higher fields. As a consequence, the mobility can decrease

49. 当测量不同温度下的电荷载流子迁移率与电场的依赖关系时,观察到 $\ln \mu \propto \sqrt{F}$ 图的斜率可以显著地减小并可能变为负值。原因是载流子迁移率除了与静态的能量无序相关,还与静态的位置无序有关。静态的位置无序改变了跃进位点之间电子耦合的强度。

with field. In the intermediate to high field range, a quantitative description of hopping transport in the presence of both energetic and positional disorder has been derived from Monte Carlo simulations in the absence of site correlation[50]. The temperature T and electric field strength F dependences of the charge carrier mobility are predicted to be GDM model:

$$\Sigma \geqslant 1.5: \mu = \mu_0 \exp\left[-\left(\frac{2\sigma}{3k_BT}\right)^2\right] \exp\left\{c \sqrt{F}\left[\left(\frac{\sigma}{k_BT}\right)^2 - \Sigma^2\right]\right\}$$

(4-32)

$$\Sigma < 1.5: \mu = \mu_0 \exp\left[-\left(\frac{2\sigma}{3k_BT}\right)^2\right] \exp\left\{c \sqrt{F}\left[\left(\frac{\sigma}{k_BT}\right)^2 - 2.25\right]\right\}$$

(4-33)

where c is numerical constant that depends on the inter-site separation (e.g., for the distance of two sites = 0.6 nm, $c = 2.9 \times 10^{-4}$ cm$^{1/2} \cdot$ V$^{-1/2}$). F is the strength of the electric field, σ is energy disorder parameter - the width of the state density of Gaussian distribution, Σ is the position disorder parameter, relating to the interaction strength between two adjacent molecules.

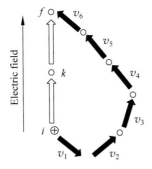

Figure 4.20 Illustration of possible pathways for a hole to get from an initial site i to a final site f that are separated by an intermediate high energy site k. The direct route (white arrows) via the high-energy site can be avoided by a detour (black arrows) over other, energetically more accessible sites. The detour can involve jumps against the direction of the electric field that become blocked when the field increases

Taking into account the long-range dipole-dipole interaction, S. V. Novikov et al. gave the Gaussian correlated model，i. e.，correlated Gaussian disorder model （CDM model）：

$$\mu = \mu_0 \exp\left\{-\left(\frac{3\sigma}{5k_B T}\right)^2 + 0.78\left[\left(\frac{\sigma}{k_B T}\right)^{3/2} - 2\right]\sqrt{\frac{qaF}{\sigma}}\right\}$$

$$(4\text{-}34)$$

where a is the distance between two hopping sites. The CDM model explains the observed Poole-Frenkel-type electric field dependence of μ over broad range of electric fields as well as the $\ln(\mu)$ versus T^{-2} type of temperature dependence. Blom and Vissenberg observed that in anisotropic disordered materials，temperature dependent mobility was CDM type and not GDM type. Heun et al. also reported a T^{-2} type of temperature dependence $\ln(\mu)$ of an 11 μm thick film of the TTB as shown in Figure 4.21[51].

51. Blom 和 Vissenberg 观测到，在各向异性无序材料中，温度依赖的迁移率是 CDM 型而不是 GDM 型。Heun 等也报道了在 11 μm 厚的 TTB 薄膜中 T^{-2} 型的温度依赖 $\ln\mu$，如图 4.21 所示。

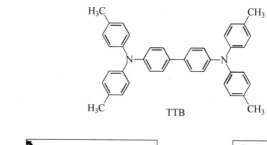

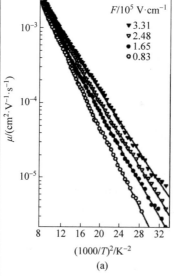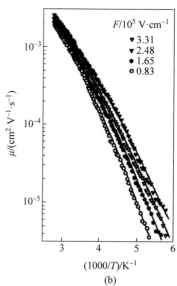

Figure 4. 21　Comparison of the temperature dependence of the hole mobility in a 11 μm thick film of the TTB shown above at different electric fields plotted on a T^{-2} (a) and a T^{-1} (b)

Figure 4.22 shows the theoretical results of field dependences of charge carrier mobilities in a doped disordered organic semiconductor at different temperatures. A Gaussian distribution width of $\sigma = 100$ meV has been used as an intrinsic DOS distribution. Although the curves follow the Poole-Frenkel-type $\ln \mu \propto F^{1/2}$ dependence at weaker fields, they tend towards saturation at stronger fields. In addition, mobility increases with increasing temperature[52].

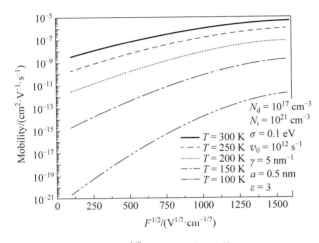
Figure 4.22　Mobility-$F^{1/2}$ curves under different temperatures

Figure 4.23 illustrates the theoretical results of temperature dependence of the mobility under different external fields. Although both the doping-induced Coulomb traps and the intrinsic DOS distribution affect this dependence, most carriers are localized in the former, which gives rise to an almost perfect Arrhenius temperature dependence with the slope affected by the external field.

Figure 4.24 illustrates the theoretical results of dopant concentration dependence of the mobility in the width of the intrinsic Gaussian DOS distribution. These dependences are strikingly different in materials with weak and strong energy disorder, i.e., with small and large width of the DOS. It can be seen that the mobility will be suppressed by concentration when doping into a weakly disordered system[53]. In a stronger disordered system, the mobility increases with doping level. It should be noted, however, that the mobility always decreases less than the doping concentration N_d increases and,

53. 这些依赖性在具有较弱和较强能量无序强度的材料中,即材料具有不同的 DOS 宽度,显著不同。可以看出,当掺杂到能量无序性不强的体系中时,迁移率会受到浓度的抑制。

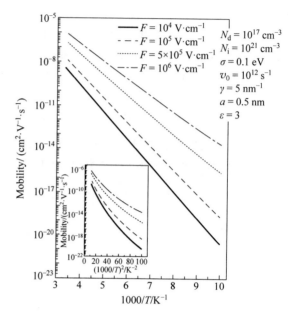

Figure 4.23　Mobility-$(1000/T)$ curves under different F

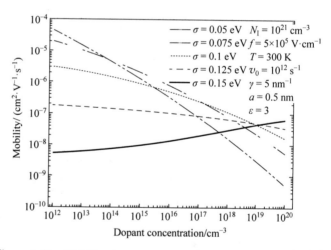

Figure 4.24　Mobility-dopant concentration curves under different σ

54. 但是要注意到，这个迁移率降低总是小于掺杂浓度 N_d 的增加，因此，与 μ 和 N_d 之积成正比的电导率随掺杂而增加，甚至在 DOS 宽度较小的材料中也如此。

therefore, the conductivity, which is proportional to the product of μ and N_d, increases upon doping even in materials with small widths of DOS[54]. It is known that dopants provide both charge carriers and deep Coulomb traps. If these traps are deeper than those states that control the mobility in the pristine material, the deep Coulomb traps will still trap majority of doping-induced carriers and their mobility has to be smaller than the carrier mobility in the undoped pristine material. However, when the effective depth of a Coulomb trap is smaller, carriers can escape from this shallow trap by jumps via localized states with energies below the maximum of the DOS distribution. Under small DOS distribution, mobility decreases with increase of doping concentration. With large DOS distribution, mobility slightly increases with increase of doping concentration. This is because that the carrier energy dispersion is large when DOS distribution is wider. Therefore, with the increase of doping concentration, the energy of carrier may form continuous band, leading to improvement of carrier mobility[55].

Figure 4.25 shows the theoretical results of the dependence of the carrier mobility on the dopant concentrations at different external fields. At weak field, mobility decreases with doping level, while at strong filed, mobility increases with doping level. Since the effective depth of Coulomb traps is controlled by the external field, one should expect different dopant-concentration dependences of the mobility at weak and strong electric fields. Indeed, at weak external fields, Coulomb potential wells are deep and ionized dopants serve as deep traps for carriers. Strong external fields reduce the barrier for carrier release from Coulomb traps, making them shallower, and, thereby, increasing the density of free carriers and the average carrier mobility[56].

55. 这是因为当 DOS 分布较宽时,载流子的能量分散性很大。因此,随着掺杂浓度的增加,载流子的能量可能形成连续的带,导致载流子迁移率的提高。

56. 由于库仑陷阱的有效深度受外电场的控制,因此可以预期在弱电场和强电场下迁移率的掺杂浓度依赖性不同。事实上,在弱的外部场中,库仑势阱很深,离化的掺杂剂是载流子的深陷阱。强的外部场降低了库仑陷阱中释放载流子的势垒,使它们变浅,从而增加了自由载流子的密度和平均载流子的迁移率。

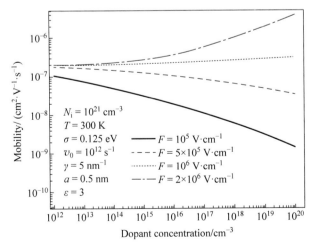

Figure 4.25 Mobility- dopant concentration under different F

4.3　Contact and Contact Potential

4.3.1　Concepts

The successful operation of some types of organic optoelectronic devices, such as OLED and OFET, requites the injection of carriers, in which contacts are the essential interface. Generally, injection contact is a heterojunction between metal and nonmetal materials, such as insulators or semiconductors, excluding those between two semiconductors, two metals, and one semiconductor and one insulator. The features of contact are dependent on the electronic structure of the nonmetal material and the metal electrode.

The most important feature relating to contact is contact barrier, which is normally determined by the energy difference between the work function of the electrode metal and the hole or electron transporting states of the nonmetal material[57]. Different from metals with carrier distribution near Femi level and adopting band like transport, charge carriers in organic materials can be both electron and hole, which distribute near the LUMO and HOMO, respectively.

In terms of contact between the injection electrode and the nonmetal material in a device, the resulted contact will be determined by the electrical properties of nonmetal materials, generally, there are three types of contact: neutral contact, Schottky contact and Ohmic contact. Different contact leads to different contact barrier and energy diagram at the contact interface.

Although the contact between two metals is not an electrical contact, there exists some physical process, which we shall discuss here. As shown in Figure 4.26, when the two metals M_1 and M_2 of different Fermi level are brought into contact with each other, an electron current flows for a short time until the two Fermi levels are equalized. The direction of electron flowing is from M_1 with higher Fermi level to M_2 with lower Fermi level[58]. Finally, M_2 is negatively and M_1 is positively charged and there will be a dipole layer between

57. 与接触有关的最重要特性是接触势垒,这通常由金属电极的功函数与非金属材料的空穴或电子输运态之间的能量差决定。

58. 如图 4.26 所示,当具有不同费米能级的两种金属 M_1 和 M_2 相互接触时,在一段极短的时间内会有电子流动出现,直到两个金属的费米能级相等。电子流动的方向是从具有较高费米能级的 M_1 到具有较低费米能级的 M_2。

the two metals with thickness of d, leading to the shift of vacuum levels (V to V_1 in M_1, and V to V_2 in M_2), and the built-in contact potential of Φ_{BI}.

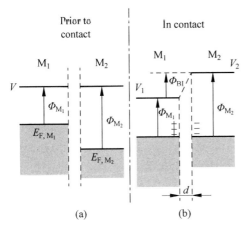

Figure 4.26   Energy levels of the (a) two un-contacted metals and (b) two contacted metals at equilibrium conduction

4.3.2   Neutral Contact

Neutral contact refers to the contact of a metal with a pure semiconductor (SC), where the interface is very pure and no electrons flow cross this interface[59]. Figure 4.27 shows the energy levels of M_1, pure semiconductor and M_2 before contact. Since E_F of M_1 is higher than that of M_2, upon the contact of the pure semiconductor with both M_1 and M_2 and under short circuit condition (Figure 4.28(a)), electrons with higher energy in M_1 flow through external circuit to M_2 until their E_F is equal. Thus, when the contact is produced via an ideal intrinsic semiconductor with thickness d, a built-in electric field $F_{BI} = \Phi_{BI}/qd$ is established within the semiconductor, where Φ_{BI} is the Fermi level difference between the two metals. Because the potential energy of the electron increases in the opposite direction along the electric potential, the built-in electric field of the semiconductor reduces the electronic potential energy of the M_1/SC interface and increases the electronic potential energy of the SC/M_2 interface, leading to the energy level tilting toward M_2 direction.

59. 中性电接触指的是一个金属和一个纯半导体(SC)的接触,其界面非常纯净,没有电子流过这个界面。

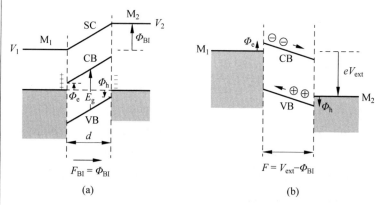

Figure 4. 27　The diagram of before contact for a pure organic SC sandwiched by two metals of M_1 and M_2

60. 在这种情况下，电子和空穴注入的接触势垒分别为 Φ_e 和 Φ_h。为了可以进行电荷传输，必须对两个金属电极施加一个反向的电压 V_{ext}，反向电压至少要足以补偿内建电场，即 $V_{ext} > -\Phi_{BI}$。

In this case, the contact barrier for electron and hole injection are Φ_e and Φ_h, respectively. To make charge transport possible, a counter-voltage V_{ext} must be applied to two metal electrodes, which is at least sufficient to compensate the built-in field, $V_{ext} > -\Phi_{BI}$[60]. The (internal) field strength in this greatly idealized case would then be $F = (V_{ext} - \Phi_{BI})/qd$ (Figure 4.28(b)).

Figure 4. 28　Neutral contacts (a) without bias and under short circuit and (b) with an applied bias

4.3.3　Schottky Contact

Neutral contact is a most ideal situation and rarely exists in practice. Practically, when the contact forms between two materials with different Fermi level, free carriers will flow from one material to the other until reaching equilibrium conditions, i. e., the Fermi levels of two materials at interface are same. As shown in Figure 4.29, the contact process between a n-type semiconductor and metal is illustrated. Before contact (Figure 4.29(a)), Fermi level of the n-type semiconductor is higher than that of the metal. When they contact with each other, electrons will flow from the n-type semiconductor to the metal, resulting in negatively charging at the metal surface and positively charging at the surface of the n-type semiconductor and finally set up an electric double layer near the contact interface (Figure 4.29 (b))[61]. Since carrier density in semiconductor is much smaller than that in metal, the distribution of positive charge in semiconductor has a thickness of W. As electron carrier density is extremely low in this region, this layer is also called as depletion region. In addition, the electric field is directed from inner semiconductor to outer semiconductor, producing an upward bending of the bands in the depletion layer, where electrons have moved from the semiconductor to the metal. The upward bended bands form a potential barrier to prevent electrons further injecting from the semiconductor to the metal (Figure 4.29(c)). The formed potential barrier (Φ_B) is equal to different between work functions of semiconductor and metal ($\Phi_m - \Phi_s$). This kind of contact is Schottky contact.

As shown in Figure 4.30, when injecting electrons to semiconductor from metal via Schottky contact, there are two situations: forward bias and reverse bias. For forward biased Schottky contact, electrons in the CB of the semiconductor can readily overcome the small potential energy barrier $\Phi_B - eV_f$ to enter the metal. However, for reverse biased Schottky junction, electrons in the metal cannot easily overcome the barrier $\Phi_B + eV_r$) to enter the semiconductor. Both in the forward and reverse biased situations, the width and bending height of depletion layer is depended on the applied biases[62]. Figure 4.30(c) shows the current-voltage curve of a Schottky contact. At reverse bias and low forward bias, the current is very small in micro-amps. However, at high forward bias, current increases rapidly. This phenomenon is the rectifying property of Schottky contact.

61. n-型半导体的费米能级高于金属的。当它们相互接触时,电子将从 n-型半导体流向金属,导致在金属表面带负电,在 n-型半导体表面带正电,最后在接触界面附近建立一个正负偶电层(图 4.29(b))。

62. 对于施加正向偏压的肖特基接触,半导体 CB 中的电子可以很容易地克服小的势能 $\Phi_B - eV_f$ 进入金属。然而,对于施加反向偏压的肖特基结,金属中的电子不能轻易地克服势能 $\Phi_B + eV_r$ 而进入半导体。在正向和反向偏压的情况下,耗尽层的宽度和弯曲高度取决于所施加的偏压大小。

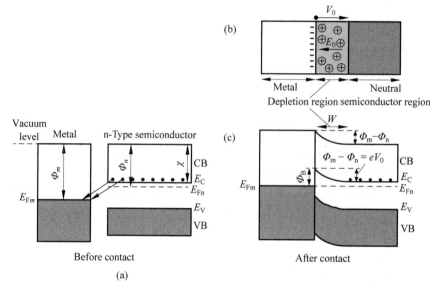

Figure 4. 29　Schottky contact and contact barrier for electron injection：（a）energy diagram before contact，（b）charge flow and the depletion region W in the semiconductor，（c）energy diagram after contact

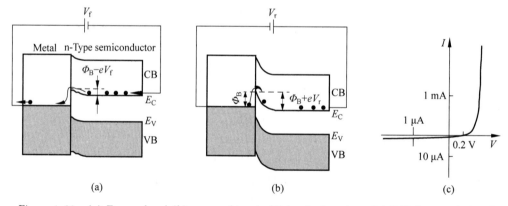

Figure 4. 30　（a）Forward and （b）reverse biased of Schottky junction. （c）I-V characteristics of Schottky junction under forward and reverse bias

4.3.4　Ohmic Contact

There is another type of contact between the metal and semiconductor，where the I-V characteristic of contact does not show rectifying property. We also take an n-type semiconductor for example，but here the Fermi level of the n-type semiconductor is lower than that of the metal[63]. Figure 4.31(a) shows the energy levels of a metal and an n-type

63. 金属和半导体之间存在另一种电接触，其 I-V 没有整流性质。我们也以 n 型半导体为例，但此时 n 型半导体的费米能级比金属的低。

semiconductor before contact. After contact, electrons in the metal flow to the semiconductor until their Fermi levels are equal. An electric double layer near the contact will also form, but the direction of the electric field is opposite compared to Schottky contact. The band bending direction is also opposite, Ohmic contact shows a downward bending of the conduction and valence bands toward the interface (Figure 4.31(b))[64]. Electron injection both in forward and reverse biased situations are permitted and there is no current limitation due to the junction. This kind of contact is called Ohmic contact. In the case of an Ohmic contact, the density of the free charge carriers at the contact and in its immediate neighborhood is much higher than the density of free charge carriers in the bulk of the semiconductor. An Ohmic contact thus provides a sufficient reservoir of charge carriers.

64. 接触后，金属中的电子流向半导体，直到它们的费米能级相等。在接触附近也会形成双层，但与肖特基接触相比，内建电场的方向是相反的。能带弯曲方向也相反，欧姆接触表现出朝着界面方向，向下弯曲的导带和价带（图4.31(b)）。

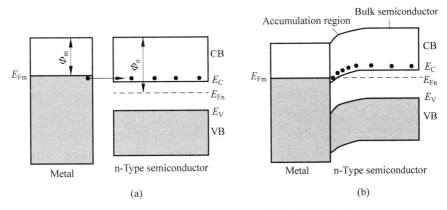

Figure 4.31   Diagrams of before (a) and after (b) Ohmic contact

The term "Ohmic contact" can be misleading in the sense that it does not necessarily mean that the current follows Ohm's law, i.e., the current depends linearly on the voltage. On the contrary, only when the voltage V is so small that the injected charge carrier density is fewer than that of intrinsic carrier in the bulk by thermal activation, the characteristic of I-V is linear. As soon as the voltage exceeds this limit, that is when the semiconductor is charged with excess charge carriers and forms zone of space charge, the characteristic of I-V becomes nonlinear in a typical manner.

4.4　Currents under Carrier Injection

To study the current-voltage characteristics of nonmetal materials, such as organic semiconductors, several factors should be considered. Due to the existence of space charge and contact problem, the characteristic of I-V in nonmetal materials is not linear; Ohm's law is no longer valid. Undoped organic semiconductors are, in fact, insulators unless charge carriers are injected from the electrodes or generated in the bulk by optical excitation. To proof this statement, one can adopt the classic semiconductor theory to estimate the concentration of charge carriers existing in the dark in an organic semiconductor under equilibrium conditions. Since the electrical gap between hole and electron transporting states is usually at least 2 eV, and the Coulomb attraction between positive and negative charges is poorly screened, the intrinsic concentration of free carriers is negligible and high electric current transport can only be achieved by introducing non-intrinsic carriers[65]. For inorganic materials, non-intrinsic carriers can be obtained by chemical doping, while for organic materials, non-intrinsic carriers are mainly obtained by injection through electrode. Non-intrinsic carriers in nonmetal materials will form excess charge carriers and space charges. These space charges can be movable or trapped charges, and they are highly depended on the temperature and field strength. Only the mobile charges contribute to the current, bound charges do not contribute to the current flow. Both mobile and bound charges will influence the distribution of electric field inside materials[66]. Additionally, there is usually uncertain potential drop at the contacts. The contact potential and space charge make the I-V characteristic of nonmetal materials is seldom linear, i.e., Ohm's law is generally not valid.

4.4.1　Injection Limited Current

An injection limited current reflects only the mechanism by which charge carriers are injected and not their subsequent transport. Any dependence of the current on the electric field

65. 由于空穴和电子传输态之间的电子能隙通常至少为 2 eV,并且正负电荷之间的库仑吸引力不能很好地被屏蔽,因此本征自由载流子浓度可以忽略不计,只有通过引入非本征载流子来实现高的电流传输。

66. 非金属材料中的非本征载流子将产生过剩的电荷载流子和空间电荷。这些空间电荷可以是可移动的或被捕获的,并且它们高度依赖于温度和电场强度。只有移动的电荷对电流有贡献,而受束缚的电荷对电流没有贡献。移动和束缚的电荷都会影响材料内部电场的分布。

and temperature is therefore a reflection of the injection process and does not yield any information on the charge carrier mobility. This is an important insight to keep in mind for the case of injection limited current.

When the contact is not Ohmic contact, the number of carriers injected from electrode is restricted, the current transport in bulk of semiconductor is injection-limited current. The crucial parameter for this is the energy difference between the work function of the electrode and the hole or electron transporting states[67]. Taken into account the case of electron injection, as shown in Figure 4.32, which describes the barrier for the injection of an electron from a metal (work function = Φ) into a semiconductor (electron affinity = EA) with an applied electric field F and electronic potential inside semiconductor, when an external electric field, F, is applied to electrode, a potential $\Phi_F = qFx$ will build up. After injecting electrons into semiconductor from electrode, the mirrored positive charges called image charges will be induced inside the electrode. These mirror charges will generate mirror force and image potential Φ_{image} inside semiconductor. The result total potential Φ_{total} inside the semiconductor will be the sum of external potential Φ_F and image potential Φ_{image}. Hence, $\Phi_e = \Phi - EA$, and $\Phi_B = \Phi_e - \Delta\Phi$ are the barrier heights without and with the image charges. The injection barrier is reduced $\Delta\Phi = \sqrt{\dfrac{qF}{4\pi\varepsilon}}\,q$ and reaching its maximum value at $x_m = \sqrt{\dfrac{q}{16\pi\varepsilon F}}$ from the contact surface. x_m is a measure of the width of the barrier. These values can be determined as follow:

$$\Phi_F = qFx \tag{4-35}$$

$$\Phi_{image} = \frac{q^2}{16\pi\varepsilon\varepsilon_0} \cdot \frac{1}{x} \tag{4-36}$$

$$\Phi_{total} = \Phi_F + \Phi_{image} = qFx + \frac{q^2}{16\pi\varepsilon\varepsilon_0} \cdot \frac{1}{x} \tag{4-37}$$

$$\Delta\Phi = \Phi_e - \text{apex of } \Phi_{total} \tag{4-38}$$

$$\Phi_B = \Phi_e - \Delta\Phi = \text{apex of } \Phi_{total} \tag{4-39}$$

The above injection barrier can be overcome by either tunneling or thermal activation, producing injection current,

67. 当接触不是欧姆接触时,从电极注入的载流子数量会受到限制,半导体中的电流传输是注入限制电流。其关键参数是电极的功函数与空穴或电子传输态之间的能量差。

68. 上述注入势垒可以通过隧穿或热激活克服,产生注入电流,如下面的讨论。

as discussed in the followings[68].

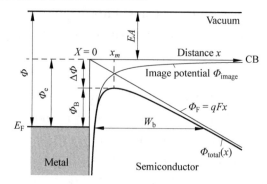

Figure 4.32　The barrier for the injection of an electron from a metal into a semiconductor

1. Current of Fowler-Nordheim Tunneling Injection

When the wavelength of carriers is not smaller than the barrier width (W_b in Figure 4.32), electron can directly tunnel through without overcoming the injection barrier. This is the tunnel effect, which is also called field emission. As shown in Figure 4.32, upon applying a strong electric field, electrons can tunnel from the Fermi level of the metal through a triangular barrier with the barrier height of Φ_e into the CB of semiconductor. The triangular barrier is determined by the superposition of the rectangular barrier due to the metal-semiconductor interface and the applied potential[69]. Fowler and Nordheim predicted that the injection current depends on the applied electric field as

69. 当载流子波长不小于势垒宽度(图 4.32 中的 W_b)时,电子可以无需克服注入势垒直接隧穿通过。这是隧穿效应,也称为场发射。如图 4.32 所示,施加强电场,电子可以从金属费米能级隧穿通过高度为 Φ_e 的三角形势垒,进入半导体 CB。这个三角形势垒,通过金属-半导体界面的矩形势垒和所施加电势的叠加来确定的。

$$j_{FN} = \frac{A^*}{\Phi_e}\left(\frac{eF}{\alpha k_B}\right)^2 \exp\left(-\frac{2\alpha\Phi_e^{\frac{3}{2}}}{3eF}\right) \quad (4\text{-}40)$$

where $\alpha = 4\pi(2m_{eff})^{1/2}/h$, A^* is effective Richardson constant, as described in next page. That is, a plot of $\ln(j/F^2)$ versus $1/F$ should yield a straight line with a slope proportional to $\Phi_e^{\frac{3}{2}}$, this is an indication of the tunnel effect. If F is determined, the value of Φ_e can be computed from the slope of the line.

2. Current of Richardson-Schottky Thermionic Injection

As shown in Figure 4.32, consider the injection of an electron that is thermally excited from the Fermi level of a planar metallic electrode with work function of Φ to a semiconductor with an electron affinity of EA. Frequently, $|EA| < |\Phi|$, so that there is an electron potential barrier $\Phi_e = \Phi - EA$ from the metal to the semiconductor. The idea behind Richardson-Schottky type injection is that the electron induces an image charge at an equal distance from the interface to that of the real charge. The total potential $\Phi_{total}(x)$ experienced by the electron is then composed of the contact barrier Φ_e, the image potential Φ_{image} determined by Equation (4-36), and the applied electrostatic potential $\Phi_F = qFx$[70].

The current density j_{RS} for this process, where electron must overcome the barrier with height of Φ_B, was originally calculated by Richardson for glow cathodes in vacuum tube and, taking the Schottky effect into account, is given by

$$j_{RS} = A^* T^2 \exp\left(-\frac{\Phi_B}{k_B T}\right) \qquad (4\text{-}41)$$

where, $A^* = 4\pi e m^* k_B^2 / h^3$, is effective Richardson constant, m^* is effective mass, e is unit charge, k_B is Boltzmann constant, h is plank constant. Suppose m^* equal to free electron mass m_0, then $A^* = 120 \ \text{A} \cdot \text{cm}^{-2} \cdot \text{K}^{-2}$.

Equation (4-41) is the expression for thermal emission electron current of electrode to vacuum, in which the Φ_B is the work function of the emission electrode. In a real Schottky contact containing a metal and a semiconductor, besides the injection barrier, the injection current density in the flow direction is also influenced by the internal electric potential (U) in the semiconductor. For $eU > 3k_B T$, j_{SR} is given by Shockley equation:

$$j_{SR} = A^* T^2 \exp\left(-\frac{\Phi_B}{k_B T}\right) \exp\left[\left(\frac{eU}{n k_B T}\right) - 1\right] \quad (4\text{-}42)$$

here, n is an "ideality factor". For ideal Schottky diodes, $n = 1$.

Another expression for injection-limited current based on Richardson-Schottky thermionic injection theory is:

$$j_{RS} = A^* T^2 \exp\left(-\frac{\Phi_B - \beta_{RS} \sqrt{F}}{k_B T}\right) \qquad (4\text{-}43)$$

here, F is the electric field intensity, $\beta_{RS} = (e^3 / 4\pi\varepsilon\varepsilon_0)^{1/2}$, ε and ε_0 are media and vacuum dielectric constant,

70. 理查德森-肖特基型注入的理论思想是,(半导体中)电子在金属内诱导出镜像电荷,它到界面的距离等于真正电荷到界面的距离。然后,电子受到的总势能 $\Phi_{total}(x)$ 由接触势垒 Φ_e、式(4-36)决定的镜像势 Φ_{image} 和施加的静电势 $\Phi_F = qFx$ 组成。

respectively. Equation (4-43) predicts the field dependence injection current as $\ln\left(j_{RS}(F)\right)\propto\sqrt{F}$. This type of field dependence is characteristic for a process in which a charge is transferred across a Coulomb potential superimposed onto an applied potential regardless if the Coulomb potential is created by a real charge or by an image charge.

The Richardson-Schottky concept is premised upon the notion that once an electron has acquired enough thermal energy to overcome the potential barrier, it traverses the potential barrier in field direction without suffering inelastic scattering. Otherwise, the electron would recombine with the image charge and be lost for the injection process[71]. Whether or not this condition is satisfied depends on the electric field and the scattering length of the charge carriers. For a representative electric field value of 2×10^{5} V · cm^{-1} and $\varepsilon_{r}=3$, the potential maximum is about 2. 5 nm away from the interface, that is, about 2. 5 times the intermolecular spacing. Therefore, one cannot expect that a thermally activated injection current obeys strictly the Richardson-Schottky model.

3. Thermally Activated Injection Current in a Disordered Organic Semiconductor

The injection currents described by either Fowler-Nordheim tunneling or Richardson-Schottky thermionic emission were originally devised with a view to crystalline inorganic semiconductors. Based upon the simulation and analytic theory, a model has meanwhile been developed that takes into account ①the existence of the image charge at the electrode, ② the hopping-type of charge carrier transport, and ③ the presence of disorder existing in a non-crystalline system[72]. It is illustrated in Figure 4. 33. The underlying idea, originally introduced by Gartstein and Conwell, is that a thermally activated jump raises an electron from the Fermi level of the electrode to a tail state of the DOS distribution of transport sites of the dielectric medium. This is subjected to the condition that this site has at least one neighboring hopping site at equal or even a lower energy. It is fulfilled by optimizing a transport parameter with regard to both jump

71. 理查德森-肖特基的概念的前提是,一旦电子获得了足够的热能来克服势垒,它就会在场方向上克服势垒,而不会遭受非弹性散射。否则,电子将与镜像电荷重新结合,产生注入过程的损失。

72. 基于模拟和理论分析,发展出了一个模型,该模型考虑了①电极处存在的镜像电荷,②电荷载流子的跃进型传输,以及③非晶体系中存在的无序性。

distance and jump energy. This condition ensures that the primarily injected carrier can continue its motion away from the interface rather than recombine with its image charge in the electrode. Subsequently, the carrier is considered to execute a diffusive random walk in the combined Coulomb potential of the image charge and the externally applied potential[73]. Simulation and analytic theory for injection barriers ranging from 0.2 eV to 0.7 eV and temperatures between 300 K and 200 K are mutually consistent. The field dependence of the injection efficiency follows a Poole-Frenkel - type of field dependence $\ln(j(F)) \propto \sqrt{F}$.

73. 这种情况保证了注入的电荷能够持续地从界面移走而不是与电极处的镜像电荷复合。然后,载流子被看成是在镜像电荷的库仑势场及外加电势的共同作用下的随机运动。

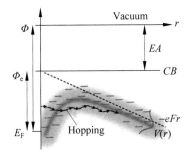

Figure 4.33   Schematic representation of electron injection from a metallic electrode into a semiconductor via hopping in a disordered organic solid. r is the distance from the electrode. The dashed line indicates the potential energy due to the applied (external) electric field F. The potential $V(r)$ in the device is shown as a gray solid line

However, when plotting an injection current on a $\ln(j)$ versus $1/T$ scale, as one does when analyzing the temperature dependence of a conventionally activated process, one finds that the apparent activation energy obtained from the slope of the $\ln(j)$ versus $1/T$ dependence is much less than the one which one would expect based upon presumed injection barriers, taken from literature values of the work function of the metal electrode and the electron affinity/ionization energy of the dielectric. The explanation of this ubiquitously observed phenomenon is related to the stochastic motion of charges in a rough energy landscape as described in section of 4.1.4. It rests on the concept of the transport energy at which charge carriers, on average, migrate within a given

74. 对这种观察到的、无处不在的现象的解释，与4.1.4节描述的在无序能量环境下电荷的随机运动有关。它是基于通过在给定的 DOS 分布中，电荷载流子的平均移动来传输能量的概念。

75. 当从电极注入电荷载流子的供应是无限的，即电流不受注入限制，而是阴极和阳极都是欧姆接触，在电极和有机半导体的界面处没有注入势垒，电流最终将受到其自身空间电荷的限制，即有机半导体的空间电荷可以屏蔽注入电极处的电场。这种类型的电流称为空间电荷限制电流（SCLC）。

DOS distribution[74]. This transport energy decreases when the temperature decreases and, thus, the energy to overcome the injection barrier is no longer constant but decreases with temperature. As a consequence, the slope of the measured $\ln(j)$ versus $1/T$ dependence is not a direct measure of the injection barrier.

4.4.2 Space Charge Limited Current

When the supply of charge carriers injected from the electrode(s) is unlimited, that is, the current is not injection limited, but instead, the cathode and the anode are Ohmic contacts and no injection barrier at interface of electrode and organic semiconductor, the current will eventually be limited by its own space charge of organic semiconductor that shields the electric field at the injection electrode. This type current is termed as space charge limited current (SCLC)[75].

Ohm's law predicts a current that varies linearly with the electric field. However, an SCLC injected from an Ohmic contact electrode does not obey Ohm's law. Considering the ideal condition for SCLC: the cathode and anode are the same, the organic material has no traps and the bandgap is large ($E_g \gg k_B T$), the contact is Ohmic contact, thickness of the semiconductor is d, area of electrodes is S, the applied bias is U. As shown in Figure 4.34, this device can be treated as a capacitor and its capacity and charge are given by:

$$C = \frac{\varepsilon \varepsilon_0 S}{d} \tag{4-44}$$

$$Q = C \cdot U \tag{4-45}$$

$$n = \frac{Q}{qSd} = \frac{CU}{qSd} \tag{4-46}$$

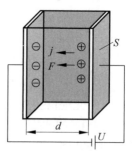

Figure 4.34   Ideal condition for SCLC

here, q is elementary charge. The mean field strength F in this ideal case is given by:

$$F = \eta \frac{U}{d} \qquad (4\text{-}47)$$

where η is a pre-front numerical factor, the value is between 1 to 2.

Using the definitions of the conductivity σ and the mobility μ as well as the charge density n ($n = Q/\text{Volume} = Q/Sd$), and with the equation of continuity, $(\partial j/\partial x) = 0$, we find the relationship of SCLC j and driving voltage U:

$$j = \sigma F = qn\mu F = q \frac{CU}{qSd}\mu\eta \frac{U}{d} = \frac{\frac{\varepsilon\varepsilon_0 S}{d}U}{Sd}\mu\eta \frac{U}{d} = \eta\varepsilon\varepsilon_0\mu \frac{U^2}{d^3}$$

$$(4\text{-}48)$$

This simple but important equation is called Child's law. Different from Ohm's law that current is proportional to voltage and inversely proportional to thickness, SCLC is proportional to the square of voltage and inversely proportional to the cube of thickness[76]. The relationship between j and U discussed above is the ideal condition for SCLC. In practice, other factors should be considered, such as traps, impurities, disorder in semiconductor. We will further discuss the relationship between j and U in four cases.

1. Ohmic Contact Current at Low Voltage

At extremely low current, carrier injection can be neglected. Current mainly originates from the thermionic current, which is linear with electric field:

$$j = \sigma F = q\mu n_0 F = q\mu n_0 \frac{U}{d} \qquad (4\text{-}49)$$

here n_0 is intrinsic carrier density, U is voltage, d is thickness.

2. Trap Free SCLC

As soon as the voltage exceeds the limit, that is when the semiconductor is charged with excess charge carriers and thus has a stationary space charge, the characteristic of current and voltage becomes nonlinear in a typical manner. When the injected carrier density is greater than the intrinsic carrier

76. 这个简单且重要的公式称为蔡尔德定律。不同于欧姆定律的电流正比于电压、反比于厚度，SCLC 正比于电压平方、反比于厚度立方。

density of the material, the space charge will become the dominant factor, which greatly affects the current density and internal electric field of the material[77]. If a material is a regular crystal, trap states are discrete and all the carriers are free, i. e., all carriers are able to move, this is trap-free space charge limited current, (TF-SCLC) and can be expressed by Mott-Gurney equation:

$$j = \frac{9}{8} \varepsilon \varepsilon_0 \mu \frac{U^2}{d^3} \qquad (4\text{-}50)$$

Equation (4-50) implies that at a given electric field, the SCLC current scales inversely with the three power of device thickness. This thickness dependence is the crucial criterium to decide whether a super-linear current field dependence of the current is caused either by space charge limitation or by field dependent charge carrier injection. The electric field (F) inside the materials varies as

$$F(x) = \frac{3U}{2d} \sqrt{\frac{x}{d}} \qquad (4\text{-}51)$$

where x is counted from the injecting electrode. The field is zero at an (ohmic) electrode and it is $\frac{3U}{2d}$ at the exit electrode which is 50% larger than the situation without space charge U/d. The charge concentration also depends on the spatial coordinate and is given by:

$$n = \frac{3\varepsilon\varepsilon_0}{4qd^2} \sqrt{\frac{d}{x}} \cdot U = \frac{1}{2} \sqrt{\frac{d}{x}} \bar{n} \qquad (4\text{-}52)$$

$\bar{n}$ is mean carrier density described as follow:

$$\bar{n} = d^{-1} \int_0^d n(x) \mathrm{d}x = \frac{3\varepsilon\varepsilon_0}{2qd^2} U \qquad (4\text{-}53)$$

that is, 3/2 times the number of capacitor charges per unit volume.

Equation (4-50) is strictly valid only if the charge carrier mobility is field-independent because a field dependence of μ would have feedback on the spatial distribution of the space charge.

If an electrode cannot supply sufficient charge carriers to maintain SCLC conditions, the current is injection limited, that is, it is determined by the number of charge carriers that an electrode is able to inject. Whether or not an electrode is able to sustain an SCLC depends on both the injection barrier and on the charge carrier mobility of the dielectric, that is, the organic semiconductor[78]. For ideal interfaces, the injection barrier is given by the energy difference between the work function of the electrode and the ionization potential or electron affinity of the organic semiconductor. Obviously, the higher the mobility, the more crucial is the height of injection barriers to reach SCLC condition. As a rough estimate for a conjugated polymer with average mobility, say, 10^{-4} cm^{-2} · V^{-1} · s^{-1}, a tolerable barrier is 0.3 eV.

3. Trap-Filled SCLC (TFL-SCLC)

A complication arises if the dielectric contains traps. If the carriers are captured by shallow traps with discrete energy levels, then they cannot play the role of conduction, but still influence internal field strength. In this case, the current under Ohmic contact is trap-filled SCLC (TFL-SCLC)[79]. The I-V curve obeys following formula:

$$j = \frac{9}{8}\varepsilon\varepsilon_0\mu \cdot \Theta \cdot \frac{U^2}{d^3} \qquad (4\text{-}54)$$

where Θ represents the ratio of free carriers (n) to total carriers ($n + n_t$):

$$\Theta = \frac{n}{n+n_t} = \frac{1}{1+n_t/n} = \frac{1}{1+\dfrac{N_t}{N}\exp\left(\dfrac{E-E_t}{k_BT}\right)} \qquad (4\text{-}55)$$

N_t and N are state density of trap and free carrier, E_t and E are energy of trap and free carrier. The above two formulas indicate that when the density of trap is zero or the traps are already fully occupied and the free carrier density is much higher than the trapped carrier density (TFL current, $n \gg n_t$), $\Theta = 1$ and Equation (4-54) is equal to Equation (4-50). Thus, Equation (4-54) can be regarded as a general expression of SCLC.

4. TFL-SCLC with Traps of Continuous Energy (TFL-SCLC-a)

In the case that the energy levels of shallow traps are not discrete or thin film is disordered, Equation (4-54) can't be applied. Using E_t to express energy of trap, N_t as the state density of trap, the distribution of state density of shallow

78. 如果电极不能提供足够的电荷载流子来维持 SCLC 条件,则电流为注入限制电流,也就是说,电流受限于电极能够注入的电荷载流子的数量。电极是否能够维持 SCLC 取决于注入势垒和电介质(即有机半导体)的电荷载流子迁移率。

79. 如果电介质中包含陷阱,则会出现更复杂的情况。如果载流子是被离散浅能级陷阱捕获,那么它们就不能起到导电的作用,但仍然会影响内部电场强度。在这种情况下,欧姆接触的电流是陷阱填充的 SCLC (TFL-SCLC)。

traps can be expressed by two modes of exponential distribution (Equation (4-56)) and Gaussian distribution (Equation (4-57)):

$$G(E) = (N_t/E_t) \exp(-E/E_t) \qquad (4-56)$$

$$G(E) = \frac{N_t}{\sqrt{2\pi} \cdot \sigma} \exp\left[-\frac{(E-E_t)^2}{2\sigma^2}\right] \qquad (4-57)$$

The above two expressions should satisfy:

$$\int_0^\infty G(E)\mathrm{d}E = N_t \qquad (4-58)$$

（1） For the case of exponential distribution, approximate solution can be obtained as follow:

Charge carrier density of filled traps:

$$n_t = N_t \left(\frac{n}{N}\right)^{1/l} \qquad (4-59)$$

Electric field inside the dielectric:

$$F(x) = \frac{2l+1}{l+1} \cdot \frac{U}{d} \left(\frac{x}{d}\right)^{l/l+1} \qquad (4-60)$$

Free charge carrier density:

$$n(x) = \frac{\bar{n}}{l+1} \cdot \left(\frac{x}{d}\right)^{l/l+1} \qquad (4-61)$$

Current density with the exponential distribution traps:

$$j = q\mu N \left(\frac{2l+1}{l+1}\right)^{l+1} \left(\frac{l}{l+1} \cdot \frac{\varepsilon\varepsilon_0}{qN_t}\right)^l \frac{U^{l+1}}{d^{2l+1}} \qquad (4-62)$$

where $l = E_t/k_B T$, $l > 1$. In the exponential distribution mode, since $(l+1) > 2$, here, j-U curve is steeper than that of TFL-SCLC; the influence of thickness d on current density is more remarkable when the trap-captured carriers present[80].

(2) For the case of traps with Gaussian distribution, the current density is

$$j = q\mu N \left(\frac{2m+1}{m+1}\right)^{m+1} \left(\frac{m}{m+1} \cdot \frac{\varepsilon\varepsilon_0}{q^m N_t'}\right)^l \frac{U^{m+1}}{d^{2m+1}} \qquad (4-63)$$

N is the state density of the free carrier, and

$$m = \left(1 + \frac{2\sigma^2}{16k_B^2 T^2}\right)^{1/2} \qquad (4-64)$$

$$N_t' = \frac{N_t}{2}\exp\left(\frac{E_t}{mk_B T}\right) \qquad (4-65)$$

Figure 4.35 schematically illustrates the variations of electric fields and I-V curves in semiconductors with different traps. As shown in Figure 4.35(a), under the cases

80. 在指数分布模型中，由于 $l+1>2$，这里 j-U 曲线比 TFL-SCLC 曲线陡；当存在陷阱捕获载流子时，厚度 d 对电流密度的影响更加显著。

of the injection limited, the space charge limited and the limitation by filled-traps with continuous energy, the electric fields at the other end of semiconductor ($x/d = 1.0$) are different, with those for the latter two cases exhibiting values of 1.5 and nearly 2.0 folds compared to that under injection-limited case. Actually, under different cases, the electric fields and the behaviors of the I-V curves are not the same. As shown in Figure 4.35(b), generally, at low voltage, injected carrier is little and the current is mainly originated from thermionic carriers, so current is linear with electric field and the slope is the smallest. When the voltage is increased, space charges present and the current is SCLC. If there are no carriers being trapped, the current increases quadratically with voltage. If there are trapped carriers, the current increases quadratically with voltage but with lower slope. When the voltage rises to a higher level, the trapped carriers can be released, slope of I-V curve increases rapidly. Further increase the voltage, slope of I-V curve decreases to the value of trap free space charge limited current, because the trapped carriers are fully released[81].

81. 如图 4.35(b)所示，一般在低电压下，注入载流子很少，电流主要来自热解离的载流子，因此电流与电场呈线性关系，斜率最小。当电压增加时，存在空间电荷，电流为 SCLC。如果没有载流子被捕获，电流随电压二次方增加。如果存在被捕获的载流子，电流随电压呈二次方的增加，但斜率较低。当电压上升到更高的水平时，被捕获的载流子可以被释放，I-V 曲线的斜率迅速增加。电压进一步升高，由于被捕获的载流子被完全释放，I-V 曲线的斜率减小到陷阱空置时的空间电荷限制电流的情况。

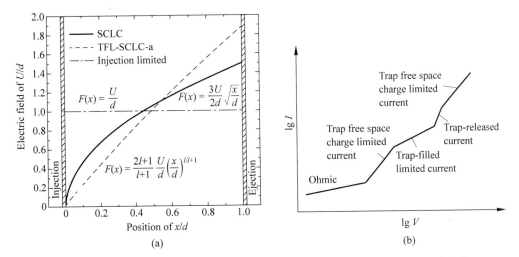

Figure 4.35　(a) Variations of electric fields and (b) I-V curves in semiconductors with different traps

Figure 4.36 shows an experimental result, the curve of $\lg(I)$ versus $\lg(V)$ in a 10 μm thick rubrene crystal, the $\lg(I)$ increases linearly with the $\lg(V)$ at the range of low voltages, then changes to quadratically increase at higher voltages，immediately following a higher voltage，the current rises sharply as the voltage increases，and when the voltage rises further，back to quadratically relationship.

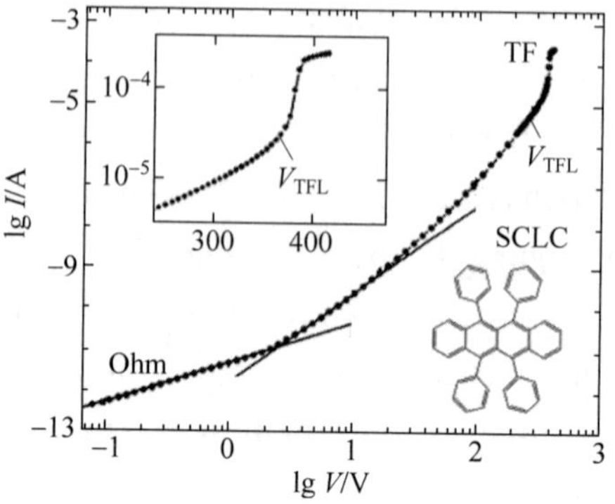

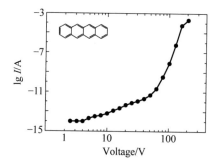

Figure 4.36　Current-voltage curve of rubrene crystal

82. 在 30 μm 厚的并四苯晶体中，如图 4.37 所示，在 4 V 到 50 V 的电压范围，电流随电压的二次方增加；在 50 V 到 150 V 之间，产生更高幂次方关系；在 150 V 以上，二次方依赖关系好像又出现了。

In a 30 μm thick tetracene crystal，as shown in Figure 4.37，the current increases quadratically with voltage in the voltage range between 4 V and 50 V and a higher power relationship occurs between 50 V to 150 V. Above $V = 150$ V, a quadratic dependence again seems to appear[82].

Figure 4.37　Current-voltage curve of tetracene crystal

4.5　Charge Carrier Recombination

The recombination of a positive and a negative charge in an optoelectronic device may occur geminately or non-geminately. The geminate recombination refers to the reaction，in which，each of the two transient species is produced from a common precursor. For example，the reaction of D + A leads to $D^+ A^-$, then recombination occurs in $D^+ A^-$ itself to produce D + A. The pair of molecular entities in close proximity resulting from a common precursor is termed a geminate pair (GP). If the recombination reaction occurs before any occurrence of separation by diffusion, this is termed primary geminate recombination. If the mutually reactive entities have been separated, and come together by diffusion, this is termed secondary geminate recombination, though this is rarely distinguished[83].

In the context of organic semiconductors, the precursor to a geminate pair would be a molecular excited state. For example, an excited donor (D) molecule, the excited donor can transfer an electron to an adjacent acceptor (A) to produce a D-A type geminate pair. The back-transfer reaction of the electron from the acceptor molecule to the positively charged donor molecule leads to geminate recombination. In contrast, if the positive and negative charges are injected at the anode and the cathode, move through the organic semiconductor film, encounter each other, and recombine, then this represents a non-geminate recombination process[84].

In optoelectronic devices, nongeminate recombination is almost an inevitable process. It is an essential process in organic light-emitting diode (OLED), however, in organic photovoltaic device it plays a negative role. As illustrated in Figure 4.38, there are two typical types of nongeminate charge recombination: bimolecular recombination and trap-involved recombination, which will be discussed below.

83. 成对复合指的是这样一种反应,即 2 个发生复合反应的物种,是来自于共同的前驱体。例如,D＋A 反应形成 D$^+$A$^-$,然后复合发生在 D$^+$A$^-$ 本身,产生了 D＋A。同一个前驱体产生的、相互邻近的一对分子,被称为双子对(GP)。如果复合反应发生在由于扩散而产生的任何分离之前,这称为初级成对复合。如果相互反应的实体已经分离,通过扩散聚集后再反应,这被称为二次成对复合,尽管这与初级成对复合的区分,很稀有。

84. 相反,如果在阳极和阴极注入的正电荷和负电荷,穿过有机半导体膜,彼此相遇并复合,则这表示非成对的复合过程。

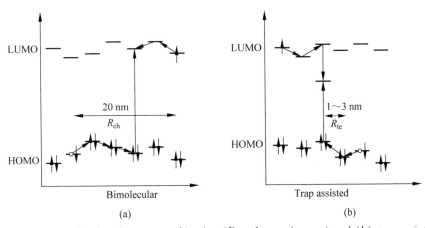

Figure 4.38　(a) Bimolecular charge recombination (R_{eh}, Langevin-type) and (b) trap-assisted (R_{te}, Shockley-Read-Hall-type or Thomson-like) recombination in OLED/OPV devices

4.5.1 Nongeminate Charge Recombination without Traps

In the operation of OLED devices, after holes and electrons are injected from the electrodes, they move oppositely under the driven of electric field; before charge recombination to form exciton, they locate at different molecules; therefore, charge recombination in OLED is a bimolecular nongeminate recombination. In organic photovoltaic devices, there are also bimolecular nongeminate recombination processes. For instance, the free charges generated by photoexcitation, e.g., electron in the acceptor (electron transport) material and hole in the donor (hole transport) material, can recombine before they are collected by electrodes, leading to the decreases of OPV efficiency. Figure 4.38(a) schematically shows the process of these bimolecular recombination i.e., Langevin-type. The rate for bimolecular recombination, k_L, is determined by the product of the concentration n of electrons and p of holes inside the materials multiplied by a rate constant γ,

$$k_L = \frac{\mathrm{d}n}{\mathrm{d}t} = \frac{\mathrm{d}p}{\mathrm{d}t} = -\gamma n p \qquad (4\text{-}66)$$

where k_L is the number of electrons and holes that react per unit time within a unit volume, i.e., the rate of recombination. γ is the product of their reaction cross-section and the velocity at which the reactants move, i.e., the rate constant for charge recombination. In a covalently bound inorganic semiconductor in which the mean scattering length of charge carriers is large, that is, much larger than inter-atomic distances, the relevant velocity to be taken is the thermal velocity of electrons and holes moving in the conduction and valence bands. In contrast, in an organic semiconductor, the mean free path of charge carriers is comparable to the intermolecular spacing and transport is of hopping type[85]. In this case, one can apply the Smoluchowski rate for a bimolecular reaction among reactants. It assumes that when the reactants A and B diffuse with diffusion coefficients D_A and D_B, the reaction occurs when they approach each other to a critical separation R_{AB}. The rate

85. 在共价结合的无机半导体中,电荷载流子的平均散射长度很大,即远大于原子间的距离,要考虑的是,在导带和价带中移动的电子和空穴的热运动速度;相反,在有机半导体中,电荷载流子的平均自由程与分子间间距相当,传输为跃进型。

constant γ is time (t) dependent, assuming $t = 0$ at the beginning of R_{AB}:

$$\gamma = 4\pi R_{AB}(D_A + D_B)\left[1 + \frac{R_{AB}}{\sqrt{(D_A + D_B)\pi t}}\right]$$

(4-67)

Usually for times exceeding a few picoseconds, the constant term dominates in Equation (4-67), so that the part of time-dependence in Equation (4-67) can be neglected. If the reactants are electrons and holes hopping among transport sites, R_{AB} is determined by their Coulomb attraction. More precisely, electron-hole recombination occurs when the Coulomb potential is equal or larger than the thermal energy $k_B T$, that is, when the electron-hole separation is smaller, then

$$R_{AB} = \frac{q^2}{4\pi\varepsilon_0\varepsilon_r k_B T}$$

(4-68)

For this reason, R_{AB} is also known as the Coulomb capture radius, or as Langevin capture radius. Using Equation (4-68) and considering that the diffusion coefficient is related to the charge carrier mobility via the Einstein relation $qD = \mu k_B T$, neglecting the time-dependence part of Equation (4-67), we get the follow relationship between recombination rate constant and mobility of organic semiconductor.

$$\gamma = \frac{q}{\varepsilon_0\varepsilon_r}(\mu_e + \mu_h)$$

(4-69)

For a molecular crystal with $\mu_h \cong \mu_e \cong 1~cm^2 \cdot V^{-1} \cdot s^{-1}$, Equation (4-69) yields $\gamma \cong 10^{-6}~cm^3 \cdot s^{-1}$.

86. 通常,被应用的有机半导体薄膜是无序的。这就提出了一个问题,即这种处理是否适用于位点能量随机分布的体系。

Usually, the organic semiconductor films used for applications are disordered. This raises the question whether or not this treatment is applicable to a system in which the site energies are distributed randomly[86]. One could argue that in a disordered solid, the critical separation at which electrons and holes recombine depends on the spread of the site energies rather than on the condition that the mutual Coulomb potential equals $k_B T$. However, Monte Carlo simulations on a system with Gaussian DOS distribution confirmed that at moderate fields the classic Langevin treatment is indeed valid. At higher electric fields, γ becomes field dependent because the charge carrier mobility becomes field dependent.

4.5.2 Nongeminate Charge Recombination with Traps

Equation (4-69) implies that when electron and hole mobilities are highly asymmetric, for example, due to trapping, the recombination is determined by the faster moving charge carrier and thus becomes a quasi-monomolecular recombination process[87]. In this case, one has to include trapping in the kinetic formalism. If trapping is associated with electrons, one has to consider that electrons are lost by both: ① by Langevin recombination (Figure 4.38(a)) and ② by capturing through electron traps (Figure 4.38(b)). If the trap depth is large, which is the ubiquitous case, thermally activated de-trapping is negligible and the only loss process for trapped electrons is by the recombination with (free) holes at a density of p. Therefore, the rate equations for the densities of free (n_f) and trapped (n_t) electrons are

$$\frac{\mathrm{d}n_f}{\mathrm{d}t} = G - n_f k_t - \gamma n_f p \tag{4-70}$$

$$\frac{\mathrm{d}n_t}{\mathrm{d}t} = n_f k_t - \gamma n_t p \tag{4-71}$$

where G is the carrier generation rate per unit volume, for example, by injection from the electrodes, k_t is the rate of trapping. It is tacitly assumed that most of the charge carriers recombine within the organic semiconductor layer rather than be discharged at the electrodes. This is granted if both anode and cathode are ohmic, that is, are able to inject an unlimited number of charge carriers[88]. We shall now consider how recombination via trapping compares to the Langevin recombination. Under a stationary condition, $\mathrm{d}n_t/\mathrm{d}t = 0$, Equation (4-71) yields the ratio of trapped to free electron densities as

$$\frac{n_t}{n_f} = \frac{k_t}{\gamma p} \tag{4-72}$$

The rate of trapping is proportional to the density N_T of traps, $k_t = \widetilde{\gamma} N_T$, so that the rate of trap induced recombination, $k_t n_f = \widetilde{\gamma} N_T n_f$, describes a bimolecular reaction between the trap site and the free electron. We can thus express the proportionality constant $\widetilde{\gamma}$ analogous to

87. 式(4-69)意味着当电子和空穴的迁移率高度不对称时,例如由于捕获,复合是由快速移动的电荷载流子决定的,从而成为准单分子复合过程。

88. 人们默认大多数电荷载流子在有机半导体层内复合,而不是在电极处被中和。如果阳极和阴极都是欧姆的,即能够注入无限数量的电荷载流子,则这个认知是成立的。

Equation（4-67）without the part of time-dependence and arrive at

$$\widetilde{\gamma} = 4\pi R_{te} D_e \qquad (4\text{-}73)$$

where R_{te} is the critical separation between trap and electron at which trapping occurs，D_e is the diffusion coefficient of the free electron. Finally，we can get the ratio of trapped to free electron density，by using Einstein relation $qD = \mu k_B T$ and combining Equations（4-69），（4-73）to（4-72）：

$$\frac{n_t}{n_f} = \frac{R_{te} N_T}{R_{eh} p} \frac{D_e}{D_e + D_h} = \frac{R_{te} N_T}{R_{eh} p} \frac{\mu_e}{\mu_e + \mu_h} \qquad (4\text{-}74)$$

R_{eh} is the critical separation between electron and hole for bimolecular recombination. If it is easier for an injected electron to directly find a hole to recombine with rather than a trap followed by a secondary recombination，the total recombination process will be kinetically largely of second order. Otherwise，it will be of the first order because the recombination is determined by the constant trap concentration[89]. This is a simple adaptation of the Shockley-Read-Hall mechanism to explain first order recombination observed in organic diodes. It explains why electron hole recombination can be a kinetic first-order process.

In contrast to Langevin recombination，trap-assisted recombination is often non-radiative or only weakly radiative. It is estimated that，for an amorphous organic semiconductor film with thickness of $d = 100$ nm，the applied voltage has to exceed 4.5 V in order that 90% of the electrons and holes recombine by the Langevin-type second order kinetics[90]. Thus，it is of advantage to limit the emission in an OLED to a thin zone at which the charge carrier concentration is high enough so that free electrons and holes can recombine in a bimolecular fashion before finding a trap.

89. 如果注入的电子更容易直接找到一个空穴来复合，而不是通过一个陷阱捕获，然后进行二次复合，那么整个复合过程在动力学上将在很大程度上是二级的。否则，它将是一级的，因为复合是由恒定的陷阱浓度决定的。

90. 与朗格文复合相比，陷阱辅助的复合，通常是非辐射的或仅有弱辐射。据估计，对于厚度为 $d = 100$ nm 的非晶态有机半导体薄膜，为使90%的电子和空穴通过朗格文型二级动力学复合，施加的电压必须超过 4.5 V。

本章小结

1. 内容提要：

本章首先对电学概念，如电导率和迁移率进行了介绍，并描述半导体中载流子能量图及载流子类型/输运；同时介绍了有机材料的电学性质。本章第二部分对载流子的迁移率特别是有机材料的载流子迁移率给出了全面的介绍；针对迁移率与温度和电场的关系，讨论了包括晶态和非晶态固体两种情况下的载流子迁移模型。在第三部分,对金属电极和有机半导体的三种有效电接触进行了介绍,并给出了在不同电接触下电压-电流满足的关系。然后,介绍了载流子注入有机半导体后的注入限制电流和空间电荷限制电流的情况。最后对载流子在有机半导体中的复合过程进行了简单的讨论。

2. 基本概念:电导率、迁移率、载流子、电荷、n/p-载流子、本征/掺杂/注入/光生载流子、带状/跃进输运、晶格/杂质散射、导电/陷阱能级、电接触、中性/欧姆/肖特基接触、接触势垒、极化子、注入限制电流、隧穿电流、热激活电流、空间电荷限制电流、电荷复合、非成对复合。

3. 主要公式:

(1) 电导率:$\sigma = \dfrac{1}{RL} = \dfrac{j}{F}$

(2) 电导率与载流子密度及迁移率的关系:$\sigma = \dfrac{j}{F} = \dfrac{qnv_{\mathrm{D}}}{v_{\mathrm{D}}/\mu} = qn\mu$

(3) 热激活载流子密度与温度的关系:$p = n = AT^{3/2}\exp\left(-\dfrac{E_{\mathrm{g}}}{2k_{\mathrm{B}}T}\right) = \sqrt{N_{\mathrm{C}}N_{\mathrm{V}}}\exp\left(-\dfrac{E_{\mathrm{g}}}{2k_{\mathrm{B}}T}\right)$

(4) 光生载流子的产率:$\eta_{\mathrm{p}} = \eta_{\mathrm{n}} = \eta_0\left(1 + \dfrac{qF}{2k_{\mathrm{B}}T}\right)\exp\left(-\dfrac{U_{\mathrm{CP}}}{k_{\mathrm{B}}T}\right)$

(5) 注入载流子浓度与场强/温度/势垒的关系:n 或 $p \propto \dfrac{1}{F}\exp\left(-\dfrac{\phi_{\mathrm{b}}}{k_{\mathrm{B}}T}\right)$

(6) 掺杂载流子浓度:$n(T) \approx \dfrac{2N_{\mathrm{D}}}{1 + \sqrt{1 + 4\dfrac{N_{\mathrm{D}}}{N_{\mathrm{C}}}\exp\left(\dfrac{E_{\mathrm{d}}}{k_{\mathrm{B}}T}\right)}}$

(7) 带状输运的载流子迁移率:$\mu = e\tau/M_{\mathrm{eff}}$

(8) 带状输运载流子迁移率与温度的关系:$\mu \propto \dfrac{1}{T^n}, 0 < n < 3,$无机物;$0 < n < 2,$有机物

(9) 跃进载流子迁移率与温度的关系:$\mu \propto \exp\left(-\dfrac{E_{\mathrm{a}}}{k_{\mathrm{B}}T}\right)$

(10) 无序材料中载流子迁移率与电场强度的关系:$\mu(F) = \mu_0\exp(\beta_{\mathrm{PF}}\sqrt{F})$

(11) 无序材料载流子迁移率的 WD 模型:
$$\mu = \mu_0\exp\left[-\dfrac{\Delta E_{\mathrm{a}}}{k_{\mathrm{B}}T}\right]\exp\left[\beta\sqrt{F}\left(\dfrac{1}{k_{\mathrm{B}}T} - \dfrac{1}{k_{\mathrm{B}}T_0}\right)\right]$$

(12) 无序材料载流子迁移率的 GDM 模型:
$$\Sigma \geqslant 1.5: \mu = \mu_0\exp\left[-\left(\dfrac{2\sigma}{3k_{\mathrm{B}}T}\right)^2\right]\exp\left\{c\sqrt{F}\left[\left(\dfrac{\sigma}{k_{\mathrm{B}}T}\right)^2 - \Sigma^2\right]\right\}$$

$$\Sigma < 1.5: \mu = \mu_0\exp\left[-\left(\dfrac{2\sigma}{3k_{\mathrm{B}}T}\right)^2\right]\exp\left\{c\sqrt{F}\left[\left(\dfrac{\sigma}{k_{\mathrm{B}}T}\right)^2 - 2.25\right]\right\}$$

（13）无序材料载流子迁移率的 CDM 模型：

$$\mu = \mu_0 \exp\left\{ -\left(\frac{3\sigma}{5k_BT}\right)^2 + 0.78\left[\left(\frac{\sigma}{k_BT}\right)^{3/2} - 2\right]\sqrt{\frac{qaF}{\sigma}} \right\}$$

（14）Fowler-Nordheim 隧穿电流：$j_{FN} = \dfrac{A^*}{\Phi_e}\left(\dfrac{eF}{\alpha k_B}\right)^2 \exp\left(-\dfrac{2\alpha\Phi_e^{3/2}}{3eF}\right)$

（15）Richardson-Schottky 热激活注入电流：

$$j_{SR} = A^* T^2 \exp\left(-\frac{\Phi_B}{k_BT}\right)\exp\left[\left(\frac{eU}{nk_BT}\right) - 1\right]$$

$$\text{或}\quad j_{RS} = A^* T^2 \exp\left(-\frac{\Phi_B - \beta_{RS}\sqrt{F}}{k_BT}\right)$$

（16）空间电荷限制电流：$j = \dfrac{9}{8}\varepsilon\varepsilon_0 \mu \cdot \Theta \cdot \dfrac{U^2}{d^3}$

（17）非成对双分子电荷复合速率常数：

$$\gamma = 4\pi R_{AB}(D_A + D_B)\left[1 + \frac{R_{AB}}{\sqrt{(D_A + D_B)\pi t}}\right]$$

（18）非成对双分子电荷复合速率常数与载流子迁移率的关系：$\gamma = \dfrac{q}{\varepsilon_0\varepsilon_r}(\mu_e + \mu_h)$

Glossary

Acoustic phonon	声子
Activation energy	活化能，激活能
Capacitor	电容器
Capacity	容量
Carrier	载流子
Charge	电荷
Conductivity	传导性，电导率
Conductor	导体
Contact	（电）接触
Contact barrier	接触势垒
Current density	电流密度
Dark current	暗电流
Dissociation	分离，解离
Doped carrier	掺杂载流子
Electrical conductivity	导电率，导电性
Electron affinity	电子亲和能
Exponential distribution	指数分布
Extrinsic carrier	本征载流子
Field emission	场发射

Filled trap	填充的陷阱
Gaussian distributed trap	高斯分布的陷阱
Heterojunction	异质结
Injected carrier	注入电荷
Injection-limited current	注入限制电流
Insulator	绝缘体
Intrinsic carrier	本征载流子
Ionization potential	电离势
Mobility	迁移率
Neutral contact	中性接触
n-type carier	n 型载流子
Ohmic contact	欧姆接触
Ohm's law	欧姆定律
Photocurrent	光电流
Photogeneration of carrier	光生载流子
Preliminary	初步的,初级的
p-type carrier	p 型载流子
Schottky contact	肖特基接触
Semiconductor	半导体
Space charge	空间电荷
Space charge limited current	空间电荷限制电流
State density	态密度
Thermionic emission	热离子发射
Transient photoconductivity	瞬态光导
Trapped charge limited current	陷阱电荷限制电流
Tunneling effect	隧穿效应
Valid	有效
Vertical bandgap	垂直能隙
Xerography	静电复印术

Problems

1. True or false questions：

（1）When we say that anthracene is a photoconductor，it means that compared to dark current，the conductivity of this material is higher with irradiation photons possessing energy larger than its bandgap.

（2）The conductivity of a material is determined by carrier mobility and does not relate to carrier density.

（3）Typically in organic semiconductors，the electron carriers for conduction locate at LUMO levels.

(4) Both producing external photoelectric effect and photoluminescence need light irradiation; and the light energy is much higher in the latter case.

(5) Irradiating 560 nm light on an organic material with bandgap of 3.0 eV, photogeneration of carrier is possible.

(6) Organic materials can be conductor, semiconductor, insulator, even superconductor.

(7) Generally speaking, in an organic material, the average energy of a hole is approximately equal to the energy of HOMO electron.

(8) When dopant with valence electron energy slightly less than those in the host semiconductor is introduced, the resulting doped system is n-type semiconductor.

(9) Longer conjunction structure in an organic material results in a larger splitting between HOMO and LUMO levels.

(10) Among different materials, such as conductors, semiconductors and insulators, the energy diagrams are generally significantly different.

(11) The carrier transport mobility in organic crystals is generally lower than that in amorphous organic materials.

(12) Bandlike carrier transport is faster than hopping carrier transport.

(13) In disordered organic films, the carrier mobility increases with increase of electric field and decreases with temperature.

(14) In organic crystals, the carrier mobility increases with increase of temperature.

(15) Increasing doping density in disordered organic materials always lowers carrier mobility.

(16) Despite doping usually lowers the mobility of carriers, it increases material conductivity.

(17) At higher temperature, the influence of electric field on mobility is relative low.

(18) Scattering will change the direction of electron transport, and it is highly dependent on temperature, so increasing temperature will decrease mobility of an organic material.

(19) The conducting of holes is generally worse than that of electrons in all kind of materials.

(20) If the charge carriers are homogeneously distributed within the solid, Ohm's law holds.

(21) It was found that in anthracene crystal, to produce hole by thermal activation, the energy needed is lower than its bandgap.

(22) When two metals contact, electrons will flow from the one with higher Fermi level to that with lower Fermi level until the two Fermi levels are equal.

(23) According to the value of contact barrier between metal and nonmetal

materials, a contact can either facilitate carrier injection or prevent carrier injection.

(24) The current in organic solids is always controlled by space charges.

(25) Injection limited current obeys Ohm's law.

(26) I - V of a Schottky junction exhibits rectifying properties.

(27) In a Schottky junction, the width of depletion region highly depends on the direction and the intensity of the applied bias.

(28) Due to space charge, electric field in semiconductors is not uniformly distributed.

(29) Traps in semiconductor generally lower current density.

(30) Current density is more dependent on voltage in disorder organic materials with traps of continuous energy, compared to those with traps of discrete energy levels.

2. Please deduce the relationship between conductivity, mobility and carrier density.

3. According to conductivity, please classify materials into four types, and abruptly give the bandgap of each type if possible; please give one material as an example for each type.

4. According to Figure 4.3, please describe three possible results in an organic material upon excitation by a photon with energy greater than the material bandgap.

5. What is photogeneration of carrier? Please explain factors that influence the yield of photogeneration of carrier.

6. Figure 4.5 shows the temperature-dependence of the conduction-electron concentration $n(T)$ and the Fermi energy $E_F(T)$ in an n-type semiconductor. E_i is the Fermi energy of the intrinsic. E_{CB}, E_{VB}, E_D are conducting band, valence band, and dopant energy level, respectively. Please explain the curve of $n(T)$ versus T, and point out the reason of temperature dependent feature of each stage.

7. According to Figure 4.10, please point out the hole transport layer in xerography, and briefly state the main principle for photocopy.

8. Classify carriers in organic materials by charge nature and material purity; give three routes for carrier generation.

9. Please explain carrier scattering and point out the influence of scattering on carrier drift.

10. Please state the three extreme carrier transport modes.

11. Please describe the dependence of the carrier mobility on temperature in organic crystals and disordered organic materials, respectively.

12. Figure 4.16(b) shows the temperature dependence of the electron mobility μ in a 370 μm thick perylene crystal. Please describe the trend of the curve change, and simply give the reason.

13. Figure 4.24 shows the theoretical results of the dependence of the carrier mobility on the concentration of dopants in materials with different variations of the intrinsic DOS distribution.

(1) Please describe the relationship between doping concentration and mobility.

(2) Please explain why at large DOS distribution, mobility slightly increases with doping concentration.

14. The following figure shows the relationship between carrier mobility and temperature for silicon at different doping density. Please briefly describe:

(1) The dependence of mobility on doping concentration.

(2) The temperature dependence of mobility both under low doping and high doping levels.

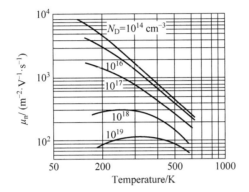

15. Figure 4.25 shows the theoretical results of the dependence of carrier mobilities on the concentration of dopants at different external fields; please briefly describe the dependence of mobility on doping concentration both under low and high external field, and give reasons.

16. According to Figure 4.13, (1) please describe the energy level changes from isolated molecule to neutral crystal, (2) please describe the energy level features of hole carrier, electron carrier and carrier-trap.

17. What is contact and what is Ohmic contact? Please enumerate the three contacts in organic electronic devices.

18. Please draw the energy band level before and after contact between metal and n-type semiconductor at the following two cases: (1) Fermi level of the n-type semiconductor is higher, (2) Fermi level of the n-type semiconductor is lower.

19. For a Schottky contact between a metal and an n-semiconductor junction, please draw the energy barrier at the interface when: (1) applied bias $V = 0$, (2) $V > 0$ and (3) $V < 0$.

20. What is space charge limited current?

21. Please deduce Child's law.

22. Based on Figure 4.35(a), please explain the abbreviated terms of TFL-SCLC-a and SCLC, then compare the curves of the electric field versus distance for the cases of SCLC, TFL-SCLC-a, and injection limited current.

CHAPTER 5 PROPERTY CHARACTERIZATION FOR ORGANIC MATERIALS

Organic materials can be applied to various optoelectronic devices depending on their optoelectronic properties, such as the ability to absorb light, conduct electricity, and emit light. To evaluate material properties for device applications, one needs to understand corresponding characterization methodologies. Thus, in this chapter, we shall discuss basic property-characterization for organic materials in terms of their optoelectronic device applications, including thermal stability, energy levels, material bandgap, carrier mobility, and photoluminescent quantum yield.

5.1 Thermal Analysis

5.1.1 Differential Scanning Calorimetry

1. 差示扫描量热法(DSC)是一种热分析技术,它检测的是温度升高时待测样品和参考样品所需要的热量差异,这里待测样品和参考样品始终保持几乎相同的温度。温度是通过程序控制逐渐增加的。在实验的温度范围内,参考样品必须没有任何相变(其比热容是常数)。

Differential scanning calorimetry (DSC) is a thermoanalytical technique, which detects the difference in the amount of heat required to increase the temperatures of a sample and reference that are always keeping at almost the same temperature. The temperature is increased gradually by a program. In the temperature range for experiment, the reference sample is required to undergo no transition (heat capacity of it is constant)[1]. The basic principle underlying in this technique is that when the sample undergoes a physical or a chemical transformation, more or less heat will need to flow to it than the reference to maintain both at the same temperature. As such, exothermic or endothermic heat flow to the sample is recorded at each temperature step.

2. 如图 5.1 所示,在 DSC 实验中,将待测样品和参考样品的坩埚置于样品架上,该样品架含有积分温度传感器,可以用来测量坩埚温度。这些装置放在一个可控气体环境的炉子内,温度是随着时间线性增加的。

As shown in Figure 5.1, in DSC experiment, sample and reference crucibles are placed on a sample holder with integrated temperature sensors for temperature measurement of the crucibles. This arrangement is located in an atmosphere-controlled oven. and the temperature is increased linearly as a function of time[2].

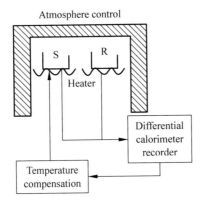

Atmosphere control

S   R

Heater

Differential
calorimeter
recorder

Temperature
compensation

Figure 5.1   Setup for DSC experiment

Figure 5.2 gives an example of DSC experiment. There are three characteristic features in the curve. The first one is an endothermic platform at about 75 ℃, which is assigned to the glass transition temperature (T_g). Glass transition may occur when the temperature of an amorphous solid is increased. This transition appears as a step in the baseline of the recorded DSC signal. For organic material of small molecules to obtain T_g using DSC experiment, usually, the sample is heated up step by step to a higher temperature relative to T_m, then quickly cooled to obtain amorphous states of the material. By thus T_g may be observed in the second heating up curve. Note that in the DSC curve, the baselines before and after T_g are different. When the sample is heated from amorphous states, after the temperature increases to T_g, some of the amorphous states start to turn to crystalline states by the assistance of heat. Hence, after T_g, the internal energy (entropy) is lowered due to more ordered molecular arrangement compared to the initial amorphous states, leading to the lowered baseline[3].

The second feature in the DSC curve of Figure 5.2 is a broad exothermic peak at about 160 ℃, which indicates that heat is released from the sample. This exothermic peak suggests that crystalline transformation occurs and a new crystalline state that is more stable in this temperature is formed. Continue increasing the temperature, there appears

3. 注意,在 DSC 曲线中,温度 T_g 前后的基线是不同的。当从非晶态样品开始加热,温度升到 T_g 时,一些非晶态开始在热能的辅助下转变为晶态。因此,高于 T_g 温度后,由于材料的有序性增加,材料的内能(即熵)相比于最初的非晶态减少了,这导致了基线的降低。

the third feature-peak at about 203 ℃, which is a strong and sharp endothermic peak and can be assigned to the melting point (T_m) of the sample[4].

4. 这个放热峰表示发生了晶型转变,在这个温度下更加稳定的新晶态形成了。继续增加温度,大约在 203 ℃,出现了第三个特征峰,这是一个又强又尖的吸热峰,是样品的熔点。

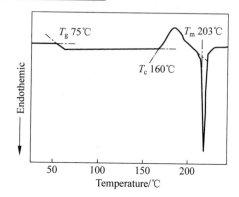

Figure 5.2　A DSC curve from an organic material

5.1.2　Thermal Gravity Analysis

Thermal gravity analysis or thermogravimetric analysis (TGA) is one thermal analysis method that provides information about physical as well as chemical phenomena related to the weight of the sample at a fixed temperature and a given atmosphere. As shown in Figure 5.3, the sample is put in a sample pan located inside a furnace containing a precision balance, an atmosphere control and a programmable temperature control. The temperature is generally increased at constant rate to incur a thermal reaction, and the sample is weighed at each temperature step[5].

5. 热重分析(TGA)是一种热分析方法,它提供给定的温度和气氛下,与样品质量相关的物理或者化学现象的信息。如图 5.3 所示,在一个包含精确天平、气氛控制和程序控温的炉腔内,将样品置于样品盘上。通常通过线性升温来引发热反应,并测量样品在每一个温度下的样品质量。

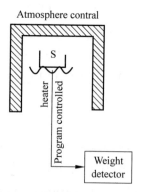

Figure 5.3　Scheme for TGA apparatus

TGA can be used to evaluate the thermal stability of a

material. In a desired temperature range, if the sample is thermally stable, there will be no observable weight loss. Figure 5.4 shows an example. At the beginning of heating, there is almost no weight loss; with the increase of temperature, weight loss occurs gradually. At 256.8 ℃, the weight loss is 5%, which is generally recognized as material decomposition temperature (T_d), often meaning that beyond this temperature the material will begin to degrade. Obviously, the higher the T_d, the better the thermal stability. Note that, when the weight loss is 5%, it doesn't definitely mean that the material starts to decompose; it might be also caused by the sublimation of the material, impurity or water in the hydrate (assuming the sample is a hydrate) at this temperature[6].

6. TGA 可以用来评价一个材料的热稳定性。在期望的温度范围内,如果这个物质是热稳定的,就没有明显的质量损失。图 5.4 是一个例子。在开始加热时,几乎没有质量损失。随着温度的升高,样品开始逐渐失重。在 256.8 ℃时,失重为 5%,一般认为这是材料的分解温度(T_d),通常意味着在这个温度之后材料将开始退化。显然,这个温度越高,材料的热稳定性越好。要注意,失重 5%,不是一定代表材料开始分解了。还有可能是材料在此温度下升华,或者材料中含有杂质,以及水合物(假定这个样品是水合物)中水的升华。

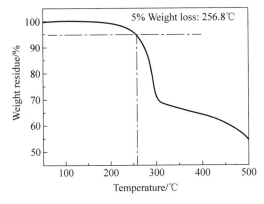

Figure 5.4　An example for TGA experiment

Both DSC and TGA are methods to analyse the thermal properties of materials; the former analyses the endothermic or exothermic heat flow and the latter studies the weight loss of the material under a certain temperature. In practice, they are often combined to probe the material thermal stability. Figure 5.5 shows an example. There are three organic functional materials, denoted by Cz_2HPS, CzHPS and HPS. As revealed in TGA data, the temperatures for their 5% weight loss are 360.4 ℃, 347.6 ℃, and 282.4 ℃, respectively. Clearly, compared to HPS without carbazole (Cz) moiety, the introduction of the Cz groups into Cz_2HPS (two Cz groups) and CzHPS (one Cz group) helps enhance material thermal-stability. The Cz groups also help increase

7. DSC 与 TGA 都是对材料热性质进行分析的方法，前者是分析材料在一定温度下的吸热或者放热情况，后者研究材料在一定温度下的失重情况。实际应用中，这两种方法通常联用来探测材料的热学性质。图 5.5 给出了一个例子。有三个有机功能材料，以 Cz_2HPS、CzHPS 和 HPS 表示。TGA 实验揭示，它们的 5% 失重温度分别为 360.4 ℃，347.6 ℃和 282.4 ℃。显然，与不含咔唑（Cz）基团的 HPS 相比，Cz 的引入，即 CzHPS 含有一个 Cz 基团，Cz_2HPS 含有 2 个 Cz 基团，有助于提高材料的热稳定性。Cz 基团也有助于增加材料的 T_g 和 T_m，如它们的 DSC 曲线所示。HPS 的 T_g 为 62.4 ℃，而 CzHPS 和 Cz_2HPS 的 T_g 分别是 95.3 ℃ 和 129.3 ℃，高出很多。类似地，HPS 在 182 ℃ 开始熔化，而 CzHPS 和 Cz_2HPS 的熔化开始于 234 ℃ 和 298 ℃。另外，HPS 的 DSC 曲线上通过加热扫描，可以观察到几个晶型的变化过程，CzHPS 的晶型变化过程发生在较高的温度，而 Cz_2HPS 体系观察不到晶型变化。这说明 Cz_2HPS 的热稳定性及形貌稳定性更好。

T_g and T_m as shown in their DSC curves. While T_g of HPS is 62.4 ℃, those for CzHPS and Cz_2HPS are much higher, being 95.3 ℃ and 129.3 ℃, respectively. Similarly, HPS starts its melting transition from 182 ℃, but CzHPS and Cz_2HPS do so at 234 ℃ and 298 ℃, respectively. Additionally, several crystallization transitions are recorded in HPS during the DSC heating scan. These transitions occur in the CzHPS system at higher temperatures, but no such transitions are detectable in the Cz_2HPS system. Cz_2HPS thus enjoys strong thermolytic resistance and high morphological stability[7].

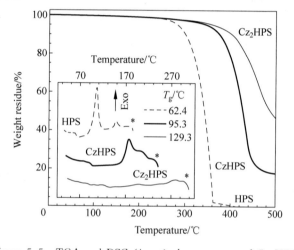

Figure 5.5　TGA and DSC (inset) thermograms of Cz_2HPS, CzHPS and HPS, taken under nitrogen at a heating rate of 10 ℃ · min^{-1}, in which the asterisk denotes the start of a melting transition

5.2　Electronic Levels and Bandgap

As discussed in previous section, for optoelectronic device applications, electronic levels and bandgap in materials are very important, which are involved in an electronic process in a device and thereby need to be considered both in device construction and device physics interpretation.

5.2.1　HOMO/LUMO Levels

HOMO and LUMO are the highest occupied molecular

orbital and the lowest unoccupied molecular orbital, respectively, which are the frontier molecular orbitals in an organic material. Typical methods to probe these levels are the ultraviolet photoemission spectroscopy (UPS) and cyclic voltammetry (CV)[8].

1. By Ultraviolet Photoemission Spectroscopy (UPS)

As we know, when light beam with suitable energies hits a substance, electrons may emit out of the material. This process is termed as photoelectric effect. The emitted photoelectrons originate from either inner atomic level of an atom or valence level of the material, depending on the energy of the incident photon beam (Figure 5.6)[9]. If X-ray is employed as incident beam, electrons from inner shells of atoms can be ejected, often carrying some kinetic energy, and can be recorded as X-ray photoelectron energy spectrum (XPS), which is usually used to probe chemical states of an atom as well as its relative quantity. When a sample is irradiated with ultraviolet photons, which possess lower energy compared to X-ray photons, only valence electrons can be emitted out from the substance. Analysis of these photoelectrons is termed as ultraviolet photoelectron spectroscopy (UPS), which is often employed to detect the valence/HOMO level in semiconductors/insulators or the Fermi level of a metal[10].

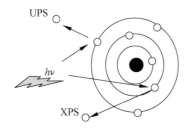

Figure 5.6   Schematic diagram for photoemission of UPS and XPS

In these techniques, a thin film of the concerned material is placed in an ultrahigh vacuum chamber. It is illuminated with UV-light or X-rays. The UV-light may result from a helium discharges source and have energies of 21 eV or 41 eV. The X-rays can result from Kα radiation of Mg at 1255 eV. The illumination promotes an electron from

8. HOMO 和 LUMO 分别是最高占据分子能级和最低空置分子能级,属于电子的前沿轨道。这些能级常用的测量方法有紫外光电能谱(UPS)法和循环伏安(CV)法。

9. 如我们所知,当具有合适能量的光束与物质相互作用时,可能会发射光电子,这个过程称为光电效应。取决于入射光子的能量,发射出来的光电子,或者来源于原子中的内层电子,或者来源于材料的价电子(图 5.6)。

10. 当样品受到能量比 X 射线低的紫外光子照射时,仅价电子可以从物质中发射出来。对这些光电子的分析称为紫外光电子能谱(UPS),通常用来检测半导体/绝缘体的价带/ HOMO 能级,或者金属的费米能级。

11. 真空中光电子的动能（E_k）相当于激发光子能量减去电子的束缚能（E_b）：

the material into the vacuum. The kinetic energy (E_k) of the photoemitted electron in the vacuum corresponds to the energy of the exciting radiation ($h\nu$) minus the binding energy of the electron (E_b)[11]:

$$E_k = h\nu - E_b \tag{5-1}$$

The binding energy of the electron is given by the energy of the orbital it occupies. Therefore, by measuring the kinetic energy of the photoemitted electrons, one obtains a spectrum that mirrors the energies of orbitals in the material.

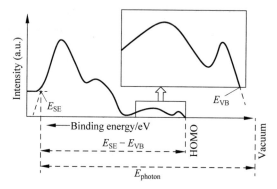

Figure 5.7　Determination of frontier level of HOMO/VB by UPS spectrum

12. 图 5.7 是一个 UPS 实验结果的示意图。x 坐标是束缚能，y 坐标是强度。当光电子的动能为零时，其束缚能最大，在曲线左边的截止位置。这时，根据式（5-1），电子束缚能（E_{SB}）等于入射光子能量（$h\nu$）。

Figure 5.7 schematically shows a result from UPS experiment. The x-axis is binding energy, and the y-axis is intensity. When the photoelectron's kinetic energy is zero, its binding energy is the highest, which corresponds to the left hand cut off point of the curve. At this point the bingding energy of electrons (E_{SB}) is equal to the energy of the incident photon ($h\nu$) according to Equation (5-1)[12]. On the other hand, when the binding energy of an electron is the smallest, its kinetic energy will be the highest. This point corresponds to the onset at right hand, and its binding energy is noted as E_{VB}. Since the electron with binding energy of E_{VB} possesses the highest energy, it belongs to HOMO level in organic materials：

$$E_{HOMO} = E_{photon} - (E_{SE} - E_{VB}) \tag{5-2}$$

CB or LUMO levels cannot be directly measured by UPS, because there are generally no electrons at these levels. However, there is a complementary technique to UPS, which is called inverse photoemission spectroscopy (IPES). In IPES, high-energy incident electron beams of a well-defined energy, which are generated by a van-der-Graff generator or a synchrotron source, is employed. Upon entering the solid, some of the incident electrons will be captured by an unoccupied state, that is, the LUMO of a molecule or the conduction band of a semiconductor. Some fraction of the dissipated energy of the electron is converted to photons (this is called Bremsstrahlung) and the leading edge of the emission spectrum is a measure of the energy of the electron-accepting states[13]. Figure 5.8 shows the determination of HOMO and LUMO levels of an organic material by UPS and IPES spectra, respectively.

13. 在翻转光电子能谱（IPES）实验中，使用的是具有明确能量的、由范德格莱夫发生器或者同步辐射装置产生的入射电子束。当进入固体时，这些入射电子束中的一部分将被材料的空置态，即分子 LUMO 能级或者半导体导带俘获。电子降低的能量（指的是入射电子能量与该电子进入 LUMO 能级的能量之差），一部分将以光子的形式释放出来（称为 Bremsstrahlung）。因此（设备记录）的发射光谱的前缘就是对电子受体位置的测量。

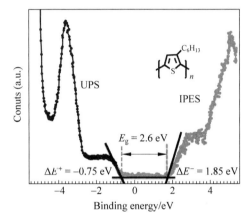

Figure 5.8　The UPS（black）and IPES（gray）spectra of 70-nm-thick P3HT film on ITO glass. The onset of HOMO and LUMO, as well as the transport gap are indicated by the lines

2.　By Cyclic Voltammetry（CV）

HOMO/LUMO level can be determined by cyclic voltammetry （CV）. Compared to UPS method, CV measurement is more convenient and popular, because of no needs for thin film preparation as well as the high vacuum and expensive equipment of UPS. The main principle of CV measurement is schematically shown in Figure 5.9. If a material is oxidized, it means that electrons are taken away from HOMO levels; while the reduction of a material means that electron-injection into the LUMO levels occurs. Therefore, the flows of oxidation and reduction currents in the CV curve relate to the HOMO and LUMO levels of the

14. 图 5.9 示意地表示了 CV 测量的主要原理。当材料被氧化时,意味着从 HOMO 位置拿走电子,而材料被还原意味着在材料的 LUMO 注入电子。因此,CV 曲线中的氧化和还原电流与材料的 HOMO 和 LUMO 相关。

15. 这个装置包含工作电极、参比电极和对电极,由此形成了 2 个电流回路:工作电极和对电极形成的电路检测样品信号和环境信号;而参比电极和对电极形成的电路,仅仅对环境信号有响应。在一定的电压下,二者之差就是该电压下来自于样品的信号。

16. 电流的开始处表明,被探测材料的相应能级与电极之间发生了能量共振。当然,这里的前提是电解质的氧化和还原不在电势扫描范围之内。一旦氧化(E_{ox})和还原(E_{red})电势获得了,HOMO/LUMO 能级相对于真空的位置,就可以确定了。

material[14]. The setup and equivalent circuit for CV measurement are shown in Figure 5.9(b) and (c). The apparatus contains working, reference and counter electrodes, which form two current loops: the circuit comprising working and counter electrodes detects signals from both the sample and the environment, while that with reference and counter electrodes only responses to environment signals. At a given voltage, the current difference between them gives signal from the sample[15].

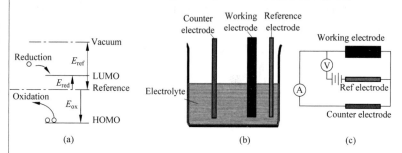

Figure 5.9　Relationship between oxidation/reduction potential and HOMO/LUMO level in an organic material (a), and the setup (b) and equivalent circuit (c) for CV measurement

By varying the applied potential, the HOMO and LUMO levels of the probed material will be swept across the Fermi level of the electrode. In the recorded current-voltage curve, the current flow of material oxidation or reduction may present in the positive and negative ranges, respectively. The onset of a current is an indication that there is energetic resonance between the respective levels of probed molecules and the electrode. This presumes, of course, that the oxidation and reduction of the electrolyte are beyond the potential scanning range. Once the oxidation (E_{ox}) and reduction (E_{red}) potentials are obtained, the position of HOMO/LUMO level (E_{HOMO} and E_{LUMO}) relative to vacuum level can be determined[16]:

$$E_{HOMO} = E_{ref} - E_{ox} (eV) \tag{5-3}$$

$$E_{LUMO} = E_{ref} - E_{red} (eV) \tag{5-4}$$

where E_{ref} is the Fermi level of reference electrode relative to vacuum: ferrocene/ferrocerium (Fc, often used as inner standard) = -5.0 eV, NHE (standard hydrogen electrode) = -4.6 eV, SCE (saturated calomel electrode) = -4.4 eV, and Ag/AgCl = −4.7 eV. Note that because all energy levels are below the vacuum level, the sign of them should be negative. However, in literature, both positive and negative signs for energy levels have been widely used, given that it is consistent in its own expression.

The principle shortcoming of CV method for energy level calculation is related to the solvation of the generated radical ions in the-inevitably-polar solvent. Since the polarization energy of an ion increases with the polarity of the (fluid) solvent, there is a systematic error in the order of a few 100 meV when using the HOMO and LUMO values inferred from cyclovoltammetry to determine the corresponding values of the molecules incorporated in a solid organic dielectric[17]. Another problem is that very often only one of the potentials, that is, either the oxidation or the reduction potential, can be determined. The reason for this is that those potentials must be located within the available potential "window" determined by the electrode/electrolyte combination. Since the electrical gap between HOMO and LUMO levels, that is, $E_{ox} - E_{red}$, is often as large as $2.5 \sim 3.0$ eV, one of the potentials-usually the reduction potential-may be outside the experimentally accessible range. Under this circumstance, it became unfortunate practice to estimate the missing reduction potential by adding the energy of the first singlet excited state, S_1, to the oxidation potential. The energy of the S_1 state, however, is reduced compared to the HOMO-LUMO difference by the Coulomb-binding energy of the electron-hole pair that ranges between 0.5 eV and 1.0 eV. To be correct, that binding energy should be subtracted. Note that reduction potentials have a negative sign[18].

Figure 5.10 shows a CV curve for a material (HPCzI). This is an oxidation process and the characteristic data are also shown in the figure. Without applied bias, no current flows due to the energy barrier between the Fermi level of the electrode and the HOMO energy of the probed molecules. As soon as an applied bias shifts the HOMO energy to match the electrode's Fermi energy, for example, around 0.37 V for HPCzI, the probed molecules are oxidized, implying the onset of a current flow[19]. Upon further increase of the

17. CV 方法的主要缺点,与在不可避免的极性溶剂中所产生的自由基离子的溶剂化有关。由于离子的极化能随着(流体)溶剂极性的增加而增加,当使用循环伏安得到的 HOMO 和 LUMO 来确定在有机固体介电材料中分子的相关数值时,大约有几百毫伏的系统误差。

18. 这种情况下,实际很不幸,需要在氧化电势上加上第一单线态能量 S_1,来估算这个丢失的还原电势。但是,由于电子-空穴对之间为 0.5 eV 到 1.0 eV 范围的库仑束缚能,S_1 的能量相比于 HOMO-LUMO 之差是减小的。作为校准,应该减去这个束缚能。要注意,这个还原电势的数值符号是负的。

19. 当没有施加电压时,由于电极费米能级与被探测分子 HOMO 之间的势垒,没有电流流动。当外加电场将其 HOMO 能量移至与电极费米能级相匹配时,例如对于 HPCzI 是在 0.37 V 附近,待测分子被氧化,表示电流流动的开始。

20. 这里存在电压位移。理想情况下，对于一个电子（氧化和还原）过程，正向峰与反向峰之间相差大约 60 mV。

voltage，the current increases until it drops eventually，limited by the insufficient diffusion of neutral molecules toward the electrode. By reversing the potential，at around 0.30 V for HPCzI⁺，the cations become charged again. The current flow is then from the electrode to molecular cations，implying a change of sign. There is usually a voltage shift，in the ideal case of about 60 mV for one electron，between the peaks obtained in the forward and backward directions[20]. Here HPCzI and HPCzI⁺ are redox couple. The half-wave potential，which is the mean value of oxidation potential and reduction potential of the redox couple，is usually used as oxidation potential to calculate HOMO level in the positive CV scan. Similarly，the half-wave potential in a negative CV scan can be used to determine LUMO level. Based on data from Figure 5.10 and according to Equation (5-3)，HOMO of HPCzI can be calculated: $E_{HOMO} = -4.7 - 0.33$ eV = -5.03 eV

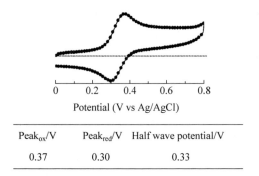

Peak$_{ox}$/V	Peak$_{red}$/V	Half wave potential/V
0.37	0.30	0.33

Figure 5.10   Current-potential curve for material HPCzI

5.2.2   Triplet Levels

Triplet states and triplet energies are usually obtained from the phosphorescence spectrum. In pure organic materials，the phosphorescence emission is too weak to be detectable at room temperature (RT) due to the forbidden nature caused by the spin conservation rule，and need to be collected at low temperature (LT)，at which the lattice vibration can be much suppressed，rendering the weak phosphorescence signal distinguishable. But if the organic material is a metal complex comprising heavy metal atoms，the spin-orbital coupling often occurs，hence breaking the spin conservation rule. As a result, its phosphorescence can be strong and detectable at room temperature[21].

As known, a light with a given wavelength, λ (nm), possesses energy (E) of

$$E = \frac{1240}{\lambda} (eV) \qquad (5\text{-}5)$$

Therefore, based on the wavelength of the phosphorescence, and using Equation (5-5), triplet energy can be calculated. Contrast to the long-wavelength cut-off point in absorption spectrum for bandgap determination, the determination of energy (level) via an emission spectrum adopts the wavelength of the highest energy peak[22]. Here is an example for the determination of triplet level/energy of a pure organic material TCPZ whose triplet level cannot be probed by emission spectrum at RT. Figure 5.11 shows its RT steady-state absorption and LT emission for fluorescence and phosphorescence. At low temperature, the peak wavelengths for fluorescence and phosphorescence are 440 nm and 482 nm, respectively. According to Equation (5-5), the singlet and triplet energy are calculated to be 2.82 eV and 2.57 eV, respectively. Hence, the singlet and triplet splitting at low temperature is 2.82 − 2.57 eV = 0.25 eV. As reported in literatures, for TCPZ, the HOMO is 6.18 eV and the LUMO is 3.23 eV. Usually, the position of LUMO is approximately treated as the first singlet excited state, hence down-shifting 0.25 eV (i.e., the singlet-triplet splitting), the triplet level is calculated as 3.23 − 0.25 eV = 2.98 eV[23].

Notice that based on the data from Figure 5.11 for triplet energy of 2.57 and triplet-singlet splitting of 0.25, combining the reference reported HOMO of 6.18, the deduced LUMO should be 3.36 eV, having 0.13 eV discrepancy compared to the reported LUMO of 3.23 eV. This may account from two aspects. Firstly, the HOMO and LUMO from the reference are obtained at RT, but the data for triplet energy and the singlet-triplet splitting based on Figure 5.11 are collected at low temperature; secondly, the approximating of LUMO to singlet state ignores the binding energy of the singlet exciton, and is another reason for the discrepancy[24].

21. 纯有机材料,自旋守恒的缘故,磷光发射是禁阻的,在室温(RT)下非常弱,不能被检测到,需要在低温(LT)下收集;低温时由于极大地抑制了晶格振动,较弱的磷光信号可以被分辨出来。但是如果有机材料是含有重原子的金属配合物,通常会发生旋-轨耦,因此打破了自旋守恒。结果使得其磷光发射可以比较强,能够在室温下探测到。

22. 与使用吸收光谱的长波截止波长来确定能隙不同,基于发射光谱的能量(能级)的确定,使用的是最高能量的发射峰位处的波长。

23. 文献报道的 TCPZ 的 HOMO 为 6.18 eV,LUMO 为 3.23 eV。通常地,LUMO 的位置近似地等于第一激发的单线态,因此下移 0.25 eV (单线态—三线态裂分),三线态能级可以计算为 3.23 − 0.25 eV = 2.98 eV。

24. 这可能来自于两个方面的原因。第一,文献中的 HOMO 和 LUMO 是室温获得的,而基于图 5.11 获得的三线态能量和单线态—三线态裂分是低温数据。第二,将 LUMO 近似地看作单线态的位置,是忽略了单线态激子的结合能,是另外一个不一致的原因。

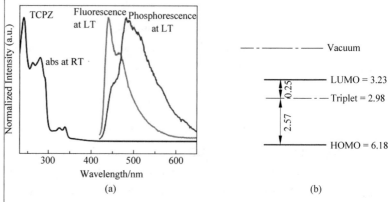

Figure 5.11 (a) Absorption of TCPZ at room temperature (left curve) and low temperature emission from TCPZ collected within the time of several nanoseconds and several microseconds after the pulse excitation turn-off for fluorescence (middle curve) and phosphorescence (right curve), respectively. (b) Schematic view for energy levels in TCPZ based on reference data for HOMO and LUMO and data from this figure for triplet energy of 2.57 eV and singlet-triplet splitting of 0.25 eV

5.2.3 Fermi Levels

The concept of Fermi level has been discussed in Chapter 2. Briefly, Fermi level refers to the energy level at which the probability of electron occupancy is equal to 0.5. Organic materials can be treated as insulators or undoped semiconductor, therefore, their Fermi level (E_F) usually locates at half of the bandgap (E_g), and can be expressed as

$$E_F = \phi_s = \frac{1}{2} E_g + E_{HOMO} \tag{5-6}$$

where ϕ_s is work function, the energy difference between the Fermi level of the solid (or liquid) and the vacuum level of the electron, E_{HOMO} is energy level for HOMO (Note: all parameters here adopt a positive sign).

Since the Fermi levels of organic materials locate at half between the HOMO and LUMO level, actually, there are no electrons at these states. But for a metal, the Fermi level contains electrons of the highest energy in the system, and hence can be detected by UPS method as described in previous section for the determination of HOMO levels in organic materials[25]. In such an experiment, one measures the kinetic

25. 由于有机材料的费米能级处于 HOMO 和 LUMO 的一半位置,这些位置上实际是没有电子的。但是对于金属,系统中能量最高的电子存在于费米能级上,因此可以通过 UPS 方法检测。该方法在前面章节确定有机材料的 HOMO 能级时讨论过。

energy spectrum of the emitted electrons at a fixed incident photon energy $h\nu$. The crucial quantity is the maximum kinetic energy (E_{max}), that is, the cut-off energy, of the emitted electrons. For a clean metal surface this is $E_{max} = h\nu - \phi_m$, where ϕ_m is the work function of the metal[26]. Besides UPS, Fermi level of metals can also be obtained by experiments of ①contact potential difference, ②currents of photoelectron emission, ③currents of thermal emission.

1. By Contact Potential Difference

The contact potential of two materials is defined as the difference of their work functions. If the work function of a material is known, and take the material as a standard surface, using kelvin method to measure the contact potential between the two material surfaces, the work function of the unknown material can be determined. When two surfaces are in contact with each other, at the condition of thermodynamic equilibrium, their Fermi energies will line up and will be at the same level throughout both materials[27]. One consequence of this is that charge will transfer from one material to the other; this charge transfer produces a potential difference at the contact, termed as contact potential difference (Δ). This contact potential is numerically equal to the differences in the work function of the two materials:

$$\Delta = | \phi_x - \phi_r | \qquad (5\text{-}7)$$

where ϕ_x and ϕ_r are the work function of the sample and the reference metal, respectively.

One of the measurements for contact potential difference is vibrating condenser method, in which the materials are not in direct contact[28]. Figure 5.12 shows the apparatus setup. This is a two electrodes system comprising working electrode of the sample and the reference electrode of the known material. The reference electrode is connected to a moveable rod enabled by a magnetic coil. An oscilloscope is integrated in the circuit for detection. During the experiment, at $V_{12} = 0$ V, exert AC current on the coil to produce a sine wave at the oscilloscope. Then, adjust the DC voltage of V_{12} between the two electrodes to cancel the sine wave. At this moment,

26. 决定性的数值是最大动能(E_{max}),即发射电子的截止能量。对于清洁金属表面有 $E_{max} = h\nu - \phi_m$,这里 ϕ_m 是金属的功函数。

27. 两种材料的接触电势定义为它们的功函数之差。如果一种材料的功函数是已知的,并将该材料作为标准表面,利用开尔文方法测量这两种材料表面间的接触电势,可以确定未知材料的功函数。当两个表面相互接触时,在热力学平衡情况下,它们的费米能级将相互对齐,贯穿两个材料(费米能级)将会在同一个能级位置。

28. 接触电势差的一种测量方法是,材料并没有真正接触的振动电容器方法。

29. 这是一个二电极系统,包含样品工作电极和已知材料的参比电极。参比电极与一个通过磁性线圈控制的移动杆相连,电路中包含一个示波器用于检测。实验过程中,在 $V_{12}=0$ V 的情况下,在线圈上施加交流电,使得示波器中产生正弦波形。之后,调节两个电极之间的直流电压 V_{12} 使得正弦波消失。此时,直流电压 V_{12} 就是两个电极之间的接触电阻,$V_{12}=\Delta$。

the voltage satisfies $V_{12}=\Delta$[29]. Based on Equation (5-7), the work function of the sample can be calculated.

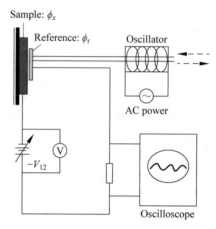

Figure 5.12 Schematic setup for work function measurement by vibrating condenser based on contact potential

Notice that Fermi levels measured by contact potential difference do not agree well with those from UPS method. It thus appears that great difficulties are encountered in making meaningful contact potential measurements, probably because of the presence of surface contamination or distortion.

2. By Photoelectron Emission

30. 在光电效应中,如果光子能量大于物质的功函数,将发生光电发射,电子将从物质的表面释放出来。

In the photoelectric effect, if the photon energy is greater than the work function of the substance, photoelectric emission may occur by liberating electrons from the substance surface[30]. The liberated electrons can be extracted into a collector and produce a detectable current. The yield of the photoelectric emission is the function of temperature, and the relationship is described by Fowler photoelectric function,

$$i \propto \alpha A T^2 F(x) \tag{5-8}$$

where i is current, α is the fraction of electrons in the metal that absorb a photon and escape, A is a universal constant, $F(x)$ is the Fowler universal exponential function of x [31] and

31. 这里,i 是电流,α 是金属中吸收了光子并逃逸出来的电子比例,A 是通用常数,$F(x)$ 是 x 的 Fowler 通用指数函数。

$$x = \frac{h\nu - \phi_m}{k_B T} \tag{5-9}$$

where ϕ_m, the work function of a metal, is the threshold for photoelectric emission, $h\nu$ is the energy of the incident photon. Thus, collect photoemission current (i) as a function of photo energy ($h\nu$), then by plotting $\ln(i/T^2)$ versus $h\nu/k_B T$, the shift parallel to the $h\nu/k_B T$ axis necessary to make the experimental and theoretical curves superimpose gives $\phi_m/k_B T$, and hence the work function of the metal[32]. The shift parallel to the $\ln(i/T^2)$ axis gives α since A is a universal constant. This is one of the most accurate ways to determine the metal work function. In most cases, the theoretical and experimental curves fit well.

3. By Thermal Electron Emission

In the theory of thermionic emission, where thermal fluctuations provide enough energy to "evaporate" electrons out of a material into the vacuum, the work function of the substance (the emitter) is important[33]. Therefore, thermionic emission can be used to measure the work function of the emitter. Generally, these measurements involve fitting to Richardson's law, and so they must be carried out in a low temperature and low current regime where space charge effects are absent[34]. According to Richardson's law, the emitted current density (per unit area of emitter), j (A · m^{-2}), is related to the absolute temperature T of the emitter by the equation:

$$j = A(1-\sigma)T^2 \exp(-\phi_m/k_B T) \text{ or } \ln\frac{j}{T^2} = \ln A(1-\sigma) - \frac{\phi_m}{k_B T}$$

$$(5\text{-}10)$$

where j is the maximum current density that can be emitted over the energy barrier ϕ_m in the absence of high field effects, A is a constant that is 120 A · cm^{-2} · K^{-2} for metals, σ is surface reflection coefficient, which is related to surface conditions, T is the absolute temperature, k_B is Boltzmann constant. Plotting $\ln(j/T^2) \sim 1/T$, a straight line will be obtained, and the value ϕ_m/k_B will be the slope[35].

5.2.4 Bandgaps

The concept of bandgap is originated from band theory, referring to the energy region between the valence band and

32. 因此，收集随光子能量变化的光电流，然后作 $\ln(i/T^2)$ 对 $h\nu/k_B T$ 图，通过将实验曲线平行于 $h\nu/k_B T$ 轴移动，使得实验曲线和理论曲线重叠，得到位移 $\phi_m/k_B T$，由此获得金属的功函数。

33. 在热电子发射理论中，热量的变化提供足够的能量，使得物质中的电子逃逸出来进入真空，物质（发射体）的功函数很重要。

34. 通常地，这些测量需要对 Richardson 规则进行拟合，因此实验必须在低温和低电流情况下进行，这时没有空间电荷效应。

35. $\ln(j/T^2)$ 对 $1/T$ 作图，将获得一条直线，ϕ_m/k_B 是直线的斜率。

the conducting band, where electron occupation is forbidden. Metals almost have no bandgap; and clear bandgap is present in semiconductors and insulators. Borrowing from that in inorganic materials, the bandgap for organic materials is the energy difference between the highest occupied electronic levels and the lowest unoccupied electronic levels in the ground state material. As discussed in Chapter 2, there are electrical gap (or transport gap) and optical gap (for light absorption), which tightly correlates to bandgap. In a direct-gap inorganic semiconductor with a dielectric constant around $\varepsilon_r \approx 11$, the energy of the absorption edge is equal to the energy difference between an electron in the valence band and a hole in the conduction band. In an organic semiconductor, poorly screened electron-electron interactions ($\varepsilon_r \approx 3$) imply that there is a significant difference between optical gap and the electrical gap[36]. The optical gap is the energy of the singlet S_1 state, inferred from either the absorption or fluorescence spectra. The electrical gap is the energy difference between the electron affinity and the ionization potential of a molecule in the condensed phase. In terms of the electrical and optical gap difference due to electron-electron interactions, the most significant contribution is the Coulomb attraction between a negative and a positive charge. Other contributions are, for example, the exchange interactions that distinguish the S_1 from the T_1 state[37].

1. Electrical Bandgap

As shown in Figure 5.8, the most direct way to get electrical bandgap of a condense material is the technique of UPS plus IPES. Another method for electrical bandgap is based on CV measurement, if both oxidation and reduction potentials are in the voltage-scanning window. The difference between the oxidation and reduction potentials gives a bandgap of material in solution. The bandgap obtained by CV method, in concept, corresponds to the electrical bandgap. However, due to the solvation effect, there exists systematic error[38]. A newly developed electrical bandgap determination method is the valence electron energy-loss

36. 在一个介电常数为 $\varepsilon_r \approx 11$ 的直接带隙无机半导体中,其吸收带边的能量等于价带电子与导带空穴之间的能量差。在有机半导体中,对电子与电子之间相互作用的较差屏蔽($\varepsilon_r \approx 3$)意味着光学能隙与电子能隙之间有显著差别。

37. 对于由于电子-电子之间作用引起的电学能隙和光学能隙的差异,其最重要的贡献是一个负电荷和一个正电荷之间的库仑吸引力。其他贡献包括,例如导致 S_1 与 T_1 差异的交换作用。

38. CV 方法获得的能隙,在定义上相当于电学能隙。但是由于溶剂效应,这里存在系统误差。

spectroscopy based on Kramers-Kronig analysis，which requires high technique instrument for both excellent spatial and energy resolutions. This method shall not be discussed here.

2．Optical Bandgap

Just as its name implies，optical bandgap is obtained from optical absorption and/or emission spectra. There are some precautions that should be paid when using UV-Vis absorption and emission spectroscopies：

For absorption measurement，one should：

（1）Take the time to have a look at the measuring beam of your spectrometer and how it is placed with respect to the cuvette. Simply set the monochromator to 500 nm，a wavelength where eye sensitivity is the highest，and check whether the beam is well centered in the cuvette and whether there might be spatial false light. Check whether the sample filling volume is sufficient[39]. If necessary，readjust the sample position.

（2）Run the spectrometer across the desired wavelength range without any sample or reference cuvette，just air against air. Is the absorbance baseline thus obtained flat or does it show offsets whenever a lamp or other switch occurs？ Adjustment of the optics may be necessary[40].

（3）Magnify this baseline to see the noise. What is the root mean square level of this noise in terms of absorbance？ For samples at low extinction，this noise-equivalent absorbance will be your detection limit.

（4）Block the light path and measure a dark spectrum. The absorbance should max out everywhere；this will give you an idea what your system does when insufficient light gets through the sample[41].

（5）Repeat the procedure in （1）with a pair of matched cuvettes both filled with the same buffer/solvent. Although this procedure is rather old style，but having a pair of matched cuvettes，i.e. cuvettes made from the same glass or quartz material and with pathlengths as similar as possible，warrants a precise subtraction of buffer absorbance. Refrain from using plastic cuvettes for quantitative measurements if

39．简单地将单色光调到 500 nm，人眼对这个波长最敏感；检查光线是否在比色皿的中央，是否可能有空间杂光，检查样品的填充体积是否足够。

40．在没有样品也没有参考比色皿的情况下，仅仅是空气对空气，在所希望的波长范围进行测量。这样获得的吸收基线是否水平，或者是否一旦光源或其他开关打开时，会发生基线偏移？ 可能需要调整光路。

41．遮挡光路并测量暗态光谱。在任何一个位置的吸光度都应该达最大值；这样可以提供给你当系统没有足够光通过时光谱的样子。

42. 如果可能,避免使用塑料比色皿进行定量测试,它们的光学品质不是很高。再一次放大这样获得的基线来观测噪声,可以告诉你光谱的可用范围。

possible; their optical quality is not very high. Again, magnify the baseline thus obtained to see the noise. This will tell you the usable spectral range[42].

（6）Use the pair of matched cuvettes for the sample and the reference buffer/solvent to run a spectrum. If you don't have such a pair, at least use the same cuvette for the measurement of the background and the sample spectrum. If absorbance exceeds a value of $1 \sim 2$, then dilute and repeat. Aiming at a maximum absorbance below 1 will prevent the fluorescence problem and guarantee a linear relationship between absorbance and concentration, i.e., complying with Lambert-Beer law.

（7）If you are concerned that the lower concentration obtained by dilution might have an impact on your sample, for example disaggregation of a protein, then use matched cuvettes with a shorter path length instead of dilution. There are standard cuvettes from quartz or glass at 1 cm, 5 mm, 2 mm, 1 mm and lower pathlengths, some even from plastic materials.

43. 如果发现散射背景,通过改变几何形状来试图降低散射,即如果可能,将样品和参考比色皿向检测器方向移动。

（8）If a scattering background is observed, try to reduce scattering by changing the geometry, i.e. move the sample and reference cuvettes closer to the detector if possible[43].

（9）Once the scattering is minimized on the experimental side, fit a $1/\lambda^4$ function to the spectrum obtained, but exclude data points at and immediately around the absorption band(s).

（10）Subtraction of this $1/\lambda^4$ function will typically result in a rather clean absorption spectrum that can be used for quantitative evaluation.

For photoluminescence（PL）measurement, precaution and related photophysics are:

（1）PL signals have lower detection limit compared to absorption, e.g., they can be detected at a concentration smaller than 10^{-5} M. Hence impurity of emission should be get rid of, by material purification, cuvette cleaning and employing spectrum scale solvent.

（2）It is best to make PL measurements with the sample in an inert atmosphere（N_2 and He）or under vacuum since the spectrum may be drastically changed by photo-oxidation.

(3) There is often a red-shift of the PL relative to the absorption for organic materials.

(4) There may exist self-absorption, which is strongly concentration dependent and can red shift the PL spectrum. Because the more molecules there are, the larger the absorption cross section becomes. Emission from the far side of the cuvette will have undergone a larger amount of self-absorption than the emission from the front[44].

(5) PL may also be intensity dependent. The intensity of the emission should scale with the intensity of the excitation light up to a saturation point. However, there may also be triplet-triplet and singlet-singlet annihilation processes at high exciton density.

(6) Decreasing the temperature has the effect of reducing the diffusion of excitons to nonradiative decay sites, so in general the intensity of the singlet PL increases and more structure can be seen. Temperature dependence of PL is particularly significant in materials which have phosphorescence from the triplet state (for example materials containing a heavy atom, which allow some spin orbit coupling - breaking the usual quantum mechanical selection rules)[45].

(7) In general the PL spectra measured for materials in solution will be different to those measured in solid state. In solution some of the moieties in a molecule may be free to take up their preferred conformation, which will also be dependent on the solvent used. Such conformational changes can alter the effective conjugation length and shift the energies of PL. In a very ordered film it is possible to have excited state species which extend over more than one molecule, which again will shift the PL spectrum in energy.

Commonly, there are three ways to determine the optical bandgap, as discussed in following.

1) By Long Wavelength Cutoff of an Absorption Spectrum

The characteristic electronic structure in organic materials is discrete energy levels, which generally endow obvious absorption peaks and hence clear cutoff positions (λ_{cutoff}, nm)[46]. According to Equation (5-5), $1240/\lambda_{cutoff}$ gives the optical gap (E_g). Here is an example. As shown in

44. 可能存在自吸收,它强烈地依赖于浓度,能够红移 PL 光谱。因为分子越多,吸收截面就越大。相比于比色皿前端的发射光,在比色皿远端的发射光将发生较大的自吸收。

45. 降低温度,将降低激子到非辐射位点的扩散,因此通常单线态 PL 强度增强,能够观测到更多的结构特性。温度依赖的 PL 在三线态的磷光发射中尤其重要(例如含有重金属原子的材料,重金属允许一定程度的自旋轨道耦合,打破通常的量子力学选择定则)。

46. 有机材料特征性的电子结构是分立的能级,这通常赋予了明显的峰位和清晰的截止位置(λ_{cutoff}, nm)。

Figure 5.13，using the cutoff wavelength of 427 nm，one can get $E_g = 1240/427$ eV $= 2.90$ eV.

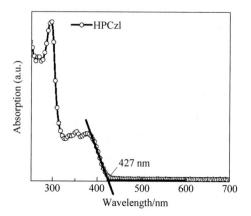

Figure 5.13　Measurement of optical gap by UV-vis absorption spectrum

2) By Crossing Point in the Normalized Absorption and Emission Spectra

If both absorption and emission spectra are available，by transforming them into normalized curves and putting in one coordinate-system，the crossing point between the two curves then indicates the energy of optical gap[47]. In the literature，the optical gap obtained from the crossing point had been correlated to zero-zero optical transition. Figure 5.14 shows normalized absorption and emission spectra with crossing point of 607 nm. Based on this the optical gap can be calculated as $1240/607$ eV $= 2.04$ eV.

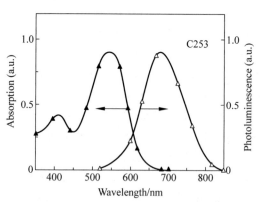

Figure 5.14　Determination of bandgap by the crossing point between the normalized absorption and emission spectra

47. 如果同时获得了吸收和发射光谱，将其转化为归一化曲线并置于同一个坐标系统，两个曲线的交点位置指示的是光隙。

3) By Tauc Plot of Absorption

If the electronic structures of a material are band like, just as those in inorganic semiconductors and insulators, it will be difficult to obtain its optical bandgap using the cutoff wavelength of the absorption spectrum with x-y coordinates of wavelength versus intensity. This arises from the fact that in materials with band structures, the optical transition is band to band, and photons with energies not smaller than the difference between the top of valence band and the bottom of the conduction band (i.e., material bandgap) get absorbed. Hence, the absorption spectra generally exhibit no clear peak as well as no clear cutoff[48]. Figure 5.15 gives such examples.

48. 这是由于这样一个事实,即在带状电子结构的材料中,光学吸收是从能带到能带的,能量不小于价带顶和导带底能量差(材料能隙)的光子都被吸收。因此,吸收光谱通常没有明显的峰位,也没有明显的截止位置。

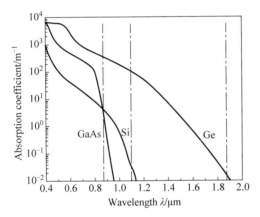

Figure 5.15　Optical absorption constants of GaAs, Si, and Ge as function of light wavelength. The dotted lines indicate the band gap wavelength

This problem can be solved by introducing the Tauc law, which describes the relationship between incident photon energy ($h\nu$) and bandgap energy (E_g):

$$(\alpha h\nu)^{\frac{1}{n}} = A(h\nu - E_g) \qquad (5\text{-}11)$$

where α is absorption coefficient, h plank constant, ν is frequency of incident photon, A is the characteristic constant of each semiconductor, E_g is the bandgap of a semiconductor, $n = \frac{1}{2}$ for direct bandgap and $n = 2$ for indirect bandgap. Transforming the absorption spectrum with coordinates of intensity versus wavelength into a Tauc plot with x-y coordinates of $h\nu$ versus $(\alpha h\nu)^{\frac{1}{n}}$, and simply by fitting a straight line to the linear (or steepest) region of the

49. 将以强度对波长为坐标的吸收光谱转换为以 $h\nu$ 对 $(\alpha h\nu)^{\frac{1}{n}}$ 为 x-y 坐标的 Tauc 曲线,对线性区域或者最陡的位置简单地拟合一条直线,外延至光子能量坐标轴(x 轴),带隙就可以获得,为 x 轴的截距。

optical spectrum and extrapolating it to the photon energy axis (x-axis), the bandgap can be achieved as the intercept of x-axis[49].

Here is an example. An absorption spectrum from an inorganic semiconductor is shown in Figure 5.16(a), and Figure 5.16(b) is its Tauc plot with y-axis of $(\alpha h\nu)^2$ and x-axis of $h\nu$. At the low energy side, a fitting straight line is extrapolated to x-axis, leading to intercept value of 1.78 eV, hence the material E_g is 1.78 eV.

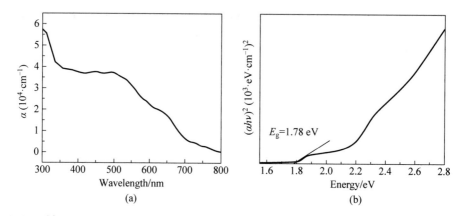

Figure 5.16　Absorption spectrum (a) and corresponding Tauc plot for a direct band semiconductor (b)

Besides the above mentioned methods for optical bandgap determination, which are more popular, other less popular methods, such as those based on photoluminescence spectroscopy, reflectance spectroscopic ellipsometry and photoacoustic spectroscopy technique are sometimes found in literatures. Here we shall not discuss them.

5.3　Carrier Mobility

50. 电荷迁移率可以通过各种技术从实验中获得。对于厚样品(约 1 mm),迁移率的值通常依赖于材料的纯度和有序性;薄样品对这些的依赖较少。

Charge mobility can be determined experimentally by various techniques. For thick samples ($\sim$1 mm), values of mobility are often dependent on the purity and order in the material; while with thin samples, they are less dependent on these characteristics[50]. Here the basic principles of some of the most widely used methods for mobility measurement are discussed.

5.3.1　By Time of Flight

The time-of-flight (TOF) technique involves the generation of carriers near one electrode with a short pulse of light and the observation of the current displaced in the external circuit by the motion of the carriers through the sample[51]. Charge mobility in organic materials was first measured with the TOF technique respectively by R. G. Kepler (1960) and O. H. Leblanc (1960). A double layer structure consisting of charge generation and transport layers (CGL and CTL, respectively), is often adopted. The electrode for pulse light irradiation should be non-injection or carrier-blocking transparent electrode, so that light can reach the sample and generate photocarriers. Copper phthalocyanine and perylenebis(dicarboximide)s can be used as CGL materials[52]. Samples are prepared using vacuum evaporation, solvent cast from solution, or by pressing melt samples with two ITO electrodes.

Figure 5.17 schematically shows the structure of the sample for TOF measurement. Either holes or electrons are produced by a pulsed light-source in the CGL with thickness of δ, then drift across the CTL to counter electrode.

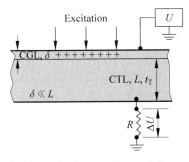

Figure 5.17　A schematic diagram for mobility measurement by TOF method

The thickness of samples (L) is usually in the range from 5 μm to 20 μm. The δ is the thickness for CGL, which should be far less than L. The irradiation time of the pulsed light source (t_e) should be far shorter than t_T, where the t_T is the flight time of the carrier (i.e. the time across the sample). Under the applied electric field $F = U/L$, carriers generated

51. 飞行时间(TOF)技术通过一束短脉冲光在一个电极附近产生载流子,然后观测外电路中由于载流子通过样品的运动而产生的位移电流。

52. 通常采用包含电荷产生层和电荷输运层的双层结构(分别为 CGL 和 CTL)。接收脉冲光辐射的电极应为非注入型或者载流子阻挡型的透明电极,以利于光达到样品并产生光生载流子。酞菁铜和苝基二亚酰胺可以作为 CGL 材料。

53. 样品的厚度（L）通常在 5 μm 到 20 μm 范围。δ 是 CGL 的厚度，它应该远远小于 L。脉冲光源的照射时间（t_e）要远远小于 t_T，其中 t_T 是载流子的飞行时间（即通过样品的时间）。在外加电场 $F = U/L$ 的作用下，薄层中产生的载流子将经过样品到达另一个电极，继而被检测器收集。

54. 在 TOF 实验中，光生载流子的运动是通过观测外电路中电阻器（图 5.17 中的 R）的电压降来监测的。串联电阻器中诱导出来的电流，是通过样品中电荷分布的变化产生的位移电流。

55. TOF 系统的特征响应时间，包括样品电容 C 和串联电阻器 R，其值为 RC。如果 RC 比飞行时间 t_T 小很多，那么当样品施加 DC 电压时，电极将产生电荷电量 $\pm Q$（$= CU$），并且在测量过程中这个电量保持不变。

in the thin layer will pass across the sample and arrive at another electrode, then be collected by detector[53]. Mobility can be calculated based on Equation (5-12):

$$\mu = \frac{v}{F} = \frac{L/t_T}{U/L} = \frac{L^2}{Ut_T} \qquad (5\text{-}12)$$

In TOF experiments, the movement of photogenerated carriers is monitored by observing the voltage drop across the external serial resistor (R in Figure 5.17). The current induced through the serial resistor is the displacement current, which is produced by the change in the distribution of charge carriers in the sample[54]. Figure 5.18(a) gives a schematic illustration for electric field distribution in the sample. Before the photogeneration of hole carriers, a homogeneous electric field, $F = U/L$, is formed when a DC voltage U is applied to a sample with a thickness of L. When N charges are photogenerated and move at position x from the illuminated electrode at a velocity of v, the electric field increases in front of moving charges and that behind the charges decreases. If the electric field in front of the charges (at the side of the counter electrode) is F_2 and that behind the charges at the side of the illuminated electrode is F_1, they are expressed by Equations (5-13) and (5-14), respectively, where d is the forward distance from the photogenerated holes, ε is the dielectric constant of the sample and S is the electrode area of the sample.

$$F_1 = F - \frac{Ne}{\varepsilon S}\left(1 - \frac{x}{d}\right) \qquad (5\text{-}13)$$

$$F_2 = F + \frac{Ne}{\varepsilon S} \cdot \frac{x}{d} \qquad (5\text{-}14)$$

The characteristic response time of the TOF system including the sample capacitance C and serial resistor R is RC. When RC is sufficiently shorter than the transit time t_T, charge $\pm Q$ ($= CU$) is induced on the electrodes when the DC voltage is applied to the sample, and the charges induced on the electrodes retain during the measurement[55]. As Equations (5-13) and (5-14) indicate, the movement of photogenerated charges attempts to change the distribution of the electric field within the sample. However, the displacement current flows through the serial resistor so that charge Q on the electrodes is constant，eliminating the change in the distribution of the electric field[56]. As shown in Figure 5.18(b)，a rectangular-shaped transient photocurrent signal is observed as the voltage drop across the serial resistor. This measurement is called the current mode.

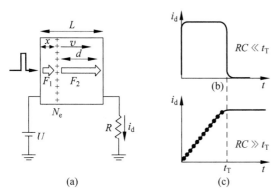

Figure 5.18　(a) Schematic distribution of an electric field in a TOF sample. Transient photocurrent signals obtained under a condition of (b) the current mode and (c) the charge mode

In contrast，if the characteristic response time RC of the sample is sufficiently longer than t_T，charges accumulated on the counter electrode are not retained during the measurement. The change in the charge on the counter electrode，ΔQ，is determined by the change in F_2. Therefore，ΔQ is described by N，e，x，and d as shown in Equation (5-15).

$$\Delta Q = \frac{Nex}{d} \qquad (5\text{-}15)$$

The voltage drop through the serial resistor，ΔU，is expressed by the carrier average velocity v，as Equation (5-16).

$$\Delta U = \frac{\Delta Q}{C} = \frac{Ne}{Cd}vt \qquad (5\text{-}16)$$

The transient photocurrent monotonically increases with time and becomes constant after t_T，as shown in Figure 5.18(c). This transient photocurrent curve indicates the charges arrived at the counter electrode. In addition，it corresponds to the integral of the curve displayed in Figure 5.18(c). This measurement is called the charge mode[57].

Figure 5.19 gives an example of the nearly ideal electron

56. 但是，由于位移电流流过串联电阻，因此电极上的电荷 Q 保持不变，这就消除了串场强度分布的变化。

57. 瞬态光电流随着时间单调上升，在 t_T 之后变为常数，如图 5.18(c) 所示。这个瞬态光电流曲线表示到达了对电极的电荷。另外，它相当于图 5.18(c) 中所表示曲线的积分。这个测量方法称为电荷模式。

transport in a perylene crystal at 40 K and at 300 K. According to Equation (5-12), with the data of F and T in the figure caption and the estimated t_T from the curves (5.5 ns for 40 K and 133 ns for 300 K), mobility can be calculated as: μ (40 K) = 1.15, μ (300 K) = 0.077 cm^2 · V^{-1} · s^{-1}, From the strong decrease in the transit time on cooling, we obtain for this example μ (40 K) / μ (300 K) = 15, indicating μ is depended on temperature.

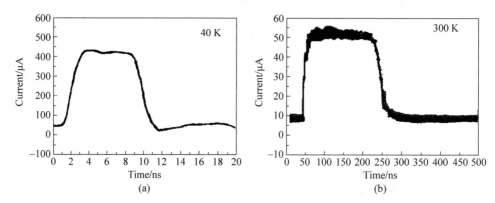

Figure 5.19　The electron TOF transients of a 225 μm thick perylene crystal disc in the b direction: (a) at $T = 40$ K and $F = 35.6$ kV · cm^{-1}; (b) at $T = 300$ K and $F = 22$ kV · cm^{-1}

5.3.2　By Space Charge Limited Current

The mobilities can also be obtained from the electrical characteristics of diodes built by sandwiching an organic layer between two electrodes (provided that carrier transport is bulk limited and not contact limited)[58]. The choice of the electrodes is generally made in such a way that only electrons or holes are injected, forming electron only or hole only device. In the absence of traps and at low electric fields, the current density J scales quadratically with applied bias U. Such behaviour is characteristic of a space-charge limited current (SCLC); it corresponds to the current obtained when the number of injected charges reaches a maximum because their electrostatic potential prevents the injection of additional charges. In that instance, the charge density is not uniform across the material and is largest close to the injecting electrode. In this regime, when neglecting diffusion

58. 通过构筑一个有机层夹在两个电极之间的二极管，测量其电学性质，也可以获得迁移率（假如输运是块体限制的而非接触限制的）。

contributions, the *J-U* characteristics can be expressed as

$$J = \frac{9}{8}\varepsilon\mu \cdot \Theta \cdot \frac{U^2}{L^3} \qquad (5\text{-}17)$$

where ε is the dielectric constant of the sample, μ is mobility, Θ is 1 for space charge limited current in the trap-free situation, U is the applied external bias to the device, L is the device thickness[59]. This equation is known as Mott-Gurney equation and applies for materials in which the mobility is independent of the electric field. Note that a field-dependent mobility has to be considered at high electric fields. The J -U curves become more complex in the presence of traps. This trap-filled SCLC is briefly discussed in Chapter 4 of this text book.

Figures 5.20(a) and (b) show examples respectively for hole and electron mobility measurement based on SCLC. In the device structure, besides the layer of the sample and the two electrodes, on both sides of the tested sample, there is an additional layer of NPB (*N*, *N* '-Di-[(1-naphthalenyl)-*N*, *N* '-diphenyl]-1, 1'-biphenyl-4, 4'-diamine) in the hole only device and BCP (2, 9-dimethyl-4, 7-diphenyl-1, 10-phenanthroline) in the electron only device. NPB promotes hole injection and prevents electron injection, while BCP helps electron injection and blocks hole injection, ensuring that only one kind of carrier is injected from electrodes[60]. Since NPB and BCP layers are thin, their thickness can be neglected. As can be seen in Figure 5.20(c), the *J-U* characteristics are linear at low drive voltages, showing ohmic behavior. At high applied voltages, the *J-U* characteristics become space-charge-limited because of the injection of charge carriers. Employing Equation (5-17) to fit the SCLC part of the curve, the prefactor, $9/8\varepsilon\mu$, can be obtained, where $\varepsilon = \varepsilon_r\varepsilon_0$, ε_0 is the vacuum dielectric constant, 为 8.85×10^{-12} F $\cdot$ m^{-1}, ε_r is the relative dielectric constant of the sample. For organic materials, ε_r is about 3.8. Combining the fitted prefactor and the value of ε, mobility, μ, can be calculated[61].

5.3.3   By Dark Injection Transient Space Charge Limited Current

The third method for mobility measurement is to record

59. 这里 ε 是样品的介电常数, μ 是迁移率, Θ 为 1, 代表陷阱空置时的空间电荷限制电流, U 是器件的外加电压, L 是器件的厚度。

60. NPB 促进空穴注入防止电子注入, 而 BCP 协助电子注入阻止空穴注入, 这样确保只有一种载流子从电极注入。

61. 使用式(5-17)对曲线的 SCLC 部分进行拟合, 就可以获得前置参数 $9/8\varepsilon\mu$ 的拟合值, 其中的 $\varepsilon = \varepsilon_r\varepsilon_0$, ε_0 是真空介电常数, 为 8.85×10^{-12} F $\cdot$ m^{-1}, ε_r 是材料的相对介电常数。对有机材料来讲, ε_r 约等于 3.8。结合拟合出来的前置因子和 ε_r, 载流子迁移率 μ 被求出。

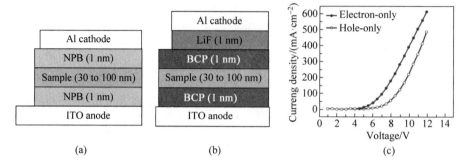

(a)　　　　　　　　　(b)　　　　　　　　　(c)

Figure 5.20　Hole (a) and electron (b) measurements based on SCLC, (c) experimental J-U curves

62. 电流随着时间而增大,在时间 t_p 时达到最大值,并逐渐减小至一个稳定值,即稳态空间电荷限制电流。时间 t_p 与 TOF 中的飞行时间 t_T 的区别在于本方法中载流子注入导致的空间电荷。在 TOF 中,载流子是通过光激发产生的,没有空间电荷。

the transient behaviour of dark injected transient space charge limited current (DI-TSCLC) after carrier injection. In the DI-SCLC method, a step voltage is applied to the sample sandwiched between two electrodes, one of which forms an Ohmic contact to enable charge injection.

An ideal transient current for trap-free materials is shown in Figure 5.21. The current increases with time, reaches the maximum at time t_p, and then gradually decreases to a constant current, which is steady-state SCLC. The difference between time t_p and carrier transient time t_T in TOF is the existence of space charge caused by carrier injection in this method. In TOF, the carrier is generated through the light excitation, without the space charge[62]. t_p is related to trap-free space-charge-limited transit time t_T as expressed by Equation (5-18), and then the mobility can be calculated from Equation (5-12).

$$t_T \sim \frac{1}{0.786} t_p \tag{5-18}$$

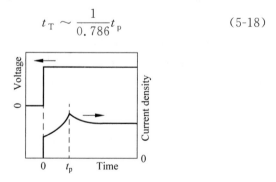

Figure 5.21　Typical DI-TSCLC measurement curves

5.3.4　By Transient Electroluminescence

In this method, organic light-emitting diode (OLED) with the concerned organic material is built. A pulse generator is used as an electrical switch, which can provide voltage pulses of 3～60 V with a rise time less than 20 ns, and a high-current output. The OLEDs are driven by the voltage pulses with a duration of 2～25 μs. The luminance response from the device is detected by a silicon photodiode attached to a high-speed storage oscilloscope[63]. The luminance delay time (t_{delay}) is the time between the application of the pulse voltage and the onset of emission, as shown in Figure 5.22. Similar to DI-TSCLC, transient EL also uses pulsed waves to generate transient voltages. But this method is to collect transient luminescence signals instead of current signals. Space charges also exist in transient EL, and carrier transport in the organic layer is restricted by them. According to Equation (5-18), carrier transport time t_T can be obtained by substituting the parameter t_p with t_{delay}. Then following Equation (5-12), mobility can be obtained[64].

63. 脉冲发生器作为电开关,可以提供上升时间小于 20 ns 的 3～60 V 电压脉冲和高电流输出。OLED 器件在 2～25 μs 的时间段内被脉冲电压驱动。器件的发光响应被一个与高速存储示波器连接的硅光电管检测。

64. 与 DI-TSCLC 类似,瞬态 EL 也是利用脉冲波产生瞬态电压。但本方法收集的是瞬态发光信号,而不是电流信号。空间电荷也存在于瞬态 EL 过程,有机层中载流子的输运受其制约。根据式(5-18),以 t_{delay} 代替 t_p,可以获得 t_T。然后根据式(5-12),迁移率就可以获得。

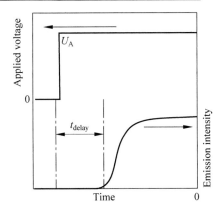

Figure 5.22　Typical transient emission behavior of OLEDs

Figure 5.23 shows single layer and double layer OLEDs for fast and low carrier mobility measurement, respectively. In the single-layer device, the carriers with lower mobility generally can be regarded as still; the generation of electroluminescence is the result of the exciton radiative decay after charge recombination when the fast carriers transport from one end of the organic layer to the other end,

65. 在单层器件中,迁移率较小的载流子通常可视为静止;电致发光的产生,是由于快速载流子从有机层的一端输运到另一端,与静止的载流子相遇时发生电荷复合之后,激子辐射衰减的结果。因此,瞬态发光时间(t_{delay})对应于快速载流子在材料中的迁移。

meeting the still carriers. Therefore, the transient luminescence time (t_{delay}) corresponds to the fast carrier drifting in the material[65].

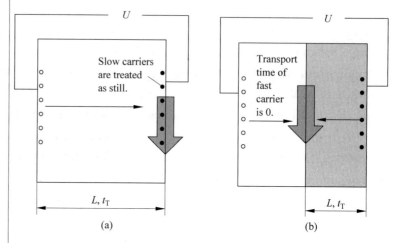

Figure 5.23 Scheme for carrier mobility measurement using transient luminescence. (a) single-layer device, (b) double-layer device. The electrodes are omitted

Since the mobility of hole is typically higher than that of electron in an organic material, the single-layer OLEDs are adopted for hole mobility measurement. In a bilayer device (Figure 5.23(b)), both hole transport layer (HTL) and electron transport layer (ETL) are included. Under applied electric fields, holes and electrons are injected from the anode and cathode, respectively. Because mobility of hole in HTL is higher, they will transport to HTL/ETL interface swiftly. The recombination of the positive (hole) and negative (electron) carriers and the subsequent luminescence occur only when the electrons, whose mobility is significantly lower, transport through ETL, and reach the HTL/ETL interface. Hence, the transient EL time (t_{delay}) represents the time of electron transport over the ETL[66].

66. 在外加电场下,空穴和电子分别从阳极和阴极注入。由于 HTL 中空穴的迁移率较大,它们将迅速地传输到 HTL/ETL 界面处。只有当迁移率较慢的电子通过 ETL 到达 HTL/ETL 之后,正负载流子的复合以及随后的发光,才可以发生。因此,瞬态电致发光时间(t_{delay})代表了电子穿越 ETL 的时间。

5.3.5 By Field Effect Transistor

The carrier mobilities can be extracted from the electrical characteristics measured in a field-effect transistor (FET) configuration. As discussed in Chapter 6, the I-V (current-voltage) curves in an OFET device can be expressed as,

$$I_D = \frac{W}{L} C_{ox} \mu \left[(V_G - V_T) V_D - \frac{1}{2} V_D^2 \right] \approx \frac{W}{L} C_{ox} \mu (V_G - V_T) V_D$$

$$(5\text{-}19)$$

$$I_D = \frac{W}{2L} C_{ox} \mu (V_G - V_T)^2 \qquad (5\text{-}20)$$

where I_D and V_D are the current and voltage between source and drain electrodes, W and L are the width and length of the conducting channel, C_{ox} is capacitance of the gate dielectric, V_G and V_T are gate and threshold voltages[67]. Equations (5-19) and (5-20) refer to the I - V expression for the linear and saturation regions, respectively. Based on the above two equations, filed-effect mobility can be determined from the slope of the linear plots of I_D versus V_G (linear region) and $(I_D)^{1/2}$ versus V_G (saturation region). Carrier mobilities extracted from the I - V curves of OFET device are generally higher in the saturated regime than those in the linear regime as a result of different electric-field distributions.

In OFETs, the charges migrate within a very narrow channel (at most a few nanometers wide) near the interface between the organic semiconductor and the dielectric. Transport is affected by structural defects within the organic layer at the interface, the surface topology and polarity of the dielectric, and/or the presence of traps at the interface (that depends on the chemical structure of the gate dielectric surface)[68]. Also, contact resistance at the source and drain metal/organic interfaces plays an important role; the contact resistance becomes increasingly important when the length of the channel is reduced and the transistor operates at low fields; its effect can be accounted for via four-probe measurements. Furthermore, the mobility can sometimes be found to be gate-voltage dependent; this observation is often related to the presence of traps due to structural defects and/ or impurities (that the charges injected first have to fill prior to the establishment of a current) and/or to the dependence of the mobility on charge carrier density (which is modulated by V_G). The dielectric constant of the gate insulator also affects the mobility. For example, measurements on rubrene single crystals and polytriarylamine chains have shown that

67. 其中 I_D 和 V_D 分别是源和漏电极之间的电流和电压，W 和 L 是传导沟道的宽度和长度，C_{ox} 是栅介电材料的电容，V_G 和 V_T 是栅电压和阈值电压。

68. 在 OFETs 中，电荷移动发生在一个邻近有机半导体和介电材料界面的、非常窄的沟道内（最宽是几纳米）。（电荷）输运受有机层内靠近界面的结构缺陷、介电材料的表面形貌和极性，或者/以及界面处的陷阱影响（即依赖于栅介电材料表面的化学结构）。

69. 例如,红荧烯单晶和三芳基聚合物链的测量表明,由于贯穿界面的极化(静电)效应,载流子迁移率随着介电常数的增加而降低;在有机半导体的导电沟道内部,靠近有机/介电材料界面的介电材料表面的电荷诱导出来的极性,与载流子运动耦合,成为 Frölich 极化子。

70. 脉冲辐解时间分辨微波电导率(PR-TRMC)是一种无电极技术,样品内部均匀分布的低密度电荷,是通过照射一个来自范德格拉夫电子加速器的、时间为 5～20 ns、能量为 3 MeV 的电子脉冲而产生的(图 5.24)。

71. 微波场的峰值振幅在 100 V·cm^{-1} 量级,相比于 TOF 实验测量迁移率通常使用的电场强度,这个值是低的。

the carrier mobility decreases with increasing dielectric constant due to polarization (electrostatic) effects across the interface; within the conducting channel of an organic semiconductor, the polarization induced by the charge carriers in the dielectric surface at organic/dielectric interface, couples to the carrier motion, which can then be cast in the form of a Frölich polaron[69].

5.3.6 By Pulse-radiolysis Time-resolved Microwave Conductivity

The pulse-radiolysis time-resolved microwave conductivity (PR-TRMC) is an electrode-less technique, with which the charge carriers with low density are generated homogeneously inside the sample by irradiation with 5～20 ns pulses of 3 MeV electrons from a van der Graaff electron accelerator (Figure 5.24)[70]. One probes their oscillatory motion by a microwave field on a short length scale yet within a nanosecond to microsecond time scale, that is, long compared to the inverse probing microwave frequency. Typical frequencies are either 10 GHz or 34 GHz. The peak amplitude of the microwave field is of the order of 100 V·cm^{-1}, being low as compared to the field strengths commonly used in TOF experiments for mobility measurement[71].

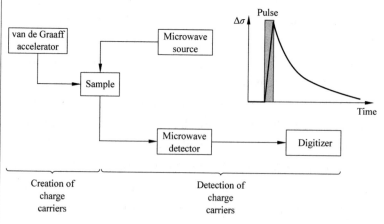

Figure 5.24　A simplified schematic representation of the pulse-radiolysis time-resolved microwave conductivity (PR-TRMC) experiment

The change in electrical conductivity $\Delta\sigma$ induced by the pulse is inferred from the decrease of the microwave power transmitted through the cell. $\Delta\sigma$ can be expressed as

$$\Delta\sigma = q \sum \mu N_{e-h} \qquad (5\text{-}21)$$

where $\Delta\sigma$ is conductivity change, q is elementary charge, $\sum \mu$ is the sum of hole and electron mobility, N_{e-h} is the density of generated electron hole pairs, which is estimated via dividing the amount of energy density of the irradiation dose to the material by the energy required to create one electron-hole pair. This value can be further multiplied by a survival probability that accounts for possible charge-recombination mechanisms during the duration of the pulse[72].

PR-TRMC is a contact-free technique that is not affected by space-charge effects and can be applied to bulk materials as well as to single polymer chains in solution. With this technique, the charges are directly generated in the bulk; their transport properties are probed on a very local spatial scale (for instance, along a portion of a single polymer chain) determined by the frequency of the microwave radiations (the lower the frequency, the larger the region that is explored); the charges trapped by impurities or structural defects are not responsive[73].

Because of its local character, PR-TRMC is considered to provide intrinsic AC mobility values for the bulk; these values should be seen as upper limits for the mobilities at low fields. TOF values are generally smaller since such DC measurements probe a macroscopic range and force the charge carriers to cross structural defects and to interact with impurities[74]. The AC and DC mobility values generally deviate about a threshold frequency that depends on the degree of order in the samples. However, there are reports in which the two techniques result in similar mobility values, for example, in the case of discotic liquid crystals based on hexathiohexyl triphenylenes, materials that have been used as reference compounds to validate the PR-TRMC technique.

PR-TRMC experiments on psolythienylenevinylene and polyparaphenylenevinylene chains provide hole/electron mobility values of $0.38/0.23 \text{ cm}^2 \cdot \text{V}^{-1} \cdot \text{s}^{-1}$ and $0.06/0.15 \text{ cm}^2 \cdot$

72. N_{e-h} 是所产生的电子－空穴对的密度，是将材料受到的辐射剂量的能量密度，除以产生每个电子-空穴对所需要的能量来估算的。这个值可以进一步地乘以考虑了脉冲过程中可能产生的电荷复合机制的生存率。

73. 通过这个技术，电荷直接产生于内部，其传输特性的探测范围依赖于微波辐射频率（频率越低，被检测的区域就越大）的非常局部的空间尺度（例如沿着一个聚合物链的一个部分）；被杂质或者结构缺陷俘获的电荷是没有响应的。

74. 基于 TOF 的数值通常较小，这是由于这个 DC 测量探测的是宏观范围，电荷被迫通过结构缺陷并且与杂质相互作用。

$V^{-1} \cdot s^{-1}$, respectively (here, one kind of charge carriers is selectively trapped for the determination of the mobility of the other type carriers). A mobility as high as $600 \text{ cm}^2 \cdot V^{-1} \cdot s^{-1}$ has been inferred from measurements in dilute solution along fully planar, ladder-type polyparaphenylene chains. In order to probe charge motion on isolated polymer chains, the polymer need to be dissolved in solvent. Monitoring the motion of electrons on the polymer chain requires thorough degassing and doping with a strong hole scavenger such as tetramethyl-p-phenylenediamine (TMPD). In order to selectively study the motion of holes, the solution can be saturated by oxygen that is an efficient scavenger for excess electrons[75].

5.4 Photoluminescence Quantum Yield

5.4.1 PLQY Definition

Photoluminescence quantum yield (PLQY) of a compound is defined as the fraction of the excited molecules that emit a photon after direct excitation by a light source. Since the PLQY concerns the quantities of the excited molecules and the emission molecules, the absorption property of material is not taken into account; it is a kind of internal quantum efficiency (IQE)[76]. PLQY can be obtained theoretically and experimentally. Here we shall discuss the experimental methods both for solution and film samples.

5.4.2 PLQY in Dilute Solution

PLQY in solution is often obtained by relative quantum yield measurement, a method by comparison with compounds of known quantum yields[77]. The initial equation is

$$PLQY_X = PLQY_R \left(\frac{K_R}{R_X}\right) \left(\frac{D_X}{D_R}\right) \left(\frac{n_X}{n_R}\right)^2 \qquad (5\text{-}22)$$

where K is the absorption intensity, D is the integrated detector response over the emission spectrum, and n is the refractive index of the solution. The X and R subscripts refer to the unknown and the standard reference solution, respectively. If optically dilute solutions are used to determine quantum yield, their absorption rests on

75. 监测电子在聚合物链上的运动，需要彻底地除气并掺杂很强的空穴清除剂，如四甲基 p-苯基二胺（TMPD）。为了选择性地研究空穴的运动，其溶液可以用氧气饱和，氧气是有效的过剩电子清除剂。

76. 一个化合物的光致发光量子产率（PLQY）定义为发射光子的分子数目占被光源直接激发的激发分子数目的比例。因为 PLQY 考虑的是激发态分子和发光分子的数量，并没有考虑材料的吸收特性，属于一种内量子效率（IQE）。

77. 溶液中的 PLQY 通常通过相对测量方式获得，一种与已知量子产率的化合物比较的方法。

Beer's law[78],
$$I_0 B = I_0 (1 - 10^{-AL}) \qquad (5\text{-}23)$$

where B is the fraction of light absorbed by the sample, I_0 is the intensity of the incident light, A is the absorbance (cm^{-1}), and L (cm) is the path length. Introducing Taylor expansion, we have,

$$B = 1 - 10^{-AL} = 1 - e^{(-AL)\ln10} = 1 - \{1 + [(-AL) \times \ln10]^1 + \frac{1}{2}[(-AL) \times \ln10]^2 + \cdots\}$$

$$\approx -(-AL) \times \ln10 = 2.303AL \qquad (5\text{-}24)$$

Substituting the K in Equation (5-22) with $I_0 B$, one gets

$$PLQY_X = PLQY_R \left(\frac{A_R}{A_X}\right)\left(\frac{D_X}{D_R}\right)\left(\frac{n_X}{n_R}\right)^2 \qquad (5\text{-}25)$$

Equation (5-25) is widely used for the determination of PLQY in solution. When the unknown and the standard compounds are excited at the wavelength of their equal absorption position, and assume the ratio of the refractive indexes for dilute standard and unknown solution is 1, Equation (5-25) reduces to[79]

$$PLQY_X = PLQY_R \left(\frac{D_X}{D_R}\right) \qquad (5\text{-}26)$$

5.4.3　PLQY in Solid State

The organic materials used in photoelectric devices usually exist as solid-state films, so it is more important to obtain the PLQY in solid state than in solution. Here is a simple and easy method to determine solid state PLQY by integrating spheres.

1. Thick Samples without Transmitted Light

An integrating sphere is a large globe, whose inner wall is coated with a highly reflective substance, such as barium sulfate or magnesium oxide, to render the sample emission reflected repeatedly in the integrating sphere, and finally detected by the fiber optical spectrometer at the outlet slit. Such a device can ensure that emissions from all directions enter into the detector, and the polarization of the light-emitting film will not affect the collection of signals[80]. The excitation light is introduced by optical fiber through the

78. 如果使用光学稀溶液来确定量子产率,其吸收符合比尔定律,

79. 当未知化合物和标准化合物被吸收相等的一个波长激发,并假设标准化合物和未知化合物的稀溶液的折射率之比为 1,式(5-25)简化为

80. 积分球是一个大的球状物,其内壁涂覆了高折射率的物质,如硫酸钡或者氧化镁,使得样品发射光在积分球内被重复地反射,最后被出口处的光纤光谱仪检测。这样的装置可以确保所有方向的光发射都进入检测器,发光薄膜的极化将不会影响信号的收集。

81. 在球内,通常含有一个反射挡板,来防止样品发射光直接进入检测器。

82. 积分球最初的目的,是通过多次漫反射来平均光发射的任何空间非均匀性。另外,折射率和极化误差也被消除。

entrance slit. The sample is usually placed at the center of the sphere and receives the excitation light from a vertical direction. There is often a reflection baffle within the sphere to shield the detector from direct sample emission[81]. A schematic view for the measurement of thick sample by integrating sphere is shown in Figure 5.25(a).

The primary purpose of the integrating sphere is to average out, by multiple diffuse reflections, any spatial anisotropy of the emission. Also, refractive index and polarization errors are eliminated[82]. If the testing sample is thick, the excitation light will be thoroughly absorbed and no transmitted light after the sample. In this case the reabsorption of the transmitted light can be ignored. Equation (5-27) gives the expression for PLQY calculation based on integrating sphere with thick samples:

$$PLQY = \frac{\text{sample emission}}{\text{sample absorption}} = \frac{\int L_X}{\int E_B - \int E_X} \quad (5\text{-}27)$$

where $\int L_X$ and $\int E_X$ are the integrals from excitation and emission signals of the sample, $\int E_B$ is the integrated signals from the excitation wavelengths of blank.

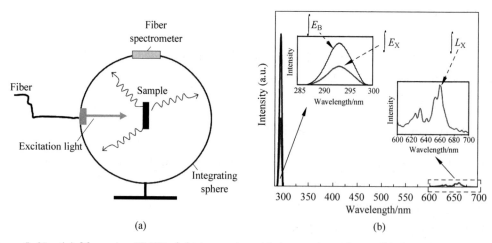

Figure 5.25 (a) Measuring PLQY of thick samples with integrating sphere. (b) Luminescence spectra recorded by integrating sphere, for both situations of blank and sample, which are excited at 293 nm. The embedded diagrams are amplifications of the signals at both excitation wavelength and sample emission wavelength

In Figure 5.25(b), an example is given. The tested sample exhibits emission at range from 600 nm to 700 nm, and the excitation is around 293 nm. The detected signals from integrating sphere comprise two components, one is the excitation information around 293 nm, and the other is the sample luminescence from 600 nm to 700 nm. In the case of blank at wavelength around sample excitation (around 293 nm), since no substance absorbs the light source, the detected signals are strong compared to the case with a sample; and their difference corresponds to the absorption quantity of the sample. In the figure, the parameters for PLQY calculation based on Equation (5-27) are depicted, with which the PLQY of the sample can be obtained[83].

2. Thin Samples with Transmitted Light

If the testing sample is thin, the excitation light will be partially transmitted through the sample, which can be further reflected by the sphere wall and irradiate the sample again[84]. In order to exclude this reabsorption, a mirror between the excitation light and the sample is introduced, as shown in Figure 5.26(a). In this case, the calculation equation is,

$$PLQY = \frac{sample\ emission}{sample\ absorption} = \frac{\int L_X - (1-A)\int L_{mirror}}{A\int E_B}$$

(5-28)

$$A = \frac{\int E_{mirror} - \int E_X}{\int E_{mirror}}$$

where $\int L_X$ and $\int E_X$ are the integrals of sample emission and excitation signals, $\int E_B$ is the integral for excitation of blank, $\int L_{mirror}$ and $\int E_{mirror}$ are the integral of emission and excitation for sample with mirror. A is the absorption intensity, which can be achieved by measuring the integral of the excitation spectra of the sample and the sample plus mirror.

83. 图 5.25(b)给出了一个例子。待测样品的发射波长在 600 nm 到 700 nm,其激发在 293 nm 左右。通过积分球检测的信号包括两个组分,一个是在 293 nm 附近的激发信息,另一个是样品在 600~700 nm 的发光。本底情况下,由于没有物质吸收光源的光,在样品激发波长位置(293 nm 附近)检测到的信号比有样品时强,二者的差异对应于样品的吸收数量。在图中,描述了基于式(5-27)计算 PLQY 的参数,由此可以获得样品的 PLQY。

84. 如果待测样品很薄,激发光将部分地透过样品,进而被积分球反射并再一次地辐照样品。

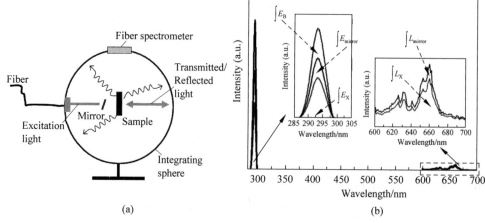

Figure 5.26　（a）Measuring PLQY of thin samples with integrating sphere.（b）Luminescence spectra recorded by integrating sphere，for the situations of blank，sample and sample with mirror. The excitation wavelength is 293 nm. The embedded diagrams are amplifications of the signals at both excitation wavelength and sample emission wavelength

85．图 5.26(b)给出了一个薄样品的例子。实验过程中，需要收集三个光谱：①样品信号，②样品加镜子的信号，③空白的信号。使用图中标示的积分参数，基于式(5-28)，PLQY 可以计算出来。

86．（积分球内壁）涂层虽然高反射，却表现出很小程度的波长依赖的反射率，这将导致积分球反射能力的很大变化。因此，积分球充当了检测装置的波长过滤器。

87．样品能够将它发射的光再次吸收，因此这部分的 PL 光谱不应该包括在效率计算中。

In Figure 5.26（b），an example of thin sample is given. During the experiment，three spectra are collected，①signals of sample，② signals of sample with mirror，and ③ signals of blank. Using the indicated parameters in the figure and based on Equation（5-28），the PLQY can be calculated[85].

3. Problematic Issues with Integrating Sphere Measurement

Besides the reabsorption of the transmitted excitation light due to thin thickness of the sample and the sphere reflection，another problem arises. The coatings，although highly reflective，show a slight wavelength dependence of the reflectance，which causes large changes in the reflectivity of the sphere. The sphere thus acts like a wavelength filter over the detector[86]. The third problem with integrating sphere measurements relates to the overlap of the PL and absorption spectra. The sample can reabsorb light it has emitted so this part of the PL spectra should not be included in the efficiency calculation[87]. Quantum yield measurements involving integrating spheres are probably more affected by reabsorption of the emission than most other methods of measurement. Although the amount of reabsorption of emitted light on a single pass through the sample might be small，the sphere causes multiple passes through the sample and the loss is multiplied.

本章小结

1. 内容概要:

取决于有机材料的光电性质,例如吸光、导电和发光的能力,它们可以应用于各种光电器件中。为了评估应用于器件的材料性能,我们需要了解相关的表征方法。因此,本章针对有机材料在光电器件中的应用,讨论其基本的性质表征,包括热稳定性、能级、能隙、载流子迁移率和光致发光量子产率。

2. 基本概念: 热分析、差分扫描量热法、玻璃转化温度、熔点、热重分析、分解温度、能级与能隙、循环伏安、光电子能谱、载流子迁移、光致发光量子产率。

3. 主要公式:

(1) 光电子动能:$E_k = h\nu - E_b$

(2) 基于 UPS 的 HOMO 能级:$E_{HOMO} = E_{photon} - (E_{SE} - E_{VB})$

(3) 基于 CV 的 HOMO 能级:$E_{HOMO} = E_{ref} - E_{ox}(\text{eV})$

(4) 基于 CV 的 LUMO 能级:$E_{LUMO} = E_{ref} - E_{red}(\text{eV})$

(5) 能量与波长的关系:$E = \dfrac{1240}{\lambda}(\text{eV})$

(6) 本征半导体的费米能级:$E_F = \phi_s = \dfrac{1}{2}E_g + \text{AC}$

(7) 接触电势:$\Delta = |\phi_x - \phi_r|$

(8) Fowler 光电流公式:$i \propto AT^2 F(x)$,$x = \dfrac{h\nu - \phi_m}{k_B T}$

(9) 热电子发射电流:$j = A(1-\sigma)T^2 \exp(-\phi_m/k_B T)$ 或 $\ln \dfrac{j}{T^2} = \ln A(1-\sigma) - \dfrac{\phi_m}{k_B T}$

(10) 光吸收的 Tauc 曲线:$(\alpha h\nu)^{\frac{1}{n}} = A(h\nu - E_g)$

(11) 迁移率公式:$\mu = \dfrac{v}{F} = \dfrac{L/t_T}{U/L} = \dfrac{L^2}{Ut_T}$

(12) 溶液的光致量子产率:$\text{PLQY}_X = \text{PLQY}_R \left(\dfrac{K_R}{R_X}\right)\left(\dfrac{D_X}{D_R}\right)\left(\dfrac{n_X}{n_R}\right)^2$

(13) 厚样品固态光致量子产率:$\text{PLQY} = \dfrac{\text{sample emission}}{\text{sample absorption}} = \dfrac{\int L_X}{\int E_B - \int E_X}$

(14) 薄样品固态光致量子产率:

$$\text{PLQY} = \dfrac{\text{sample emission}}{\text{sample absorption}} = \dfrac{\int L_X - (1-A)\int L_{mirror}}{A \int E_B}, \quad A = \dfrac{\int E_{mirror} - \int E_X}{\int E_{mirror}}$$

Glossary

Cyclic voltammetry   循环伏安

Dark injection transient space charge limited current	注入型瞬时暗电流法
Decompose temperature	分解温度
Differential scanning calorimeter	差分扫描量热法
Field effect transistor	场效应晶体管
Glass transition temperature	玻璃转化温度,玻璃态转变温度
Melting point	熔点
Normal hydrogen electrode	标准氢电极
Photoluminescence quantum yield	光致发光量子产率
Pulse radiolysis	脉冲射频分解
Saturated calomel electrode (SCE)	饱和甘汞电极
Secondary electron	次级电子
Sublimation temperature	升华温度
Thermal gravimetric analysis	热重分析
Transient electroluminescence	瞬态电致发光法
Ultraviolet photoemission spectroscopy	紫外光电子能谱

Problems

1. True or false questions

(1) In DSC analysis, the glass transition point is an exothermal process.

(2) Generally speaking, the 8% weight loss in thermal gravity analysis is considered as decomposition temperature.

(3) Melting point can be observed as a sharp endothermic peak in a DSC curve.

(4) Sometimes, exothermal peaks can be observed in DSC analysis, which could be crystalline transformation temperatures.

(5) The redox means a pair of oxidation and reduction species.

(6) Triplet energy state (E_T) can be probed by phosphorescent spectrum, where the highest energy peak is assigned to E_T.

(7) In intrinsic semiconductor and insulator, the Fermi level locates at the half position of material energy gap.

(8) UPS refers to ultraviolet photoemission spectrum, and can be used to detect HOMO/VB level of materials.

(9) Material bandgap can be obtained by absorption spectrum.

(10) In a material, if the electronic structure is band like and the band width is wide, its bandgap can be easily determined by long wavelength cutoff in the absorption spectrum.

(11) In carrier mobility measurement, TOF is the abbreviation of "taken on free".

(12) In TOF based mobility measurement, beside the layer for mobility measurement, there must be a thin layer of light-pumped charge generation layer.

(13) Compared to a material with nondispersive carrier transport, the TOF signals

from materials with dispersive carrier transport are clearer.

(14) Using the formula of space charge limited current, carrier mobility can be obtained by constructing a device capable of both hole and electron injections.

(15) In the method of dark injection transient space charge limited current for carrier-mobility measurement, the carriers are obtained by photon generation.

(16) In mobility measurement for organic materials by transient electroluminescence with organic light-emitting diodes (OLEDs), double-layer OLEDs often determine the mobility of electron.

(17) Based on the devices of organic field effect transistor, mobility of organic materials can not be obtained.

(18) PLQY only refers to the luminescence quantum yield of a material in solution.

(19) PLQY refers to the luminescence quantum yield of a material excited by electrical energy.

(20) PLQY refers to the percentage of the number of emission photons to the number of the absorbed photons.

2. Please explain why thermal stability, bandgap, carrier mobility, energy levels, absorption coefficient and photoluminescent quantum yield are important parameters when evaluating an organic material for optoelectronic applications.

3. For thermal analysis, what does DSC mean? And what can be obtained from DSC curve?

4. According to Figure 5.5, please answer questions: (1)What kinds of experiments are done to get these curves? (2)What material parameters are obtained? please point out their values. (3)Please discuss the findings based on these data.

5. Please state two methods for measuring HOMO/LUMO, and give detail equations.

6. Please enumerate the methods for determining the bandgap of a material.

7. (1)How to measure the energy of a triple state? (2)According to the following figures, please give the triplet energy of materials CBP, TPD, BCP, Ir(ppy)$_3$ and PtOEP.

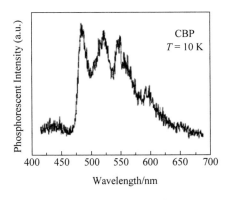 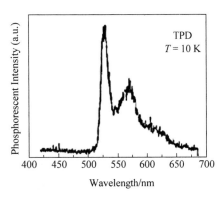

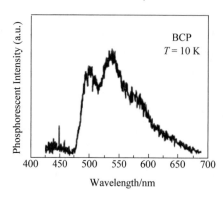
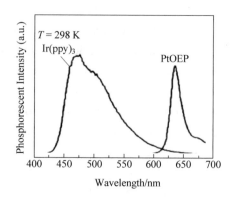

8. The following figure shows the absorption and emission spectra of three phosphorescent materials at room temperature, the absorption peaks and the long wavelength cutoff of the singlet absorption band are 242/396.1 nm for 1, 273/402.6 nm for 2, 261/423.2 nm for 3. The peaks of emission are 506 nm,516 nm and 544 nm for 1, 2 and 3, respectively. The HOMOs of them are known as 6.04 eV, 6.10 eV, and 5.71 eV for 1, 2 and 3, respectively. Please deduce: (1) triplet energy of the three materials, (2) singlet-triplet splitting of each material, (3) give schematic energy level diagram (including HOMO, LUMO, and triplet levels) for each material.

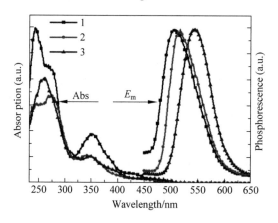

9. Please give three methods for Fermi level measurement and discuss them in detail.

10. Please explain the methodology for mobility measurement based on TOF.

11. Employing space charge limited current (SCLC), mobility of materials can be obtained, please explain this methodology in detail, combining with equation and device structure.

12. State the experimental method for mobility measurement based on dark injection transient space charge limited current.

13. Give schematic drawings together with sentences to explain how to obtain mobility of electron and hole by transient electroluminescence.

14. How to obtain the mobility of an organic material by field effect transistor?

15. (1) Please explain the concept of PLQY in detail, and (2) tell that based on PLQY what material property can be evaluated, and for what kind of electronic application?

16. Suppose both for absorption and emission, the influence of instrument to different wavelength is negligible. We know: (1) the PLQY of a standard material is 100%; (2) the excitation wavelength for both the standard material and the unknown material is 285 nm. And at this wavelength, the intensity of absorption for the standard and the unknown materials is 0.06 and 0.03, respectively. The emission area of the standard and the unknown materials are 271.40, 53.20 respectively. According to these information, please calculate the photoluminescence quantum yield (PLQY) of the unknown material.

17. Please describe the method for film PLQY measurement, and give an example.

CHAPTER 6   ORGANIC FIELD EFFECT TRANSISTORS

In this chapter, after the comprehensive introduction about the structure and working principle for OFETs, device characterization parameters, such as transfer and output curves, threshold voltage, and subthreshold swing, current on-off ratio, etc, are discussed in detail. Finally, the applications of OFET devices in display, sensor, radio-frequency labels, and logical circuit are presented.

6.1   Introduction

1. 虽然电子时代开始于 K. Braun 的阴极射线管(1897年)和 A. Fleming 的真空整流器(1904年),但实际上,是从 L. Forest 的基于真空管的、通过在阳极和阴极之间加入一个"网格"的"三极管"(1906年)开始的;三极管使得整流器转变为放大器,从而使无线电通信和长途电话成为现实。

Although the age of electronics was started with K. Braun's cathode ray tube (1897) and A. Fleming's vacuum rectifier (1904), it was actually launched by L. Forest's vacuum-tube based "triode" (1906); by including a "grid" between the anode and the cathode, the triode transformed the rectifier into an amplifier, thus making radio communications and long-distance telephone a reality[1]. The vacuum triode has its limitations, e. g., it is fragile, rather slow, difficult to miniaturize, consumes too much energy and produces too much heat. The idea of replacing the triode with a solid-state device offering an alternative to the thermionic principle can be traced back to the mid-1920s. In October 1926, J. E. Lilienfeld filled a patent describing an "apparatus for controlling the flow of an electric current between two terminals of an electronically conducting solid by establishing a third potential between said terminals". He probably never got his device to work, and his patent went into obscurity. It was not until thirty years later (1956) that this early concept, i. e., a voltage-controlled electronic switch which is termed field effect transistor (FET), could be successfully demonstrated. This was not with the celebrated Bardeen and Brattain's "point-contact" transistor (1947), nor with Shockley's bipolar transistor (1948) - both devices were based on different principles[2].

2. 直到 30 年后(1956年),这个早期的概念,即称为场效应晶体管(FET)的通过电压控制的电子开关,才得以成功展示。这不是著名的 Bardeen 和 Brattain 的"点接触"晶体管(1947年),也不是肖克利的双极晶体管(1948年)——这两种器件(与 FET)的原理是不同的。

In 1960s, silicon-silicon dioxide based metal oxide semiconductor FET (MOSFET) appeared. Actually, nearly fifteen more years of material technology research were needed to finalize the MOSFET. Today, MOSFET dominates our environment; there are millions of them in the processors used in personal computers, cellular phones, and many other microelectronic devices. The success of MOSFET actually rests on a continuous improvement in the handling of one semiconducting material, silicon.

Besides their numerous technological applications, FETs have also been used as tools for studying charge transport in solid materials; this is because the device gives direct acccss to charge-carrier mobility. A celebrated example of such a concept is with hydrogenated amorphous silicon (a-Si:H). For this, an alternative architecture was employed, the thin-film transistor (TFT), which differs from the MOSFET in that the conducting channel is induced in the accumulation regime rather than through the formation of an inversion layer[3]. The first a-Si:H TFTs were actually designed to measure the mobility of the material, which was at that time difficult to access by other techniques. It was only later that the technological importance of the device was recognized in applications in which large area is required and where single crystalline silicon can no longer be used. Today, a-Si:H TFTs play a crucial role in active-matrix liquid-crystal displays (AM-LCD).

The first metal-oxide-organic semiconductor device was demonstrated by Ebisawa, Kurokawa, and Nara at NTT (Nippon Telegraph and Telephone public corporation) in 1982. The device deposited polyacetylene on a polysiloxane gate dielectric and used aluminum for the gate and gold for the source and drain electrodes. While the device operated in depletion mode and showed only a few percent current modulation when analyzed for trans-conductance, the researchers clearly recognized the concept's potential and ended their paper suggesting that this type of device "appears promising" for TFTs. The next significant milestone was the development of the first organic field effect transistor (OFET) with recognizable current gain by an in-situ

3. 这种概念的一个著名例子是氢化非晶硅（a-Si:H）。在此，采用了另一种结构，即薄膜晶体管（TFT），其与 MOSFET 的不同之处在于，导电沟道是在聚集积状态下诱导产生的，而不是通过形成反转层。

4. 下一个重要的里程碑是 1986 年由三菱化学的津村、小冢和安藤开发的具有可观电流增益的第一个有机场效应晶体管(OFET)。

polymerized polythipohene transistor by Tsumura, Koezuka, and Ando of Mitsubishi Chemical in 1986[4]. Because of the poor performance of these initial devices, interest in OFETs remained limited to a small number of academic groups for nearly ten more years. During that period, much research effort was devoted to improving the charge-carrier mobility. A soluble form of polythiophene, developed by Jen et al., and applied to OFETs by Assadi et al. ignited excitement about the possibility of printable semiconductor systems which could be made with the same economies of scale as printed paper media. Soluble polymers, dispersible polymers, and a range of small molecule systems have been developed for compatibility with a wide range of deposition and patterning technologies. Specialized materials for electrodes and insulators also continue to be developed and refined. The availability of OFET devices may open the way to completely new set-ups, fabrication processes, and applications. Thus, one can envisage processing of organic materials by printing, which enables high-volume, low-cost production[5]. New products include radio-frequency identification (RFID) tags, that might replace the optical bar code found on nearly all consumer goods today, single-use electronics, low-cost sensors, biocompatible products, and flexible displays, etc.

5. OFET 器件的出现可能为全新的装备、制造工艺和应用开辟道路。为此,人们可以设想通过印刷工艺来加工有机材料,从而实现大批量、低成本的生产。

OFETs stand at a crossroad today. Performance has, by many metrics, exceeded that of amorphous silicon. Several industrially scalable processing strategies have been developed, and it is quite likely that the cost and energetic input of such processes are significantly less than that incurred for amorphous silicon. Industrial acceptance, however, has been limited[6]. It remains one of the primary goals of the OFET research community to develop applications which uniquely exploit the properties of OFETs and translate them into killer applications which will cement OFETs' relevance and exploit their unique characteristics.

6. 现如今 OFET 站在一个十字路口。从许多指标来看,性能已超过非晶硅,已经开发出了几种可进行工业化大规模制备的工艺,并且这些工艺过程的成本和能量需求似乎明显低于非晶硅工艺的。然而,这些工艺在工业界的接受程度还有限。

6.2　Device Structure of OFETs

A FET is a three-terminal device: gate, source, and drain. Three electrodes contacted to a double-layer structure including a dielectric layer and a semiconductor layer，where gate electrode is contacted with the dielectric layer，and source and drain electrodes are contacted with the semiconductor conducting channel. The semiconductor itself is separated from the gate electrode by a thin layer of gate dielectric material[7]. The depth of semiconductor conducting channel is very thin，no more than a few nanometers at the surface of the semiconductor nearing the semiconductor-dielectric interface. Although the source and drain are physically identical，the source electrode is conventionally considered as the source of charge carriers. Transistors are primarily described by the type of their conduction channels：p-type transistors are those with hole carriers in transporting channels，while n-type transistors are those that operate with electron carriers in transporting channels. The structure of the FET is not only material-dependent，but also related to the sequence of thin-film deposition when the device is fabricated. Depending on the order in which the thin films are stacked，the devices can be classified into top-gate（TG）FETs and bottom-gate（BG）FETs（Figure 6.1）.

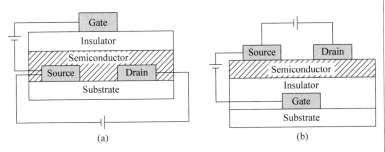

Figure 6.1　Top（a）and bottom（b）gate structures for FETs

When the organic semiconductor was used as the channel component，the transistors are called as organic field effect transistor（OFET）. An organic TFT（OTFT）is a special case of OFET，in which the organic semiconductor and dielectric are deposited as thin films on top of an inactive substrate that plays no part in the transistor behavior. For convenience，we will use OFET to describe both OFET and OTFT in this chapter[8].

For p-type OFETs，as holes are the transporting charge carriers，the source is the most positive terminal，and is often

7. FET 是一种具有三个端点的器件：栅极、源极和漏极。三个电极分别与一个包括一个薄层电介质和一个半导体层的双层结构接触，其中栅极与电介质层接触，源极和漏极与半导体沟道接触。半导体层通过一薄层电介质与栅极隔开。

8. 有机薄膜场效应晶体管（OTFT）是 OFET 的一种特殊情况，其中有机半导体和电介质作为薄膜沉积在非活性衬底的顶部，该衬底在晶体管工作过程中不起作用。为方便起见，我们将在本章使用 OFET 来描述 OFET 和 OTFT 二者。

9. 对于 p-OFETs,由于空穴是电荷传输载流子,因此源极是最正的端点,并且通常由高功函数金属制成,例如金(5.1 eV)。相反,对于 n-OFETs,电子需要从负的源极端点注入,理想的源极是由低功函数金属制成,如钙(2.9 eV),尽管这些可能不具有环境稳定性。

10. 相反,源和漏电极是通过掩模板形成的。因此,器件的尺寸将受到很大限制。另外,TC 结构器件的接触电阻小于 BC 结构器件的。因此,TC 结构的器件表现出较高的场效应迁移率。

made of a high-work-function metal such as gold (5.1 eV). Conversely, for n-type OFETs, electrons are injected from the negative source terminal, which is ideally a low-work-function metal such as calcium (2.9 eV), though these may not be environmentally stable[9]. Due to the poor thermal resistance of organic materials, the structure of OFET is widely adopted as the bottom-gate structure, which can avoid the damage of the organic semiconductor layer by afterwards deposition of the insulating layer and the gate electrode with too high temperatures, which are almost unendurable to organic materials.

As shown in Figure 6.2, bottom-gate OFETs can be further divided into top contact (TC) and bottom contact (BC) devices. The TC device is formed by depositing the source and drain electrodes on the top of organic semiconductor layer through a mask. The BC device is formed by depositing organic semiconductor material on the pre-fabricated source and drain electrodes. Both structures have their advantages and disadvantages. In the BC structure, the source and drain electrodes directly contact with the insulating layer, and if the insulating layer is an inorganic oxide such as silicon dioxide, the gate, source and drain electrodes can be prepared by a well-established micro-etching process, with guaranteed quality and convenience. In the meantime, since the preparation of the organic semiconductor layer in the BC device is the last process, avoiding any damage due to afterwards film-deposition. Comparatively, in TC devices, it can't be possibly to utilize a micro-etching process because we need to make source and drain electrodes on the top of a relatively weak organic semiconductor film. Instead, masks are used to form the source and drain electrodes. As a result, the size of the device will be greatly limited. On the other hand, the contact resistance of the TC structure device is smaller than that of the BC structure device. Therefore, the TC device exhibits higher field effect mobility[10]. It is also hypothesized that in TC devices, the thermal evaporation of metal can cause the source/drain atoms to migrate into the organic semiconductor, increasing the carrier mobility, but the mechanism is not fully understood.

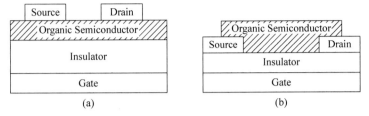

Figure 6.2　Bottom-gate structure: (a) top contact and (b) bottom contact of OFETs

6.3　Working Principle of OFETs

FET is the active devices that modulate the gate voltage to control the channel current between the source and drain electrodes. During a FET operation, a certain voltage is exerted between the source and the drain electrode. If gate voltage is zero or very small, the channel current between the source and the drain is usually small and the characteristics of current are similar to those of the insulator[11]. Thus, the device is in the "off" state. However, if a sufficiently voltage is applied to the gate, a conductive channel will appear inside the semiconductor near the insulator, and the current between the source and drain will rapidly increase to turn the device in the "on" state. This is one of the basic working principles of FET.

In the case of OFETs, when a gate voltage is applied, a field is induced at the organic semiconductor-insulator interface, which causes the highest occupied molecular orbital (HOMO) and lowest unoccupied molecular orbit (LUMO) to shift relative to the source and drain Fermi levels which are held at a fixed value by external voltages. If the gate voltage is large enough, mobile charge carriers will flow from the source contact into the organic semiconductor. When a potential difference is then applied to the source and drain, charge carriers will flow to the drain contact, completing the electrical circuit. We shall discuss the detail device physics in this section.

11. 如果栅极电压为零或非常小,则源极和漏极之间的沟道电流通常很小,此时器件的电流特性与绝缘体相似。

6.3.1　Accumulation or Depletion Mode

A FET device can work either at accumulation or depletion mode. As shown in Figure 6.3(a), in the accumulation mode, there is a moveable charge in the semiconductor channel if a gate voltage is applied. Then, when a certain voltage is applied between the source and drain electrodes, a current will occur in the semiconductor channel, the intensity of which depends on the source-drain voltage. In depletion mode (Figure 6.3(b)), when gate voltages are exerted, the original existing charge carriers are depleted, thus there is no mobile charge in the semiconductor channel[12]. As a result, with the applying of voltages between the source and drain electrodes, almost no current can flow. Therefore, the load of gate voltage in the FET operating in the accumulation mode turns the device from "off" to "on", whereas in the depletion mode-based FET, the loading of the gate voltage changes the device from "on" to "off". In inorganic FETs, since free carriers exist in the inorganic semiconductor, the operating mode of the device can be either accumulation mode or depletion mode; in the absence of a gate voltage, both p- and n-type accumulation-mode FETs have no conductive channel while there is always a conduction channel in depletion-mode FETs which require an opposite-polarity gate voltage to turn off the devices. Because organic semiconductors are normally undoped with no intrinsic carriers, in OFET, the carrier sources rely on the injection through the source electrode, the OFETs usually work at an accumulation mode[13].

12.　在耗尽模式下（图 6.3(b)），当施加栅极电压时，半导体中原本存在的电荷载流子被耗尽，因此半导体沟道中没有可移动电荷。

13.　在无机 FET 中，由于无机半导体中存在自由载流子，器件既可以工作在聚集模式也可以工作在耗尽模式。在没有栅极电压的情况下，p-型和 n-型聚集模式的 FET 都没有导电沟道，而耗尽模式的FET 中总是存在一个传导沟道，这需要施加一个相反极性的栅极电压来关闭器件。由于有机半导体通常未掺杂不存在本征载流子，因此在 OFET 中载流子依赖于源极注入，OFET 通常在聚集模式下工作。

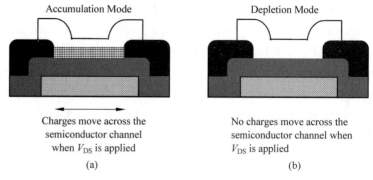

(a)　　　　　　　　　　　(b)

Figure 6.3　Accumulation mode and (a) depletion mode of FET (b)

6.3.2 Band/Energy-level Bending

In FET devices, the applied gate voltage induces an electric field perpendicular to the conduction channel, which leads to dipole moments in the dielectric layer[14]. As shown in Figure 6.4, taking p-type semiconductors as an example, a description for the change of electrical properties of semiconductors by applying a vertical gate voltage is present. There are usually three energy levels: vacuum level, conduction band (CB or LUMO), and valence band (VB or HOMO), at the interfaces between the metal electrode, the dielectric layer and the semiconductor in a FET. When the gate voltage is zero (Figure 6.4(a)), there is no charge transfer between the metal electrode and the semiconductor because no electric field is built. In this case, their vacuum levels are the same and a straight energy bands/levels, i.e., flat bands/levels, are present. When a negative voltage is applied to the gate electrode (Figure 6.4(b)), an electric field will be generated with the direction towards to gate electrode and perpendicular to the channel of the source and drain. At the same time, negative charges will form in dielectric layer near the interface of semiconductor/dielectric, which attract positive charge in p-type semiconductor. Hence, hole accumulation occurs in the semiconductor at a distance of about 5 nm from the interface of dielectric/semiconductor, resulting in the VB/HOMO and CB/LUMO bands/levels of semiconductor upward-bending toward the interface of semiconductor/dielectric. These accumulated holes can be driven to produce current when a voltage between the source and drain is applied. As shown in Figure 6.4(c), when the gate voltage is positive, the direction of generated electric field points from the gate to the semiconductor. At this time, the positive charges in the dielectric layer locate near the interface of semiconductor/dielectric, which, then induce the negative charges in the semiconductor, leading to the depletion of hole in the p-semiconductor and the downward-bending of the CB/LUMO and VB/HOMO band/levels of the semiconductor toward the interface[15]. Since negative charge induced in the

14. 在 FET 器件中,施加的栅极电压感应出垂直于导电沟道的电场,在电介质层中诱导出偶极矩。

15. 如图 6.4(c)所示,当栅极电压为正时,器件内部产生的电场从栅极指向半导体。此时,电介质层中的正电荷位于半导体/介电层的界面附近,诱导出半导体层中的负电荷,导致 p 型半导体层内的空穴耗尽,半导体的导带/LUMO 轨道和价带/HOMO 的能带/轨道的能量朝向界面向下弯曲。

16. 应该注意到,在图4.29和图4.31中,对 n 型半导体的能带弯曲进行了讨论,那是电极和半导体之间电接触的情形,与这里的栅极在半导体中诱导出电场的情形不同。但是,结果是一致的:在界面处的半导体内,指向界面的有效电场导致向下弯曲,离开界面的有效电场导致能级向上弯曲。

semiconductor results in the depletion of the main carrier holes in the p-type semiconductor, the current between source and drain will be blocked by the applied gate voltage. It should be noticed that in Figure 4.29 and Figure 4.31, cases of band bending in an n-semiconductor are discussed, the situation of which is the electrical contact between the electrode and the semiconductor, in contrast to this situation of gate-electrode induced electrical field in the semiconductor. However, the results are consistent: In the semiconductor, the effective electric field pointing to the interface causes the down-bending, and the effective electric field pointing away from the interface leads to up-bending of the energy levels in the semiconductor at the interface[16].

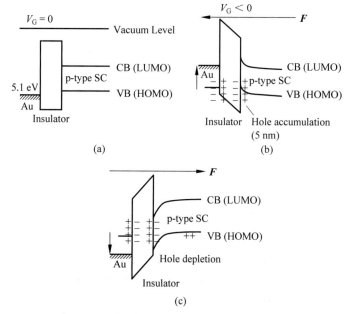

Figure 6.4　The energy levels of the p-type semiconductor material in the OFET device with different gate voltage (V_G): (a) flat levels at $V_G = 0$, (b) upward bending of energy levels due to hole accumulation at $V_G < 0$, (c) downward bending of energy levels due to hole depletion at $V_G > 0$

6.3.3　Carrier Distribution

The distribution of charge carriers in the semiconductor channel of an OFET, which greatly influences the output characteristic of the device, varies with the gate and source-drain voltages. Here three cases will be discussed (Figure 6.5).

At very lower source-drain voltages （V_D）, and the gate voltage （V_G） is much larger than threshold voltage （V_T） (Figure 6.5(a)), there is only a vertical electric field in the channel induced by the gate voltage, which is distributed evenly in the source-to-drain direction. Carrier distribution from the source to the drain is almost uniform. In this case, the channel current obeys the Ohm's law, the current characteristic of the device is similar to a resistor[17]. The carrier density in the channel is related to the gate voltage and the channel current is directly proportional to the drain voltage, and the output characteristic of the device is within the linear region. With the increase of drain voltages, the electric fields in the channel are no longer uniform and are affected by both the gate voltage and the drain voltage. Near the drain electrode, the electric field is small due to the mutual offset of the gate-drain voltage and the source-drain voltage. This uneven electric field results in an uneven distribution of carriers in the channel, e.g., decreasing from the source to the drain. The output current is no longer a linear relationship with the drain voltage[18].

17. 从源极到漏极的载流子分布几乎是均匀的。在这种情况下,沟道电流遵循欧姆定律,器件的电流特性类似于电阻。

18. 随着漏极电压的增加,沟道中的电场不再均匀,同时受栅极电压和漏极电压的影响。在漏极附近,由于栅-漏电压和源-漏电压的相互抵消,电场很小。这种不均匀的电场导致沟道中载流子分布的不均匀,即从源极到漏极递减。器件的输出电流不再与漏极电压呈线性关系。

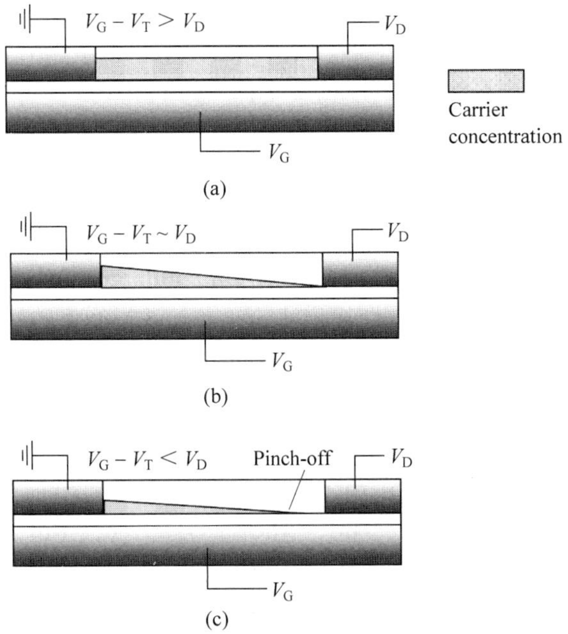

Figure 6.5　Carrier distribution in OFETs: (a) the linear regions, (b) the occurrence of pinch points, (c) the saturated region

Increasing source-drain voltage to such a situation that V_D is almost equal to $V_G - V_T$, in the channel, at the point near drain electrode, there is now no longer an effective V_G to maintain the charge, the electric field provided by the gate electrode at the drain electrode completely cancels the electric field provided by the drain electrode. Therefore, the voltage drops as well as the number of charges at the drain electrode is zero, the channel is pinched off, and there is a pinch-off point near the drain electrode of the channel (Figure 6.5(b)). Further increases in V_D will widen the depletion region next to the drain and move the pinch-off point towards the source electrode (Figure 6.5(c)), resulting in a shorter accumulated channel length, less resistance, and a small increase in I_D. At this time, as long as the channel length is much longer than that of the pinch-off area, the increase of drain voltage will not further increase the output current. The channel current shows saturation feature, that is, at a fixed gate voltage, the device's leakage current remains essentially unchanged. The drain voltage that is required to begin forming the pinch-off region is called pinch-off voltage[19]. The pinch-off voltage at different gate voltages is different.

19. 此时,只要沟道长度比夹断区域的长度大得多,漏极电压的增加就不会进一步增加输出电流。沟道电流显示出饱和特性,即在固定栅极电压下,器件的漏电流基本保持不变。开始形成夹断区域所需的漏极电压,称为夹断电压。

6.3.4 Generation of Output Currents

According to the polarity of the majority carrier in the device, the OFET can be divided into the p-channel that conducts holes and the n-channel that conducts electrons. In very limited cases, if the organic semiconductor is ambipolar, the device may conduct both electrons and holes[20].

In Figure 6.6, the work diagrams of the p-channel and n-channel OFETs in accumulation mode are shown. In the n-channel device (Figure 6.6(a)), when the gate is loaded with a large enough forward voltage, electrons will be injected from source into the LUMO of the organic semiconductor. If a forward voltage is applied to the drain electrode, electron current from source to drain is formed. In the p-channel device (Figure 6.6(b)), when the gate is applied a sufficient negative voltage, holes will be injected from source into the HOMO of the organic semiconductor. If

20. 根据器件中多数载流子的极性,OFET 可分为传导空穴的 p 沟道和传导电子的 n 沟道。在非常有限的情况下,如果有机半导体是双极性的,OFET 器件可能同时传导电子和空穴。

a negative potential is applied to the drain electrode, a hole current from source to drain will be produced.

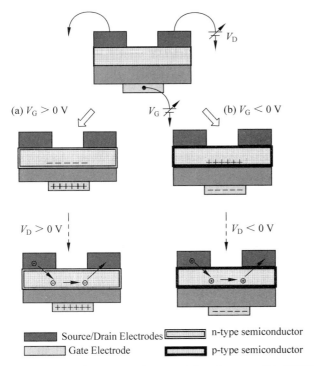

Figure 6.6　Work diagram of n-type (a) and p-type (b) OFETs

6.4　Performance Parameters for OFETs

FET devices have been widely used in various optoelectronic applications. The diverse applications require the different parameters. For a battery powered device, the threshold voltage is required to be a few volts because the power consumption is key factor[21]. For some circuits, such as active liquid crystal displays, to better distinguish between transistors On and Off states for good visual effect, fast switch speed is required, i.e., the on/off ratio of current is higher than 10^6 and the field effect mobility is greater than $0.1 \text{ cm}^2 \cdot \text{V}^{-1} \cdot \text{s}^{-1}$; for other applications, the mobility is usually required not less than $0.01 \text{ cm}^2 \cdot \text{V}^{-1} \cdot \text{s}^{-1}$, and the switching ratio greater than 10^3. The following content will describe various parameters for OFET characterization.

21. 对于电池供电设备，因为功耗是关键因素，阈值电压要求为几伏。

6.4.1 Transfer and Output Curves

Conventionally, the conductivity of OFETs is controlled by a vertical electric field from the gate terminal acting upon a horizontal conduction channel that has been formed in the organic semiconductor between the source and drain electrodes. When the device is working, the constant voltage loaded between the source and the drain is called the source-drain voltage (or drain voltage, V_D). The corresponding current is called source-drain current (or leakage current, I_D). It is also called channel current, because it is confined in the conductive channel between the source and drain, and it is determined by the density and mobility of majority carriers in the channel. The variable DC voltage applied to the gate is called gate voltage (V_G).

To describe the current-voltage characteristics in OFET devices, basically, there are transfer and output curves (Figure 6.7). The former tells the transfer characteristics that show the variation of the leakage current (I_D) with the gate-source voltage (V_G), taking V_D as the fixed parameter; while the latter gives the main static characteristics of the OFET that show the variation of the leakage current (I_D) with the source-drain voltage (V_D), taking V_G as the fixed parameter[22]. From these characteristic curves, we can deduce various parameters that define the performance of OFETs, including the on-off ratio of current, threshold voltage, field effect mobility, subthreshold drift (swing), and contact resistance, etc. The on-off ratio of current is defined as the ratio of the device output current in both on and off states; the threshold voltage is the gate voltage at which the channel current begins to appear; the mobility defines as the average migration speed of charge carriers under the unit electric field; subthreshold swing (drift) describes the switching speed of the device transitions from the off state to the on state; and the contact resistance refers to the interfacial resistance between the semiconductor and the source-drain electrode.

22. 基本上,描述 OFET 器件的电流-电压特性的,有转移曲线和输出曲线(图 6.7)。前者以 V_D 为固定参数,表示漏电流(I_D)随栅极电压(V_G)变化的转移特性;后者给出了 OFET 的主要静态特性,以 V_G 为固定参数,表示漏电流(I_D)随漏电压(V_D)的变化。

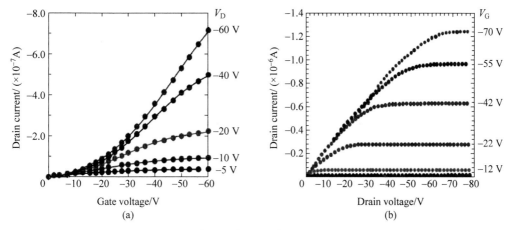

Figure 6.7　Under a certain drain voltage, the transfer characteristics of the OFET (a) ; under a certain gate voltage, the output characteristics of the OFET (b)

6.4.2　Linear/Saturation Region and Pinch-off Voltage

In the output curves of an OFET, there are three features: linear region, saturation region and pinch-off voltage, as shown in Figure 6.8. There are three output curves in the figure, and as can be seen, with the increase of the gate voltage, the output curve shifts upward, i.e., the current increases. In each curve, there are two regions, namely, the linear and the saturated regions. In the linear region, the drain voltage is smaller than the gate voltage, and the current is proportional to the drain voltage. The $I_D - V_D$ curve is linear[23]. In the saturated region, the drain voltage exceeds the gate voltage, the current shows saturation, i.e., no more increase with the drain voltage. The current in the saturation region is usually referred as the output current of an OFET device at a fixed gate voltage; and it increases with the gate voltage. The pinch-off voltage refers to the turning point between the linear region and the saturated region in the output curve at a fixed gate voltage, which generally increases with the increase of the gate voltage[24].

23. 在线性区域,漏电压小于栅电压,电流与漏电压成正比,I_D-V_D 曲线是线性的。

24. 夹断电压是指在固定栅极电压下,输出曲线中线性区域与饱和区域之间的转折点,一般随着栅极电压的增加而增大。

6.4.3　Threshold Voltage and Subthreshold Swing

Based on the transfer curve, the threshold voltage and subthreshold drift of the device can be obtained. As mentioned

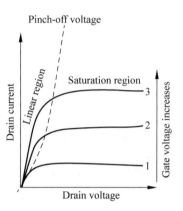

Figure 6.8   Linear and saturation regions, and pinch-off
point in output curves of an OFET

25. 如上所述,当漏极电压较小时,器件的电流特性处于线性区域;当漏极电压较大,超过夹断电压时,器件的电流特性将进入饱和区。

26. 与曲线中斜率最大的点相对应的电压表示为 V_T'(图6.9的曲线1)。阈值电压可通过 $V_T = V_T' - 1/2V_D$ 计算得出。得到阈值电压的另一种方法,是在饱和区域使用 $(I_D)^{1/2}$-V_G 曲线(图6.9的曲线2)。把 $(I_D)^{1/2}$-V_G 曲线中的直线部分外推到 V_G 轴,直线与 V_G 轴交点处的电压为阈值电压(图6.9中的 V_T)。

above, when the drain voltage is small, the current characteristics of the device are in a linear region; when the drain voltage is large, exceeding the pinch-off voltage, the current characteristics of the device will be in the saturation region[25]. In order to facilitate the analysis of device transfer characteristics, the transfer curves of saturated and linear regions are put in the same figure, as shown in Figure 6.9. The leakage current in the linear region (smaller V_D) varying with the gate voltage ($I_D - V_G$) is drawn as curve 1; the square root of the leakage current in the saturated region (larger V_D) changing with the gate voltage $[(I_D)^{1/2} - V_G]$ is drawn as curve 2. It can be seen from curve 1 that when the gate voltage is small, the leakage current of the device is small and can be neglected basically. As the gate voltage increases, the leakage current increases rapidly. The voltage corresponding to the point in the curve with the largest slope is denoted as V_T' (curve 1 in Figure 6.9). The threshold voltage can be calculated from equation $V_T = V_T' - 1/2V_D$. Another way to extract the threshold voltage is to use the $(I_D)^{1/2} - V_G$ curve in the saturated region (curve 2 in Figure 6.9). The portion of straight line in the $(I_D)^{1/2} - V_G$ curve is extrapolated to the V_G axis and the voltage value at the intersection point is the threshold voltage (V_T in Figure 6.9)[26]. The device threshold voltage obtained by different methods is usually inconsistent, with larger value obtained from the transfer curve of the linear region. It is noteworthy that the method for extracting the threshold voltage is not unique indeed, even if using the same region, because the approximation of the line location is different, the value is not the same. It is proved that the threshold voltage from the saturation region is more accurate.

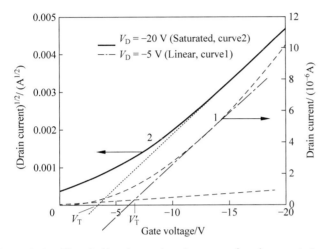

Figure 6.9   Threshold voltages based on transfer characteristics of
OFETs at both linear and saturation regions

An OFET generally operates in accumulation mode, with no intrinsic carriers and no depletion layer to isolate the conduction channel from the bulk. From a theory point of view, the threshold voltage of OFET can be thought as the gate voltage when a depleted channel becomes one in charge accumulation[27]. Usually, the energy barrier between the Fermi level of the source and drain electrodes and the HOMO/LUMO level of the organic semiconductor is exist, resulting in the movement of charges at the interface, creating an electric dipole layer and band bending; preventing charge injection into organic layer from the electrode. In order to generate mobile carriers in the organic semiconductor, a gate voltage needs to be applied to overcome the potential barrier between the electrode and the organic semiconductor to eliminate the dipole layer and the band bending[28]. In addition, if a large number of carrier traps exist in the organic semiconductor, the traps must be filled first before movable carriers can be generated. Considering various factors mentioned above, a certain gate

27. OFET 通常是在聚集模式下工作的,没有本征载流子,也没有可将传导沟道与半导体本体隔离开来的耗尽层。从理论的角度来看,OFET 的阈值电压可以认为是当耗尽沟道变成聚集沟道时的栅极电压。

28. 为了在有机半导体中产生可移动的载流子,需要施加栅极电压来克服电极和有机半导体之间的势垒,以消除偶极层和能带弯曲。

voltage must be added into the device channel to result in additional free carries. This gate voltage that turns on the device is the threshold voltage. Below the threshold voltage, the leakage voltage will not produce leakage current; above the threshold voltage, the output of channel current occurs.

With the $I_D - V_G$ curve, we can also determine another device parameter within a range of less than the threshold voltage: the subthreshold swing (S), which is defined as follows:

$$S = \frac{dV_G}{d\left[\lg(I_D)\right]}, \quad V_G < V_T \qquad (6\text{-}1)$$

where, V_G is the gate voltage, I_D is the leakage current. The S can be deduced from the $\lg(I_D) - V_G$ curve in the linear region of the device below the threshold voltage, as shown in Figure 6.10. To determine the S, firstly, find the threshold voltage point on the logarithmic curve $\left[\lg(I_D) - V_G\right]$, and then through this point, make a straight line approaching the $\lg(I_D) - V_G$ curve from the low gate voltage. The reciprocal of the slope of resulting straight line is subthreshold drift S. S is a measure of the speed of the device transitions from off to on and the unit is $V \cdot decade^{-1}$, or $mV \cdot decade^{-1}$; the slope is usually the maximum one of $\lg(I_D) - V_G$ curve within the range of the below threshold voltage. This parameter generally reflects the carrier leakage of gate electrode or relates to the bipolar of material, the second area of charge accumulation and the higher concentration of shallow traps and so on[29]. The smaller the subthreshold drift value (e. g., 60 mV $\cdot$ decade^{-1}), the steeper the $\lg(I_D) - V_G$ curve, and the better the device performance. Larger subthreshold drift usually means that the above problems occur in the device, the turn-on condition of device will drift. Multiplying S by the capacitance parameter (C_{diel}) of the device dielectric layer, a normalized subthreshold drift can be obtained, namely: $S_n = C_{diel} \times S$, in units of V $\cdot$ nF $\cdot$ cm^{-2} $\cdot$ decade^{-1}, which can be used to compare the performance of device with different dielectric materials.

29. 该参数一般反映栅极的载流子泄漏,或者与材料双极性、第二个电荷聚集区域以及较高浓度的浅陷阱等有关。

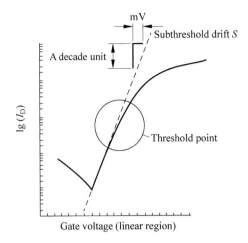

Figure 6.10   Method for extracting the subthreshold drift S

6.4.4 Current On-off Ratio

Combining with the output and transfer characteristics of an OFET device, the current switching ratio can be obtained. The current switch ratio is defined as the ratio of the device's on-state current to the off-state current, usually expressed in 10^x form. The off-state current can be obtained from the current value corresponding to the threshold voltage in the output curve of the device under a given gate voltage, and the on-state current value corresponds to the saturation current in the device output curve under the same given gate voltage[30]. Obviously, high switching ratio requires high on-state current and small off-state current, which are related to the mobility and doping state of the semiconductor, respectively. Generally, high mobility semiconductor ensures a device with a faster switching response and a higher on-off ratio of current. In addition, the on-off ratio of current has a strong dependence on the gate voltage, thus the on-off ratio of the two devices is usually compared at the same gate voltage[31].

6.4.5 Hysteresis

If the gate voltage is swept in both directions, a cyclic $I_D - V_G$ curve is obtained that can be used to study the hysteresis of the device. Hysteresis refers to the phenomenon that the backward leakage current is different with the forward leakage current as shown in Figure 6.11[32]. There are

30. 器件的关状态电流可以从给定栅极电压下器件输出曲线中阈值电压对应的电流获得,导通状态电流对应于同一给定栅极电压下器件输出曲线中的饱和电流。

31. 通常,高迁移率半导体可确保器件具有更快的开关响应和更高的电流开关比。此外,电流的开关比对栅极电压有很强的依赖性,因此两个器件的开关比通常在相同的栅极电压下进行比较。

32. 迟滞指的是在一定电压下,OFET 反向漏电流与正向漏电流不同,如图 6.11 所示。

a few precautions that can be taken to monitor the effect of hysteresis: ① Questionable *I-V* characteristics should be measured in both directions（double sweeps）at the same sampling speed; this serves as a qualitative measure of bias-history hysteresis，② Hysteresis can be measured using sweeps which are at the frequency of interest. DC hysteresis may be immaterial for AC circuits，and ③ Device measurement can be performed after the device is held at a fixed bias for some time; this will expose the direction and nature of the hysteresis.

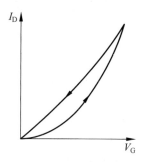

Figure 6.11　A schematic transfer characteristic exhibiting hysteresis

33. 迟滞在驱动特性方面是一个问题。由于它们的本质,迟滞响应对(模拟)模型构成了挑战,因为晶体管保留了驱动历史,且响应依赖于那个历史。

Hysteresis in drive characteristics is problematic. Hysteretic responses are，by their nature，challenging to model since the transistor retains a history of drive and the response depends on that history[33]. Hysteretic device characteristics also combine the transport of carriers，which are induced by the applied bias，and carriers which are released by traps or dielectric relaxation，which can lead to errors in parameter extraction. While it is generally preferable to have devices with as little hysteresis as possible（with the exception of devices designed to store state）.

Several studies have identified three major contributions to short term hysteretic effects: ①Slow relaxation of the gate dielectric is usually strongest in polymer-based gate dielectrics which have polar groups or absorb water. The relaxation of the gate dielectric invariably increases the dipole in the gate dielectric as time passes. This causes a decrease in current after a bias change which increases the current flow; ②Dielectric charge storage is caused by injection of charges into the gate dielectric. The driving force for the relaxation increases the field drop across the gate dielectric and decreases the field which reaches the semiconductor. This generally leads to a decrease in current after a bias change which increases the current flow[34]; and ③ Semiconductor traps have been directly observed in transistors and organic semiconductor crystals by a variety of techniques including temperature dependent current analysis, photoluminescence, and transient current spectroscopy, among others. These traps can increase or decrease the observed current depending on their charge; if they trap majority carriers the current will decrease over time, but if they emit majority carriers (and therefore hold a charge which reinforces accumulation) current can increase over time. These traps are often extrinsic (water and oxygen often form these centers) and their formation is sometimes reversible. Understanding their nature can lead to insight on how to process and encapsulate devices to prevent their formation.

In addition to reversible hysteretic effects, irreversible storage and bias-stress can also be observed in many OFET systems. Repeated characterization through the application of the stress will expose these effects.

6.4.6 Mobility of OFETs

1. Source-drain currents in an OFET

The gate- and drain- voltage dependent current of an OFET device can be expressed by equations, which can be used to curve-fitting of the experiment data for mobility calculation. Figure 6.12 shows the structure of the FET and its geometric dimensions is also marked on the graph. The X and Y directions in the figure represent the direction of the channel current and the direction of the vertical electric field respectively. The thickness of the dielectric layer, the channel length, and the channel width are represented by d_i, L, and W, respectively. The conducting channel in the organic semiconductor of an OFET is only a few nanometers thick and locates adjacent to the dielectric, hence the current will be affected by the interface scattering. Besides the dependence of current on the properties of device components

34. ①基于聚合物的栅电介质，它含有极性基团或者吸水，栅介电的缓慢松弛通常最强。随着时间推移，栅介电的松弛不可避免地增加栅电介质内的偶极。这在使电流流动增加的电压变化之后，引起电流降低。②栅介电材料中电荷的注入，引起介电层电荷存储。释放介电存储电荷的驱动力，增加了通过栅电介质的电场，减少了到达半导体的电场。这通常在一个提高电流流动的偏压变化之后，导致电流降低。

35. 除了对器件组分性质和界面状况的依赖,电流还取决于栅极和漏极电压。

and the interface situation, the current also depends on both gate and drain voltages[35]. Since an OFET device operates in distinct regimes or regions, the carriers have different concentration/distribution (Section 6. 3. 3), leading to different device current.

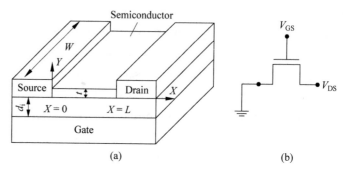

Figure 6. 12　Structure of a FET device with geometrical size (a) and symbol for FET devices (b)

1) Off State

At this stage, gate voltage is far small than the threshold voltage, there is no accumulated conducting channel. The source-drain current, I_D, is very small and is determined by the intrinsic conductivity of the bulk semiconductor:

$$I_D(\text{cutoff}) = I_{\text{leak}}, \quad V_G \ll V_T \qquad (6\text{-}2)$$

where I_{leak} is a small intrinsic leakage current in the bulk semiconductor.

2) Subthreshold Region

36. 在亚阈值区域,栅电压仍不大于阈值电压,源-漏电流随栅电压指数增加,类似于正向偏压下的二极管电流。

In the subthreshold region, the gate voltage is still not greater than the threshold voltage, and the source-drain current increases exponentially with gate voltage, similarly to the current due to a forward-biased diode[36]. This exponential behavior is often attributed to deep trap states:

$$I_D(\text{subthreshold}) = I_0 \exp\left(\frac{V_G}{SV_{\text{th}}}\right), \quad V_G \leqslant V_T \quad (6\text{-}3)$$

where $V_{\text{th}} = k_B T/q$ is the thermal voltage (~26 mV at room temperature), S is the subthreshold factor, and I_0 is a process-dependent parameter that also has some dependence on V_T. In organic devices, owing to the large values of V_T, subthreshold behavior is important and accounts for a significant amount of current. Performance of a device in the subthreshold region is measured by the subthreshold parameter S, which is defined as the change in V_G needed to affect a decade increase in I_D current (Section 6.4.3).

3) Linear Region

In the linear region, the voltages have relationships of $V_G > V_T$ and $V_D < V_G - V_T$. Typically, a conducting channel at the semiconductor-dielectric interface is only formed when V_G is greater than V_T. An equal amount of charge, but opposite in sign, appears on either side of the dielectric. The V_T is the result of either shallow traps in the semiconductor, which need to be filled, or charged dipoles already in the channel, which need additional gate voltage to form the channel[37]. The current at this region can be expressed as:

$$I_D(\text{linear}) = \frac{W}{L}\mu C_{\text{diel}}\left[(V_G - V_T)V_D - \frac{1}{2}V_D^2\right], \quad V_D < V_G - V_T$$

$$(6\text{-}4)$$

where μ is the surface mobility of the channel, C_{diel} is the capacitance per unit area of the gate dielectric.

4) Saturation Region

In the saturation region, where $V_G \geq V_T$ and $V_D \geq V_G - V_T$, I_D is now substantially independent of V_D and is mainly controlled by V_G. The current can be derived as

$$I_D(\text{sat}) = \frac{1}{2}\frac{W}{L}\mu C_{\text{diel}}(V_G - V_T)^2(1 + \lambda V_D), \quad V_D < V_G - V_T$$

$$(6\text{-}5)$$

where λ is the channel length modulation parameter.

For convenience, in Equations (6-4) and (6-5), a transistor gain factor, β, may be defined:

$$\beta = (\mu C_{\text{diel}})\frac{W}{L} = \left(\mu\frac{\varepsilon_r\varepsilon_0}{d_i}\right)\frac{W}{L} \qquad (6\text{-}6)$$

Gain factor β can be further separated into two other factors, a geometry gain factor (W/L) and a process gain factor ($\kappa' = \mu C_{\text{diel}}$) or $\left(\kappa' = \mu\frac{\varepsilon_r\varepsilon_0}{d_i}\right)$, where ε_r is the relative permittivity of the gate dielectric and d_i is the thickness of the gate dielectric.

2. Device Mobility

Field effect mobility μ (mobility) is the term to describe

37. V_T 是由半导体中需要填充的浅陷阱或沟道中已有的电偶极子导致的结果,这些陷阱和电偶极子使得沟道的形成需要额外的栅极电压。

the degree of carrier movement inside the active layer under electric field in FETs. It is the most important parameter to evaluate the different active materials and preparation methods of FETs. Generally, the field-effect mobility of single-crystal silicon (c-Si), polycrystalline silicon (p-Si) and amorphous silicon (a-Si) FETs are in the ranges of $10^2 \sim 10^3$ cm^2 · V^{-1} · s^{-1}, $10^1 \sim 10^2$ cm^2 · V^{-1} · s^{-1}, and $\sim 10^0$ cm^2 · V^{-1} · s^{-1} respectively; the mobility of the preferred OFET is in the range of 1 to 10 cm^2 · V^{-1} · s^{-1}, closing to the level of amorphous silicon. One of the exceptions of OFET mobility is from single crystal rubrene based device, achieving value up to 20 cm^2 · V^{-1} · s^{-1}.

Mobility in OFET devices is affected by various factors, such as the electrical property of the organic semiconductor, the roughness of the organic semiconductor-dielectric interface, and the dipoles in the dielectric[38]. For organic semiconductors, there is a gate-voltage-dependent mobility, usually modeled as

$$\mu = \mu_0 \left(\frac{V_G - V_T}{V_{aa}} \right)^\gamma \qquad (6\text{-}7)$$

where γ and V_{aa} are empirical fitting parameters, and μ_0 is normally taken to be the mobility in crystals (the band mobility). Parameter γ is related to the conduction mechanism and can describe both an increase and a decrease in mobility with V_G. With an increase in mobility, $\gamma > 0$, which is typical of amorphous and nanocrystalline devices and is related to a trap conduction mechanism. A decrease in mobility with gate voltage, $\gamma < 0$, appears in polycrystalline OFET when surface scattering starts to be important.

Generally, the mobility of an FET device can be obtained by curve-fitting based on either the transfer curve or the output curve. There are three different types of mobility in OFETs, which can be calculated to a first order as follows:

1) Effective Mobility

It refers to the mobility value that is calculated from the drain conductance (g_d is the slope of the output curve) in the linear region of the output curve (Equation (6-4) with V_G as fixed parameter):

38. OFET 器件的迁移率受很多因素影响,例如有机半导体的电学性质、有机半导体—电介质界面粗糙度以及电介质中偶极子等。

$$g_d = \left. \frac{\partial I_D}{\partial V_D} \right|_{V_{G=const}} = \frac{W}{L} C_{diel} \mu_{eff} (V_G - V_T) \qquad (6\text{-}8)$$

$$\mu_{eff} = \frac{L}{W} \cdot \frac{g_d}{C_{diel}(V_G - V_T)} \qquad (6\text{-}9)$$

2) Field Effect Mobility

It refers to the mobility of that is calculated from the transistor transconductance (g_m is the slope of the transfer curve) in the linear region (Equation (6-4) with V_D as fixed parameter):

$$g_m = \left. \frac{\partial I_D}{\partial V_G} \right|_{V_{D=const}} = \frac{W}{L} C_{diel} \mu_{FET} V_D \qquad (6\text{-}10)$$

$$\mu_{FET} = \frac{L}{W} \frac{g_m}{C_{diel} V_D} \qquad (6\text{-}11)$$

Any discrepancy between μ_{FET} and μ_{eff} is due to the neglect of any gate-voltage-dependent mobility effect[39].

3) Saturation Mobility

It is the mobility derived from plotting $\sqrt{I_{D(sat)}}$ against ($V_G - V_T$) from the output curve at saturation region (Equation (6-5) with V_D as fixed parameter):

$$m = \left. \frac{\partial \sqrt{I_D}}{\partial (V_G - V_T)} \right|_{V_{D=const}} = \sqrt{\frac{W}{2L} C_{diel} \mu_{sat}} \qquad (6\text{-}12)$$

$$\mu_{sat} = \frac{2L}{W} \frac{m^2}{C_{diel}} \qquad (6\text{-}13)$$

As with μ_{FET}, that with μ_{sat} also neglects any gate-dependent effects. Also, by definition, in the pinch-off region there are no charge carriers at the pinched-off drain electrode, but there are plenty of charge carriers at the source electrode. As the density of charge carriers varies considerably along the saturated channel, μ_{sat} can therefore only measure the mean mobility along the channel. For this reason, it is often better to extract mobility when the transistor is in the linear region, where the charge carrier distribution is more uniform. The calculated mobility in the saturated region is generally larger than the mobility in the linear region[40].

6.5　Applications of OFETs

6.5.1　Requirements for OFET Applications

OFET devices have a number of potential applications

39. μ_{FET} 和 μ_{eff} 之间的任何差异都是由于忽略了栅极电压依赖的迁移率效应。

40. 此外,根据定义,在夹断区域,夹断的漏极上没有电荷载流子,但在源极处有大量的电荷载流子。由于此时电荷载流子的密度沿饱和沟道变化很大,因此 μ_{sat} 只能测量沿着饱和沟道的平均迁移率。由于这个原因,通常地,最好在晶体管处于线性区域时测量迁移率,此时沟道中的电荷载流子分布比较均匀。计算出的饱和区域的迁移率通常大于线性区域的迁移率。

which can take the advantage of their properties. Because organic semiconductors are fully satisfied van der Waals solids which do not require any epitaxial templating and high substrate temperature process，most OFET processes have a low thermal budget，simple manufacturing processes (including partially or totally solution printing-based process flows)，and are compatible with a range of substrates. It is these features that drive interest in OFETs，and the technology's longevity will be determined by its ability to address the challenges of applications of interest.

41. 期望的 OFET 质量/特性包括：高的场效应迁移率、高的电流开关比、低的泄放电流，小的阈值电压偏移和陡的亚阈值斜率。具体的应用对某些 OFET 参数的要求会更为严格。

Desirable qualities/characteristics of an OFET include：a high field-effect mobility, high on-off current ratio, low leakage current, minimal shift of threshold voltage, and sharp subthreshold slope. Depending on the specific application, the demands imposed on certain OFET parameters are more stringent[41]. For active-matrix liquid crystal displays (AMLCDs)，the OFET behaves as a switch，and thus，minimal leakage is the most critical requirement. On the other hand，low current drive and high on-off current ratio are of great importance in active-matrix (AM) organic light-emitting diode (OLED) displays，since the OFET must provide sufficient current output to drive the OLED for light emission. For radio frequency identification (RFID) tags，speed is the most critical parameter. In most cases, the threshold voltage (V_T) should be as low as possible for large dynamic range (in displays) and for low voltage operation. Good device stability is demanded in all applications. These application-specific requirements can be addressed via choice of material, device structure, and/or the associated fabrication method[42].

42. 在大多数情况下，阈值电压（V_T）应尽可能低，以实现大的动态范围（在显示应用中）和低电压工作。所有应用都需要良好的器件稳定性。这些应用的特定要求可以通过选择材料、器件结构和/或相关的制造方法来解决。

6.5.2　Display Applications

Direct view flat panel displays are currently the leading application for large area electronic devices. One of the most promising applications of OFETs is their use as an on-pixel switching element in active-matrix displays，similar to the functions that are currently fulfilled by a-Si：H TFTs in AMLCD and AMOLED display applications. Active-matrix backplane electronics based on OFETs are attractive because of the large-area capability and low-cost advantage of organic technology, as well as the opportunity to realize a new generation of flexible, lightweight displays, and e-paper[43]. The incorporation of mechanically flexible active matrix displays using OFETs on plastic substrates has recently been demonstrated including a mechanically flexible polymer dispersed liquid crystal display (LCD) and a display containing electrophoretic material (e-Ink). Development of OFET-driven AMOLED displays is also in progress. OFET-OLED integration is appealing because it suggests the potential of manufacturing inexpensive and flexible display modules with completely functional organic materials.

Table 6.1 summarizes some of the display prototypes using OFETs; most of the existing prototypes were built using pentacene OFETs. The e-paper and liquid crystal display (LCD) were made with an OFET matrix array and the AMOLED with dot patterns. Since the OFET serves as an active element, the AMOLED is very sensitive to non-uniformity in the OFET performance across the array, which can lead to non-uniformity in the display's brightness. The non-uniformity often arises from the grain size distribution of polycrystalline organic semiconductors. On the other hand, OFETs are more accommodating for LCD or e-paper because they act mainly as switches and the key performance requirement is a high on-off current ratio[44].

43. 基于 OFET 的有源矩阵基板电路具有吸引力,因为它具有有机电子技术的大面积和低成本的优势,以及实现新一代柔性、轻型显示器和电子纸的机会。

44. 不均匀性通常源于多晶有机半导体的晶粒尺寸分布。另外,OFET 更适合 LCD 或电子纸,因为它们主要充当开关,关键的性能要求是高的开关电流比。

Table 6.1　Display prototypes using OFETs

Application	Semiconductor	Specification	Organization
E-paper	Polyfluorene-based polymer (inkjet printing)	60 pixel×80 pixel on PET	Plastic Logic (UK) and E-ink (USA)
E-paper	Pentacene(solution-process)	QVGA on PEN	Philips (Netherlands)
LCD	Pentacene	1.4 inch 80×80 RGB on glass	Hitachi (Japan)
LCD	Pentacene	12 inch full color XGA on glass	Samsung Elec. (Korea)
OLED	Pentacene	8 pixel×8 pixel on glass	Pioneer (Japan)
OLED	Pentacene	4 pixel×4 pixel on PC	NHK (Japan)

6.5.3 Sensor Applications

Among the exciting developments of organic optoelectronics during the last three decades, organic sensor applications in the fields of medical diagnosis, industrial safety, household security, food safety, and environmental observations have drawn tremendous attention[45]. Researchers foresee the tremendous capability of organic devices, especially, OFETs made up of organic semiconducting materials with virtually endless combinations and compounds to embed system-on-chips using low cost, low temperature, and efficient process techniques.

OFET sensors were proposed for the first time in the late 1980s, just a few years after the first OFET device was proposed. These preliminary studies reported OFET sensor based on the changes of source-drain current upon exposure to different volatile/odor molecules. Although the idea of using a bottom gate organic thin-film transistor as a sensing device was put forward by these early works, it was not clarified why a three-terminal device structure should have been beneficial to the sensor performance[46]. As time goes by, OFET-based sensors, specifically with an upside-down structure, have drawn immense research interest and emerged as a most promising candidate for sensing applications. Generally, OFET sensors have device structures and detection mechanisms different from those of commercial inorganic FET sensors. In the case of OFET sensors, the analyte detection is performed employing a bottom gate device structure where the active layer is directly exposed to the analyte and acts as both transistor channel material and sensing membrane[47].

OFET sensors are capable of producing amplification of sensed quantity; hence they are suitable for direct sensing application as the active organic material directly reacts with the analytes depending on adsorption, diffusion, or absorption properties of the organic semiconductor[48]. Furthermore, the reactive properties of organic semiconductors can be modified with the inception of receptor molecules as pre-requisition during the synthesis

45. 在过去 30 年中有机光电子令人激动的发展中,有机传感器在医疗诊断、工业安全、家庭安全、食品安全和环境监测等领域的应用引起了广泛的关注。

46. 这些初步研究报道了 OFET 传感器暴露于不同挥发性/气味的分子时源漏电流的变化,尽管这些早期工作提出了使用底栅有机薄膜晶体管作为传感器件的想法,但没有明确说明为什么三端器件结构有利于传感器性能。

47. 在 OFET 传感器的情况下,待检测物的检测是采用底栅结构进行的,其中有机活性层直接暴露于分析物中,同时充当晶体管沟道材料和传感膜。

48. 基于 OFET 的传感器,能够产生感知数量的放大。因此,它们适用于直接传感应用,因为活性有机材料可根据有机半导体的吸附、扩散或吸收等特性直接与分析物发生反应。

process to enhance the selectivity and sensitivity of the sensor, which would be rather difficult to achieve using their inorganic counterpart. <u>The OFET sensor holds all the essential properties as a sensing element such as selectivity, sensitivity, disposability, and reproducibility at very low cost and low operating temperature. The foremost advantage of the OFET sensor is the multipara-metric sensing operation, wherein various electrical parameters such as the bulk conductivity, two-dimensional field-induced conductivity, threshold voltage, and field-effect mobility of the OFET device, are employed to characterize the sensing output</u>[49].

1. Mechanical Sensors

<u>OFETs, because of their mechanical flexibility and large size, are well suited to switch and amplify mechanical actuations</u>[50]. The first large scale application of OFETs in a mechanical sensor was demonstrated by T Someya, et al. A flexible OFET backplane was laminated together with a flexible conductor loaded elastomer whose resistance changed in response to an applied pressure. The schematic application circuit diagram is shown in Figure 6. 13(a). By switching through the transistor matrix and observing the current flow from a common power supply through the variable resistor, it is possible to create a force map for use as a flexible sensor skin.

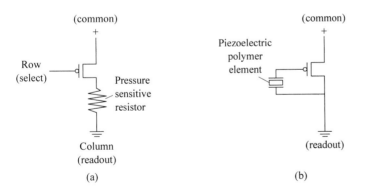

(a)　　　　(b)

Figure 6. 13　(a) A pressure sensitive circuit which uses OFETs to switch an otherwise linear pressure sensitive resistor sheet. (b) Diagram of a piezoelectric sheet film (typically PVDF or its derivatives) locally amplified using an OFET

49. OFET 传感器在非常低的成本和低的工作温度下,具有作为传感元件的所有基本特性,例如选择性、灵敏度、弃置性和再现性。OFET 传感器的首要优点是多参数传感工作,它可采用各种电学参数,如体电导率、二维场效应电导率、阈值电压和 OFET 器件的场效应迁移率来表征传感输出。

50. OFET 由于其机械柔韧性和大尺寸,非常适合对机械驱动进行转换和放大。

In addition to producing the sensing array using OFETs, this team also fabricated much of the peripheral drive circuitry using OFET-based logic elements and introduced a creative architecture in which the unit dimensions can be customized by cutting several elements with scissors and attaching them with a pressure sensitive adhesive.

Another application of OFETs to measuring mechanical stimuli is the buffering of charge signals from large piezoelectric polymer sheets, such as PVDF. Extracting spatially localized information about the stimuli applied to the sheet can be challenging because the charge signal is dissipated across the parasitic capacitances between the stimulus and the location of the sensing circuitry. OFETs can serve as local transimpedance amplifiers which convert the charge signal into a current signal that can overcome this capacitance. This amplification can be achieved across heterogeneous substrates or by building the OFET directly onto the piezoelectric polymer sheet[51]. Figure 6.13(b) shows the schematic circuit diagram of this architecture.

2. Chemical Sensors

The typical OFET chemical sensor structure is shown in Figure 6.14. It consists of a conductive substrate covered by a thin gate dielectric film that is interfaced to the organic active layer, which is generally a film of a few tens of nanometers of conducting polymers or oligomers, such as region-regular polythiophenes or pentacene molecules. It is deposited by solution processing (such as solution casting, spin coating, and Langmuir-Shäfer or Langmuir-Blodgett techniques) or thermal evaporation. Organic active layers are generally polycrystalline, with a granular morphology characterized by grains having linear dimensions of, at most, hundreds of nanometers and a crystalline-type degree of structural order. Polycrystalline organic active layers are generally described as contiguous grains having a crystalline core and amorphous grain boundaries[52]. Gold source and drain contacts are deposited by thermal evaporation through a shadow mask directly over the organic active layer, while a contact on the conducting substrate is used to apply the gate bias.

51. 另一个关于 OFET 在测量机械力方面的应用是，缓冲来自大片压电聚合物材料（如 PVDF）受压后产生的电荷。获得有关施加到压电材料片上力的空间局部信息可能具有挑战性，因为电荷信号可在作用力点和传感电路之间的寄生电容器上消散。OFET 可用作本地的跨阻放大器，将电荷信号转换为可以克服寄生电容的电流信号。这种放大可以在非均质基板上，或者通过将 OFET 直接制备在压电聚合物片上来实现。

52. 多晶有机活性层通常被描述为是由结晶核和无定形晶界组成的毗连晶粒。

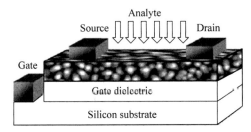

Figure 6. 14　Structure of a sensing OFET

Generally, there are two detection modes in OFET chemical sensors - no gate bias and applied gate bias. At no gate bias and applied a fixed V_D bias, the OFET measures I_D variation caused by the interaction and/or permeation of the analyte in the active layer. In this regime, a permeation of the analyte down to the gate dielectric will result in a three dimensional conductivity variation. The response acquired in this case is exactly that of an equivalent chemiresistor. Upon application of a gate bias, on the other hand, a much higher two-dimensional I_D, confined near the gate dielectric/organic interface, is induced. In this regime the "on" I_D change can be recorded and different device parameters, such as V_T and μ_{FET}, can also be simultaneously influenced during the exposure to the analyte. Since the two conduction regimes are completely independent, two orthogonal OFET I_D responses can be recorded and the simultaneous variation of different field effect parameters can be monitored in the "on" regime.

The conduction mechanisms of organic semiconductor in OFETs play a crucial role in determining the sensing mechanism of an OFET sensor[53]. V_T and μ_{FET} of OFETs depend on the volume density of trapped charges and on the potential barrier between contiguous grains, respectively. Several reactive species cause charge trapping-de-trapping processes to occur, enhancing or lowering barriers between grains hence influencing carrier transport from one grain to another. Therefore, V_T and μ_{FET} can be greatly influenced by the interaction of the transistor active layer with a chemical species, which results in a change of the device on-current and the two-dimensional conductivity.

Many active layers such as substituted thiophene-based

53. OFET 中有机半导体的传导机制在决定 OFET 传感器的传感机制中起着至关重要的作用。

54. 在所有观察到的情况下,活性层与分析物的相互作用已被建立模型,描述为分析物分子在活性层晶粒表面被吸附甚至捕获。

55. 此外,晶界起着至关重要的作用,因为当晶粒尺寸减小或沟道长度 L 增加时,OFET 传感器响应会增加。这意味着与分析物相互作用的强度,随着单位体积内暴露于分析物分子的活性层晶界的增加而增加。

56. 如文献所展示的,原则上可以通过选择适当的电介质/有机半导体界面来调制 OFET 的选择特性。

57. OFET 传感器的主要缺点之一,是场感应的载流子可落入能量更深的局域态,并相应地降低载流子的迁移率,导致直流电流随着时间的变化。因此,特别是当变化很小时,很难获得由气味分子引起的电流变化。

polymers and oligomers, naphthalenes, copper phthalocyanines, and pentacene have been investigated in OFET sensors, and different analytes such as alcohols, ketones, thiols, nitriles, esters, and ring compounds have been sensed with these systems. In all the observed cases, the active layer-analyte interaction has been modeled as the analyte molecules being adsorbed, or even trapped, at the surface of the grains[54]. This is supported by the fact that no film swelling has been detected upon exposure, suggesting that the analyte is not able to permeate the crystalline grains.

In addition, grain boundaries play a critical role because the OFET sensor response increases when the grain size is reduced or channel length L is raised. This means that the strength of the interaction with the analyte increases with increasing the grain boundaries exposed to the odor molecule per volume unit[55]. Moreover, upon exposure to analytes, a mass uptake is recorded along with an increase or decrease in I_D. The OFET's sensing mechanism can be plausibly depicted as the analyte molecules being trapped at the grain boundaries and changing the height of energy barrier and, eventually, the film mobility of the organic conducting channel. Associated charge trapping can have an effect on V_T. A minor effect of doping has been observed in specifically designed systems, and this has resulted in chemical modulation of the off and on currents of the OFET. As demonstrated in literatures, it is in principle possible to tailor the recognition properties of OFETs also by means of a proper choice of the dielectric/organic semiconductor interface[56]. The ability to design selectivity properties of OFET sensors, in addition to the repeatability and gate bias enhancement of their response, will make OFET sensors a much more flexible option than existing portable sensor systems for a large number of sensing applications.

One of the main OFET sensor drawbacks is the temporal change in the DC current caused by field-induced carriers falling into energetically deeper localized states with a corresponding decrease in mobility. For this reason, it is difficult to extract the change in current caused by an odorant, especially when the change is small[57].

3. Imagers

The large and flexible backplanes available via OFET technologies can also benefit image sensing applications，many of which can benefit from large area（e. g. X-ray sensors），mechanical flexibility（contact scanners for non-planar objects）or both characteristics.

The basic architectures which have been proposed for image sensing using organic semiconductors couple a single transistor per cell OFET architecture with an organic photodiode（Figure 6. 15（a））or a photoconductor material such as titanyl pthalocyaninie（Figure 6. 15（b））. These architectures allow the creation of fully additive photodetector elements on essentially arbitrary substrates including plastic sheet，injection molded objects，or non-planar surfaces.

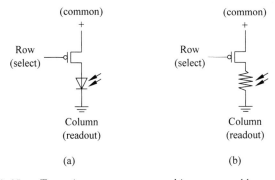

Figure 6. 15   Two imager sensor architectures addressed using OFETs：（a）an organic photodiode as the light sensitive element，and（b）an organic photoconductor

6.5.4 RFID and Logic Applications

In recent years，there has been significant interest in the development of RFID tags for their versatile detecting and tracking capabilities. RFID technology provides an automatic way to collect product，place，time，or transaction data quickly and easily without human intervention or error[58]. This technology is expected to dramatically improve automation，inventory control，distribution，shipment，tracking，and purchasing operations，provided it is cheap enough to be widely deployed. Depending on the intended

58. RFID 技术提供了一种自动的方式来快速轻松地收集产品、地点、时间或交易数据的方法,而无需人为干预和避免人为错误。

59. 通常地，较高系统频率，可实现较长距离的读取范围，但要与较高价格进行协调。对于低端应用，RFID 标签是当今条形码技术的潜在替代品。RFID 标签能提供更全面的数据收集存储，并可以从远处读取；相比之下，条形码提供了有限的识别信息和必须在线检测。

application，the RFID system will operate at different frequency bands，as listed in Table 6.2. <u>In general，a higher system frequency allows for a longer read range，but with the trade-off of higher cost. For lower-end applications，RFID tags are viewed as a promising alternative to today's barcode technology. RFID tags can provide more comprehensive data collection/storage and can be read from a distance；in contrast，a barcode provides limited identification information and requires in-line detection</u>[59].

Table 6.2　Frequency bands and applications of RFID systems

Frequency band	RFID system characteristics	Example application
Low： 100~500 kHz	Short read range，inexpensive	Access control，animal ID，inventory control
Intermediate： 10~15 MHz (13.56 MHz)	Medium read range	Access control，smart cards
High： 850~950 MHz，2.4~5.0 GHz	Long read range，high read speed，line of sight required，expensive	Railroad car monitoring，toll collection systems

Recently，RFID tags based on the OFET elements have received substantial attention as a potential application for printed organic transistors. The primary driver for consideration of this application is cost；it is expected that the cost of an RFID tag fabricated using printed transistors will be substantially lower than that achievable using conventional technologies.

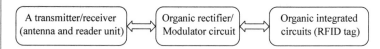

Figure 6.16　Signal flow setup of RFID tag with a transmitter/receiver unit

A block diagram of a typical organic RFID tag is shown in Figure 6.16. It consists of three different functional blocks：a transmitter/receiver（antenna and reader unit），rectifier/modulator unit，and an organic integrated circuit（RFID tag）. A schematic of a capacitive coupled organic rectifier-modulator circuit is shown in Figure 6.17，which consists of a transmitter/receiver，a rectifier-modulator unit and an RFID tag. <u>In an RFID system，the signal flows in two directions：① from a reader to the tag，when transmitter</u>

sends the command to read a code from the tag, and ②from the tag to the reader, when the tag sends the code back to the receiver[60]. As shown in Figure 6.17, for reading data from the tag, the antenna receives the voltage signal of the RF frequency from the reader through coupling capacitors, C_1 and C_2. This high-frequency AC signal is rectified (full-wave) by two half-wave rectifying diodes realized using p-type organic transistors, $OFET_1$ and $OFET_2$. These transistors are configured in diode load logic, wherein, drain and gate terminals are shorted. The capacitor C_3 filters out the redundant AC ripples, thereby, producing a smooth DC signal to be fed to the tag.

60. 在 RFID 系统中,信号是双向流动的:① 当发射器发送命令从标签读取代码时,信号从读取器到标签,以及②当标签将代码发送回接收器时,从标签到读取器。

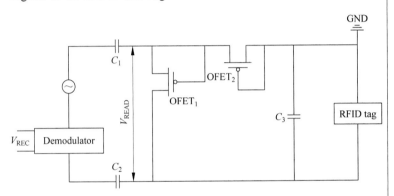

Figure 6.17 Schematic diagram of a capacitive coupled organic rectifier-modulator

To send back the code from an RFID tag to the reader, $OFET_2$ performs a modulation of electric signal generated by the tag. $OFET_1$ works as a feedback unit providing a return path to the modulated signal[61]. Using an antenna, this signal is transmitted to the reader section, where it is demodulated to regenerate the signal.

Figure 6.18 describes the major functional blocks inside an RFID tag. Organic RFID tags are still in their early stage of development, with some initial progresses. They operate in the low to medium frequency ranges (Table 6.2). For these lower-end applications, the logic circuitry of an RFID tag usually operates in the vicinity of 100 kHz. These data rates are expected to be deliverable by OFETs. However, the RFID tag has a front end that must handle rectification and operate at the frequency of the incoming RF signal (e.g.,

61. 为了将代码从 RFID 标签发回给读取器,OFET_2 对标签产生的电信号进行调制。OFET_1 作为反馈单元工作,提供调制信号的返回路径。

62. 对于这些低端应用，RFID 标签的逻辑电路一般运行在 100 kHz 附近，这样的传输速率，OFET 有望达成。但是，RFID 标签有一个前端必须处理整流，需要工作在 RF 信号输入的频率（13.56 MHz）。在整流阶段，界面需要将吸收的 RF 能量转换成直流，以便整个 RFID 的运行。

13.56 MHz）. The rectification stage at the interface needs to convert the absorbed RF energy into DC power to run to the entire RFID tag[62]. Designing organic circuits to operate at this high frequency is a major challenge; this will require clever and innovative designs to overcome the speed limitations of OFETs. Figure 6.19 outlines the development of OFET-based RFID circuitry, in which the key circuit components (as identified in Figure 6.18) are included. The progress shall take it step by step from simple circuit to the more complex logic circuit. Integration of these components with an appropriate RF interface and antenna constitutes the next step to realizing all organic-based RFID tags.

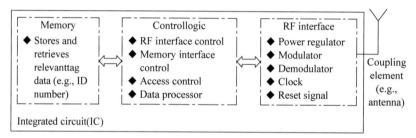

Figure 6.18　Key modules/components of a RFID tag

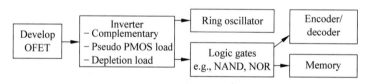

Figure 6.19　Development strategies for OFET-based RFID tag circuitry

本章小结

1. 内容概要：

本章首先对 OFET 的结构和工作原理进行了详细的介绍，然后对 OFET 器件的性能参数进行了仔细的讲解，包括转移和输出曲线、阈值电压和亚阈值漂移、电流开关比等。最后对 OFET 在显示技术、传感应用、射频标签及逻辑电路领域的应用进行了讨论。

2. 基本概念： 场效应晶体管、源极、漏极、栅极、沟道、聚集模式、耗尽模式、顶接触、底接触、转移曲线、输出曲线、线性区域、饱和区域、阈值电压、亚阈值漂移、电流开关比、夹断电压、场效应迁移率、接触电阻。

3. 主要公式：

（1）亚阈值漂移：$S = \dfrac{dV_G}{d\left[\lg(I_D)\right]}, V_G < V_T$

（2）关状态电流：$I_D(\text{cutoff}) = I_{\text{leak}}, V_G \ll V_T$

（3）亚阈值电压区域电流：$I_D(\text{subthreshold}) = I_0 \exp\left(\dfrac{V_G}{SV_{\text{th}}}\right), V_G \leqslant V_T$

（4）线性区域电流：$I_D(\text{linear}) = \dfrac{W}{L} \mu C_{\text{diel}} \left[(V_G - V_T)V_D - \dfrac{1}{2}V_D^2\right], V_D < V_G - V_T$

（5）饱和区域电流：$I_D(\text{sat}) = \dfrac{1}{2}\dfrac{W}{L} \mu C_{\text{diel}}(V_G - V_T)^2(1 + \lambda V_D), V_D < V_G - V_T$

（6）晶体管增益参数：$\beta = (\mu C_{\text{diel}})\dfrac{W}{L} = \left(\mu \dfrac{\varepsilon_r \varepsilon_0}{d_i}\right)\dfrac{W}{L}$

（7）栅电压依赖迁移率：$\mu = \mu_0 \left(\dfrac{V_G - V_T}{V_{\text{aa}}}\right)^{\gamma}$

（8）有效迁移率：$\mu_{\text{eff}} = \dfrac{L}{W}\dfrac{g_d}{C_{\text{diel}}(V_G - V_T)}, g_d = \dfrac{\partial I_D}{\partial V_D}\bigg|_{V_G = \text{const}} = \dfrac{W}{L}C_{\text{diel}}\mu_{\text{eff}}(V_G - V_T)$

（9）场效应迁移率：$\mu_{\text{FET}} = \dfrac{L}{W}\dfrac{g_m}{C_{\text{diel}}V_D}, g_m = \dfrac{\partial I_D}{\partial V_G}\bigg|_{V_D = \text{const}} = \dfrac{W}{L}C_{\text{diel}}\mu_{\text{FET}}V_D$

（10）场效应迁移率：$\mu_{\text{sat}} = \dfrac{2L}{W}\dfrac{m^2}{C_{\text{diel}}}, m = \dfrac{\partial \sqrt{I_D}}{\partial(V_G - V_T)}\bigg|_{V_D = \text{const}} = \sqrt{\dfrac{W}{2L}C_{\text{diel}}\mu_{\text{sat}}}$

Glossary

"Always on" transistor	"常开"器件
Ambipolar	双极性的
Amplifying effect	放大效应
Antenna	天线
Bottom contact	底接触
Capacitor	电容器
Channel length	沟道长度
Channel width	沟道宽度
Charge transfer	电荷转移
Charge transport	电荷传输
Collection electrode	集电极
Complementary metal-oxide-semiconductor	互补型金属氧化物半导体场效应晶体管
Conductivity	传导率,电导率
Contact resistance	接触电阻
Detection	侦查,探测
Device structure	器件结构

Directionality	定向性；指向性
Dominate	在……中占首要地位
Drain	漏极
Driving voltage	驱动电压
Emission electrode	发射极
Exhaust	用尽，耗尽，排出
Flexible	柔性的
Gate electrode	栅极
Glass substrate	玻璃基底
Large area	大面积
Linear region	线性区域
Low cost	低成本
Mobility	载流子迁移率
Net gain coefficient	净增益系数
On-off current ratio	电流开关比
Organic field effect transistor	有机场效应晶体管
Oscillation	振动
Output power	输出功率
Pentacene	并五苯
Performance characterization	性能表征
Pinch-off voltage	夹断电压
Relay	传递，传达，继电器
Resistor	电阻器
Resolution	分辨率
Response	反应
Saturation region	饱和区域
Semiconductor channel	半导体沟道
Sensitive	敏感的，灵敏的
Sensor	传感器
Signal processing	信号处理
Source	源极
Subthreshold swing	亚阈值漂移
Threshold voltage	阈值电压
Top contact	顶接触
Transfer and output curves	转移特性曲线和输出特性曲线
Transistor	晶体管
Vacuum tube	真空管

Problems

1. True or false questions:

(1) OFETs refer to field effect transistors based on organic semiconductors.

(2) The commonly used structure for OFET is top gate type.

(3) Top contact OFETs generally havw the advantage of higher mobility compared to bottom contact OFETs.

(4) OFETs generally work under depletion mode.

(5) Without gate and drain voltages, the electronic band of the semiconductor in an OFET is generally flat.

(6) In linear region, the current of an OFET is not proportional to gate voltage.

(7) The majority carrier in p-type OFETs is electron.

(8) Generally speaking, in OFETs, the mobility of those with ordered organic film is higher than those with disordered organic film.

(9) Transfer curve of an OFET refers to the curve of gate-voltage dependent currents under fixed drain voltage.

(10) Generally, an output curve of an OFET contains both linear and saturation regions.

2. Currently, in most electronic devices, what is the dominating electronic component for logical and digital circuits? Please compare this electronic component between the one based on inorganic semiconductor and the one with organic semiconductor.

3. Under bottom gate, please discuss the two structures of OFETs.

4. Please explain the electronic level changes in the organic semiconductor layer of an OFET by the loading of gate voltage, please also give schematic figures.

5. (1)At given gate voltages of -42 V and -20 V, please schematically draw the output curves of an OFET. (2)At given drain voltages of -10 V and -60 V, please schematically draw the transfer curves of an OFET.

6. Please explain the threshold of an OFET.

7. Please explain the on-off ratio of an OFET.

8. Please explain the subthreshold swing of an OFET.

9. Please deduce the expression for currents in an OFET.

CHAPTER 7　SOLAR CELLS CONTAINING ORGANIC MATERIALS

This chapter is devoted to solar cells containing organic materials，which include organic solar cell，dye sensitized solar cell and perovskite solar cell. After a brief introduction to solar radiation and the history of solar cell research，device structures and working principles for the above three kinds of photovoltaic technologies are discussed comprehensively. Finally，device characterization parameters are studied，with theoretical interpretation to the feature of the current-voltage curve from a solar cell.

7.1　Introduction

Nowadays，energy and environment have become two of the most critical subjects of wide concern，and these two topics also correlate to each other. An estimated 80% or more of today's world energy supplies are from the burning of fossil fuels such as coal，gas，or oil. Carbon dioxide and toxic gases released from fossil fuel burning contribute significantly to environmental degradation，such as global warming，acid rains，smog，etc. In addition，fossil fuel deposits on Earth are not unlimited. Due to today's increased demands for energy supplies coupled with increased concerns of environmental pollution and the exhausting of traditional energy resources，alternative renewable，environmentally friendly as well as sustainable energy sources become desirable.

Sunlight is an unlimited（renewable and thus sustainable），clean（non-polluting），and readily available energy source，which can be exploited even at remote sites where the generation and distribution of electric power present a challenge[1]. Today，any crude oil supply crisis or environmental degradation concern resulting from fossil fuel burning has prompted both people and government to consider solar energy resource more seriously[2]. The technique

1. 太阳光是无限（可再生因此可持续）、清洁（无污染）且便捷可得的能源，它甚至在发电和配电具有挑战性的遥远地点也可以被利用。

2. 目前，原油供应危机或者化石燃料燃烧造成的环境劣化问题，都促使人们和政府更加认真地考虑应用太阳能资源。

of converting sunlight directly into electric power by means of photovoltaic (PV) materials has already been widely used in spacecraft power supply systems, and is increasingly extended for terrestrial applications to supply autonomous customers (portable apparatus, houses etc.) with electric power.

7.1.1 Solar Radiation

Solar radiation is electromagnetic radiation given off by the sun, which composes infrared, visible, and ultraviolet light. The visible light from the sun is called photosphere. Most of the radiation reaching the earth originates from this layer. Although the sun is a gaseous body, the photosphere is usually referred to as the surface of the sun. The temperature of this layer varies between 4000 K and 8000 K. The sun has traditionally been treated as a blackbody (thermal equilibrium) radiator with surface temperature of 5800 K and distance of 1.5×10^{11} m from the Earth.

According to Planck, blackbody radiation implies a universal dependence of the energy density per photon energy interval dE_p. This results in an energy current density dJ_E per photon energy interval dE_p, emitted from a hole in the blackbody cavity into the solid angle Ω, perpendicularly to the area of the hole[3].

$$\frac{dJ_E}{dE_p} = \frac{2\Omega}{h^3 c^2} \frac{(h\nu)^3}{\exp\left(\dfrac{h\nu}{k_B T}\right) - 1} \tag{7-1}$$

The cavity temperature T is the only variable, by which the energy current density of the radiation is controlled. h is Planck constant, c is speed of light in vacuum, and k_B is Boltzmann constant. Radiation described by Equation (7-1) is thus called thermal radiation.

Any other body which has absorptivity $\alpha(h\nu) = 1$ for photons with energy $h\nu$ will emit radiation according to Equation (7-1). Although the sun consists mainly of protons, alpha particles and electrons, its absorptivity is $\alpha(h\nu) = 1$ for all photon energies $h\nu$ by virtue of its enormous size, hence satisfying Equation (7-1). Its temperature is not

3. 根据普朗克的研究，黑体辐射隐含着一个关于单位光子能量 dE_p 的能量密度的普适关系。这就得出了从黑体的空洞辐射到垂直于空洞面积的立体角 Ω 中的单位光子能量 dE_p 下的能量流密度 dJ_E。

4. 虽然太阳主要由质子、α粒子和电子组成，由于其巨大的尺寸，其对所有光子能量 $h\nu$ 的吸收率为 $\alpha(h\nu)=1$，因此满足方程(7-1)。太阳的温度不是均匀的，但其发射的光子是来自太阳表面几百千米厚的相对薄的表面层，表面层的温度恒定，并且所有入射光子都被吸收。相反地，只有在这个表层内发射的光子才能到达太阳表面。

5. 衰减程度取决于大气组成和光子通过大气的路径。当光子以倾斜角度到达地球表面时，后者（光子通过大气的路径）比大气的径向厚度长。从技术上，空气质量系数定义了光子穿过地球大气层的实际光程长度，表示为相对于大气径向厚度的比率。空气质量系数可用于表征太阳辐射穿过大气层后的太阳光谱。空气质量系数通常用于表征太阳能电池(SCs)在标准化条件下的性能，通常使用"AM"后跟数字表示。

homogeneous, but the emitted photons originate from a relatively thin surface layer of a few hundred kilometers thick, in which the temperature is constant and in which all incident photons are absorbed. Conversely, only photons emitted within this surface layer may reach the surface of sun[4]. The solar spectrum observed just outside the Earth atmosphere agrees well with Equation (7-1) for a temperature $T_S = 5800$ K, taking the solid angle subtended by the sun as $\Omega_S = 6.8 \times 10^{-5}$.

The spectrum of solar radiation observed at the surface of the Earth is modified by scattering and absorption in the atmosphere. In particular, it is attenuated in the ultraviolet and infrared regions. The degree of attenuation depends on the composition of the atmosphere and the photon path through it. The latter is longer than the radial thickness of the atmosphere when photons arrive obliquely. Technologically, the air mass coefficient defines the real optical path length through the atmosphere of the Earth, expressed as a ratio relative to the radial thickness of the atmosphere. The air mass coefficient can be used to characterize the solar spectrum after solar radiation has travelled through the atmosphere. The air mass coefficient is commonly used to characterize the performance of solar cells (SCs) under standardized conditions, and is often referred to using the syntax "AM" followed by a number[5]. As shown in Figure 7.1, the spectrum passing through the atmosphere to sea level with the sun directly overhead is referred to AM 1.0. As a standard spectrum for which solar cell efficiencies are rated, a distance of 1.5 times the thickness of the atmosphere, corresponding to a solar zenith angle of $48.2°$, is chosen and the spectrum is designated as AM 1.5. The solar spectrum outside the atmosphere is accordingly AM 0. The spectra of AM 0 and AM 1.5 are shown in Figure 7.2. The total energy intensity obtained by integrating over the spectrum amounts to 1.35 kW $\cdot$ m^{-2} for the AM 0 spectrum and 1.0 kW $\cdot$ m^{-2} for the AM 1.5 spectrum.

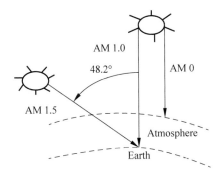

Figure 7.1   Schematic illustration for AM 0, AM 1.0 and AM 1.5 of solar irradiation to the Earth

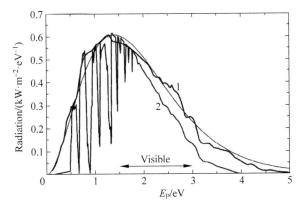

Figure 7.2   Energy current densities per photon energy for AM 0 (1) and AM 1.5 (2) solar radiations. The thin solid line is the spectrum of a 5800 K blackbody emitted into the solid angle 6.8×10^{-5}

7.1.2   History of Solar Cells

The generation of a voltage or electric current in a material or a device upon illumination of light is known as the photovoltaic effect. The devices that exhibit photovoltaic effect are known as photovoltaic devices or solar cells. In general, this effect takes place in semiconductor devices, where exposure to light causes the photons to get absorbed in the semiconducting material that excites electrons from the valance band (VB) to the conduction band (CB). Such electrons in the CB are known as photo-generated electrons, and they leave behind the corresponding holes in the VB[6]. These electrons and holes need to be extracted out to get

6. 光照下,在一个材料或器件中产生了电压或电流,被称为光伏效应。表现出光伏效应的器件被称为光伏器件或者太阳能电池。一般地,这种效应产生于半导体器件,该器件暴露于光下,引起光子被半导体吸收,导致电子从价带(VB)被激发到导带(CB)。这些 CB 电子被称为光电子,它们在 VB 留下相应的空穴。

7. 在半导体材料两侧的电极上,收集光生载流子会产生穿过器件的电动势或光电压。

electricity. Due to opposite charge on electrons (e) and holes (h), they have the tendency of recombination, which should be prevented. The solar cells are designed and prepared in a way that the photo-generated electrons and holes move in opposite directions via drift and diffusion processes and get collected at electrodes. A solar cell possesses two electrodes with the light-absorbing medium sandwiched between them. <u>Collection of photo-generated charge carriers on the electrodes in opposite sides of the semiconductor causes an electromotive force, or a photovoltage, to develop across the device</u>[7]. If an electronic circuit is connected to the device, an electric current would pass through the circuit and by this way the light energy gets converted into electricity. The drift of carriers is provided by the built-in electric field, whereas carrier diffusion is caused by the concentration gradient of the photo-generated charge carriers. If the energy of incident photons is less than the band gap of the semiconductor, the photons will not be absorbed and no photovoltaic effect will be observed.

Photovoltaic effect was firstly observed in 1839 by French physicist A. H. Becquerel when he found that shining light on an electrode submerged in a conductive solution would create an electric current. Later in 1873, W. Smith discovered photoconductivity in selenium and then this gave birth to photovoltaic technology. The first solid-state solar cell was prepared by F. Charles in 1883, when he coated a thin film of gold over the selenium semiconductor to form a junction (Figure 7.3) and got around 1% power conversion efficiency (PCE).

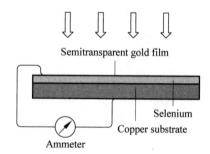

Semitransparent gold film

Selenium
Copper substrate

Ammeter

Figure 7.3　Selenium solar cells

Although these selenium solar cells had been investigated

for the next several decades, nothing appreciable was achieved. Therefore, experiments were performed on other materials and this gave birth to the next generations of solar cells. The first generation of solar cells is based on crystalline and polycrystalline silicon (c-Si). These solar cells are also known as conventional, traditional, c-Si or wafer-based solar cells. c-Si solar cells had reasonably high performance (PCE ~ 25%), which made them a reliable source of electricity generation for satellites and space vehicles, but their cost was very high compared to their power output[8].

For cost-effective solar cell technology, new processes and materials were developed, which include amorphous Si (a-Si), copper indium gallium selenide (CIGS) and cadmium telluride (CdTe). These materials are processed in thin films and the solar cells made of them are also known as thin-film solar cells or second generation solar cells. These solar cells incorporate a small amount of active materials and are processed on inexpensive substrates like glass[9]. Thin-film solar cells are comparatively less efficient than c-Si based one, due to inferior film quality and more recombination losses. Nevertheless, to date, the best CIGS solar cell can reach PCE of 23%; and they promise to be highly cost effective. The reduced cost of thin film solar cells comes out from little usage of materials, cheaper substrates, and high throughput in production. However, incorporation of some toxic elements and photo induced degradation are their main drawbacks.

8. 第一代太阳能电池是基于晶体或多晶硅材料（c-Si）。这些太阳能电池也被称为传统的、c-Si 或基于晶片的太阳能电池。c-Si 太阳能电池具有相当高的性能,这使它们成为卫星和航天器的可靠发电来源,但与其输出功率相比,它们的成本非常高。

9. 为了获得具有成本效益的太阳能电池技术,开发了新的工艺和材料,包括非晶硅（a-Si）,铜铟镓硒化物（CIGS）和碲化镉（CdTe）。这些材料是以薄膜形式加工的,由这些材料制成的太阳能电池也被称为薄膜太阳能电池或第二代太阳能电池。这些太阳能电池需要的活性材料少,并且可在玻璃等廉价基板上制造。

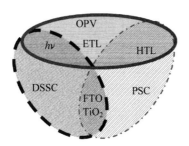

Figure 7.4 Comparison of functional components in organic based solar cells

The third generation of solar cells is multi-junction solar cells with emerging photovoltaic technologies like dye-sensitized solar cells (DSSCs), organic solar cells (OSCs, also termed as

OPV-organic photovoltaic), and quantum dot solar cells (QDSCs). The third generation of solar cells is quite futuristic and is expected to have very high PCE. They can be prepared in large area with great ease like OPVs; in addition, high PCE would make them potential candidates for future power generation. The fourth generation of solar cells includes the organic-inorganic hybrid solar cell and organometal halide perovskite solar cell (PSC). Organic-material-contained solar cells are among the third and fourth generations of solar cells, i. e., OPV, DSSC and PSC. As shown in Figure 7.4, in OPV devices, all active components, i. e., light harvesting, electron and hole transports, are organic materials; DSSC device contains light harvesting of dyes that belong to organic material, and in PSC devices, the hole transport component can be organic materials. The comparison among these devices is summarized in Table 7.1[10].

10. 含有有机材料的太阳能电池包括在第三代和第四代太阳能电池中,它们是 OPV、DSSC 和 PSC。如图 7.4 所示,在 OPV 器件中,所有活性组分,即光吸收、电子和空穴传输,都是有机材料;DSSC 器件中吸收光的染料分子是有机材料,而在 PSC 器件中,它们的空穴传输组分可以是有机材料。表 7.1 给出了这些器件的比较。

Table 7.1　Differences and similarities in OPV, DSSC and PSC

	OPV	DSSC	PSC
Photoactive material	organic	organic	perovskite
Exciton	Frenkel or CT	Frenkel or CT	Wannier
Separation	organic-organic heterojunction	organic-inorganic heterojunction	$k_B T$
Solid/liquid	solid	liquid/solid	solid
Flexibility	yes	possible	yes
Best PCE (year 2022)	~17%	14%	24%
Challenge	efficiency and stability	efficiency and stability	stability

One major challenge for large-scale application of photovoltaic technique at present is the high cost of the commercially available inorganic semiconductor-based solar cells. In contrast, recently developed organic and polymeric conjugated semiconducting materials appear very promising for photovoltaic applications due to several reasons: ①Ultrafast optoelectronic response and charge carrier generation at organic donor-acceptor interface (this makes organic photovoltaic materials also attractive for developing potential fast photo detectors); ②Continuous tunability of optical (energy) band gaps of materials via molecular design, synthesis, and processing; ③ Possibility of lightweight, flexible shape，versatile device fabrication schemes，and low cost on large-scale industrial production；and ④ Integrability of plastic devices into other products such as textiles，packaging systems，consumption goods，etc. This chapter mainly deals with organic solar cell（i. e.，OPV）；the other two organic containing solar cells（i. e.，DSSC and PSC）will also be discussed briefly.

7. 2　Device Structures

7. 2. 1　OPV Structures

OPV devices are solid state thin-film solar cells based on organic semiconductors. Organic semiconductors are a special class of aromatic hydrocarbons where carbon atoms are sp^2 hybridized and have alternate single and double bonds. Due to alteration in single and double bonds among C atoms these materials are also known as conjugated molecules. In principle，these materials are insulators in pure form，but they are called semiconductors because their electrical conductivity increases exponentially with temperature and some other properties like band gap are similar to those of the inorganic semiconductors. Most of the organic semiconductors are amorphous in nature，which gives mechanical flexibility to devices[11].

Depending on the molecular weight，the organic semiconductors can be classified into small molecules（low molecular weight materials with no repeating units），oligomers（low molecular weight materials with some repeating units），and conjugated polymers（high molecular weight materials with repeating units in the polymer chain）. Usually，thin films of small molecules and oligomers are prepared by thermal evaporation in vacuum，whereas the thin films of conjugated polymers are prepared by spin coating from their solutions. Therefore，there are two subcategories of OPVs：those incorporating small molecules are known as small molecular solar cells，whereas those based on conjugated polymers are known as polymer solar cells. No matter what kind of OPVs，their device structures are typically characterized with the type of junctions for the

11. 原理上，这些材料的纯粹形式是绝缘体，但是它们被称为半导体，因为其电导率随温度指数增加，而且它们的一些其他性质如带隙与无机半导体相似。大多数有机半导体本质上是无定形的，使其器件具有机械柔性。

12. 无论哪种OPV,其器件结构通常以用于解离的异质结的类型为特征：平面异质结（PHJ）、本体异质结（BHJ）和分子结,它们决定器件的结构。

purpose of excition dissociation: planar heterojunction (PHJ), bulk heterojunction (BHJ) and molecular junction, which determine the device structure[12].

1. Planar and Bulk Heterojunction OPVs

Generally, as shown in Figure 7.5(a), the structure of OPV devices comprises organic active layer sandwiched between two electrodes, and beneath the electrodes, cathode buffer and anode buffer are often inserted to modify the energy-level for matching. Based on the structure of the active-layer, OPV devices can be classified into planar heterojunction OPV (PHJ OPV) and bulk heterojunction OPV (BHJ OPV), with the former containing individual donor and acceptor layers and the latter containing the mixture of donors and acceptors in one layer (Figure 7.5(b)).

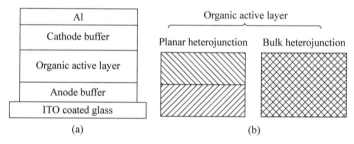

Figure 7.5　Structure of a typical OPV (a), and the organic active layer with different form of the donor and acceptor (b)

13. 一个传统平面异质结OPV(也称为双层OPV),根据制备顺序,依次包含阳极、空穴收集层、由给体层和受体层组成的活性层、电子收集层和阴极。空穴收集层和电子收集层用于修饰电极的功函数,以形成欧姆接触。在给体层和受体层之间存在一个单独、明确的界面,激子在这个界面上进行解离。

A typical PHJ OPV (also termed as bilayer OPV) consists of an anode, hole collection layer, active layer composed of donor layer and acceptor layer, electron collection layer, and cathode fabricated sequentially. The hole collection layer and electron collection layer are used to modify the work function of the electrodes to form an ohmic contact. A single, well-defined interface exists between the donor layer and acceptor layer at which excitons dissociate[13]. A significant drawback for PHJ OPV is that the short exciton diffusion length of organic materials limits the thickness of the donor and acceptor layers. If the donor or acceptor layer is too thick, the excitons generated far away from the heterojunction may decay before reaching the heterojunction.

In addition, the donor and acceptor layers are limited to tens of nanometers, which lead to weak absorption. To ensure that excitons are generated near the heterojunction, interference effects have to be considered fully during the design of PHJ OPV. These mutual trade off factors lead to low PCE and impose challenges in the design of bilayer OPV.

One of the most important breakthroughs in the field of OPVs is arguably the discovery of the bulk heterojunction (BHJ) structure. The BHJ devices consist of blends of a conjugated polymer and a molecular sensitizer, or blends of two different molecular or conjugated polymers. Such blends have the donor-acceptor interfaces throughout the whole active layer[14]. Although thermal co-deposition methods can be used to fabricate a BHJ layer, the junction is commonly formed by intermixing donor and acceptor materials in a solution, then forming the active layer by spin-coating of the mixed solution on a substrate. The resulting film is an interpenetrating nanoscale network of donor and acceptor materials. The phase separation within the film is commonly $10 \sim 20$ nm, which is within the exciton diffusion length of many organic semiconductors. Consequently, nearly unity internal quantum efficiency for excitons dissociation has been achieved for BHJ OPV, which means that nearly all photogenerated excitons are dissociated. Carriers are then transported through percolated pathways within the active layer toward the respective contact for collection. Due to the small nanoscale phase separation in BHJ OPV, a thicker active layer can be fabricated in these cells when compared to PHJ OPV. However, as the spin-coating process is inherently less controlled than the vapor deposition process commonly used in PHJ OPV, the performance of BHJ OPVs is susceptible to various parameter changes. Their device efficiency is strongly dependent on the morphology of the BHJ and various processing methods, such as thermal annealing and solvent annealing[15].

2. Molecular Junction OPVs

Currently, the mainstream of OPV devices contains two materials as active component, i.e., the electron donor and

14. OPV 领域最重要的突破之一无疑是体异质结(BHJ)结构的发现。BHJ 器件含有共轭聚合物和分子敏化剂的混合层,或者两种不同分子或共轭聚合物的混合层。这种混合使得给体-受体界面存在于在整个活性层中。

15. 它们的器件效率强烈地依赖于 BHJ 的形貌以及各种加工方法,例如热退火和溶剂退火。

the electron acceptor. Nevertheless，there are researches on OPV devices with only one active material，i. e.，the molecular junction OPV（Figure 7. 6）. Researches on this topic are not as much as the PHJ and BHJ OPVs，due to their low efficiencies now.

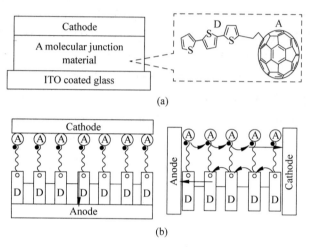

(a)

(b)

Figure 7. 6　Structure of a molecular junction OPV（a）and the schematic view of double cables in molecular junction OPVs（b）

16. 分子结 OPV 中所含的有机活性材料需要同时具有电子给体和电子受体的特征结构(D-A 分子)，这两部分之间形成分子结。理想条件下，光吸收形成激子后，分子结可以解离激子产生电荷，然后电子和空穴分别沿受体部分和给体部分传输。

The organic active material in molecular junction OPVs needs to carry structural features of both electron-donor and electron-acceptor moieties（D-A molecule），between which molecular junction is formed. In an ideal condition，after light absorption to form exciton，the molecular junction can dissociate the exciton to charge，and then，electron and hole transport along the acceptor moiety and the donor moiety，respectively[16]. If these molecules stack in such that the donor and the acceptor moieties are segregated，respectively，the material will behave like a double cable，endowing this simplest OPV with high efficiency（Figure 7. 6（b））. This topic will be further discussed in Section 7. 3. 2.

3. Inverted OPV Structures

The structure of OPV devices discussed in previous section is conventional（Figure 7. 5（a）），that is the high work-function anode locates at the bottom glass substrate，while the low work-function cathode，which is more chemically active，is deposited as the top layer，possibly bringing stability trouble to the device. To solve this problem, inverted OPVs are constructed (Figure 7.7). Because the top layer of the device is a high work-function anode, which possesses more stable chemical nature, hence being quite inert to air and moisture, the inverted OPV is expected more stable than conventional one. Inverted OPVs can have PHJ, BHJ or molecular junction in the active layer.

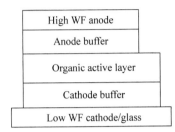

Figure 7.7   Device structure for inverted OPVs (WF = work function)

7.2.2   DSSC Structures

Dye-sensitized solar cell (DSSC) is another class of solar cells that contain organic materials; it has light harvester of organic dyes. Contrast to inorganic photovoltaics, such as Si, CdTe, and CIGS, that demand high-purity materials, and require costly, high-energy and complex fabrication, as well as place more restrictions on where they can be installed, DSSC devices offer many unique advantages, such as low cost, offering a short payback period of less than 1 year, due to less stringent demands on materials, use of more accessible feedstock, and flexibility in processing that can be scaled from a simple lab environment to an industrial process without significant investment cost[17]. These make DSSCs attractive in a broad range of applications that are not solely confined to terrestrial deployment but also in areas from low-power consumer products to large-scale architectural structures.

A typical DSSC possesses a TCO-coated substrate as electron collecting electrode, which is subsequently coated with a thin porous layer of large band gap semiconductor like TiO_2. The porous structure of TiO_2 film (TiO_2 nanoparticles connected with each other to form a porous network) is the basic element for DSSCs. Due to the large band gap of TiO_2, it absorbs very little light in UV range and works as a window

17. 与无机光伏,如 Si、CdTe 和 CIGS,需要高纯材料、昂贵、高能耗、复杂制备、及安装场地备受限制形成对比,DSSC 提供许多独特优点,如:低成本,小于一年的短投资回报期。这些缘于 DSSC 对材料不那么严格的要求、比较易得的原料和加工的弹性,即可以无需重大投资成本,就可从简单实验室环境放大到工业加工。

18. 由于 TiO$_2$ 的带隙很大，它对 UV 范围内的光吸收得非常少，可以用作窗口材料。有时 ZnO 也用于此目的，但 TiO$_2$ 显示出优越的性能。多孔 TiO$_2$ 层化学吸附了一层有机染料，作为光吸收和 TiO$_2$ 敏化剂。

material. Sometime ZnO is also used for this purpose, but TiO$_2$ has shown superior properties. The porous TiO$_2$ layer chemically adsorbs a layer of organic dye, which works as a light absorber and sensitizer for TiO$_2$[18]. The top electrode, also known as the counter electrode, is again a TCO coated on other substrate, which is typically a hole collecting electrode. Usually, FTO is used as TCO in DSSCs. The counter electrode is coated with a thin layer of catalytic platinum or lead, and to formulate a solar cell, the counter electrode is coupled with an electrolyte medium (usually nitrile derivatives) between them. Generally, the electrolyte medium possesses an I^-/I_3^- redox couple, which is achieved by addition of I_2 to some metal iodide salt. Structures of DSSC devices are described in following.

1. n-DSSCs and p-DSSCs

Typical DSSC device can be regarded as a liquid-containing micro-box with thickness of several to ten μm, which consists of six main components (Figure 7.8(a)): ① a conductive mechanical support (cathode) that is a glass or plastic substrate coated with a transparent conductive oxide (TCO), most commonly, fluorine-doped tin oxide (FTO) with low sheet resistance; ② a compact semiconductor film composed of a transparent wide bandgap semiconductor, most commonly, TiO$_2$ in the anatase crystalline form; ③ a mesoporous, sintered network of $10 \sim 30$ nm nanoparticles with a layer thickness of around 10 μm and a porosity of $50\% \sim 60\%$ (i.e. a photoanode); ④ a dye sensitizer, most commonly a ruthenium complex, that is adsorbed on the semiconductor nanoparticle surface as a molecular monolayer; ⑤ an electrolyte, typically an organic solvent containing a redox mediator, such as the iodide-triiodide (I^-/I_3^-) couple; and ⑥ a counter electrode made of a transparent conductive support with a thin layer of catalyst, most commonly, platinum (i.e., anode). Actually, this kind of DSSCs is a n-type DSSC.

The conventional DSSC is n-type, which was developed by B. O'Regan and M. Grätzel in 1991; and the p-DSSC (Figure 7.8(b)) was developed afterwards, in the year of

1999. As shown in Figure 7.8(b), the six components of a p-DSSC are: ① an anode on a glass or plastic substrate; ② a compact layer; ③ a photocathode: a mesoporous, sintered network of 10~30 nm nanoparticles with a layer thickness of around 10 μm and a porosity of 50%~60% (e.g., NiO); ④ a dye sensitizer that is chemically adsorbed on the photocathode nanoparticle surface as a molecular monolayer; ⑤ an electrolyte, typically an organic solvent containing a redox mediator, such as the iodide-triiodide (I^-/I_3^-) couple; and ⑥ a counter electrode (cathode).

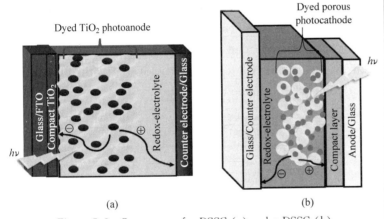

Figure 7.8　Structures of n-DSSC (a) and p-DSSC (b)

The main difference between n-DSSC and p-DSSC is the photoelectrode: n-DSSC has photoanode while p-DSSC contains photocathode. This means that in n-DSSC, after light absorption by the dye, electron injection into the photoanode occurs, while the corresponding process in p-DSSC is the hole injection to the photocathode. Compared to the well-developed n-DSSC with PCE up to 14%, p-DSSC is still in its infant stage, with good PCE around 1%. The reasons for this bad performance in p-DSSC include unstable charge-separated state in the dye, unsatisfied photocathode and the small open circuit voltage due to the small level difference between the often used redox (I^-/I_3^-) and the VB of the often used photocathode (NiO)[19].

2. Solid State DSSCs (ss-DSSCs)

DSSCs comprising liquid electrolyte often have

19. n-DSSC 和 p-DSSC 之间的主要区别在于光电极：n-DSSC 具有光阳极，而 p-DSSC 含有光阴极。这意味着,n-DSSC 中,在染料吸收光子后,发生的过程是电子注入光阳极,而 p-DSSC 中的相应过程是空穴注入光阴极。与高度发展的 n-DSSC 相比,其 PCE 高达 14%,p-DSSC 仍处于初级阶段,较好的 PCE 只有约 1%。这个 p-DSSC 性能不佳的原因包括不稳定的染料电荷分离状态,由于通常使用的氧化还原对(I^-/I_3^-)与通常使用的光阴极(NiO)VB 之间的能级差太小导致的较小的开路电压。

disadvantages of liquid leak, solvent volatile, dye-desorption, electrode corrosion and difficult to be flexible. Therefore, solid state DSSCs (ss-DSSCs) with feature of solid electrolyte were proposed in the year of 1996. Like OPV and PSC devices, ss-DSSCs are made of multiple layers in solid state; Figure 7.9 shows the device structure, which contains the following layers in sequence: glass substrate, FTO electrode, TiO$_2$ compact layer, dyed porous photoelectrode, solid electrolyte, and counter electrode. In a n-type ss-DSSC, the <u>electrolyte layer is a redox-activated hole transport layer;</u> <u>while that in a p-type ss-DSSC is a redox-activated electron</u> <u>transport layer.</u> The main problems need to be solved in ss-DSSCs are the bad contact between electrolyte and the dyed photoelectrode due to illiquidity of the electrolyte; as well as the unsatisfied conductivity of the solid electrolyte layer compared to liquid electrolyte[20].

20. 在 n 型的 ss-DSSC 中,电解质层是具有氧化还原活性的空穴传输层;而在 p 型的 ss-DSSC 中则是具有氧化还原活性的电子传递层。ss-DSSC 中需要解决的主要问题是,由于固态电解质的流动性差所引起的,电解质与吸附染料的光电极之间的不良接触,以及与液体电解质相比,较差的固体电解质层的电导率。

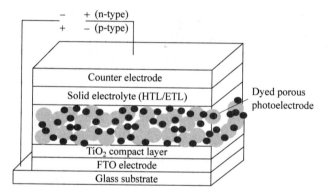

Figure 7.9   Structure of ss-DSSCs

7.2.3   PSC Structures

The perovskite solar cell (PSC) is a recently emerged technology, which has shown great potential and promise to be highly cost effective and more efficient than any other thin-film solar cell technology known so far. Generally, perovskite materials possess advantages of strong/broad absorption, low exciton binding energy, long diffusion length for exciton/carrier, and defect tolerant structure. Perovskite materials do not incorporate any rare elements hence are low cost. These devices can be easily prepared on flexible substrates at room temperature with spin coating process on a small scale, whereas printing processes are appliable to PSC devices on a large scale. Due to its high efficiency and high cost effectiveness, around 2020s, the commercialization of PSCs has been a worldwide focus for the third generation of solar cells.

Similar to OPV, PSC is also thin solid film devices. A typical perovskite solar cell possesses several hundred nanometer-thick perovskite film sandwiched between two electrodes along with electron and hole transport layers with (Figure 7.10(a)) or without mesoporous scaffold (Figure 7.10(b)). As shown in Figure 7.10(a), the prototype PSC is actually an ss-DSSC with perovskite as light-absorbing dye that is adsorbed on the surface of porous TiO_2 (bulk PSC). Planar PSC came after the bulk PSC and became the main device structure afterwards. The structure for planar PSC is: glass substrate, ITO or FTO, electron transport layer (ETL), perovskite layer, hole transport layer (HTL), and anode. Since perovskites often possess good electron transport ability, in some cases there is no ETL in the device structure; while the HTL is always included. The drawbacks of PSC include: not environmentally friendly in lead based PSC, suspectable stability in organic HTL based PSC, and expensive of Au for anode[21].

21. 由于钙钛矿材料通常具有良好的电子传输能力，在某些情况下器件结构中没有 ETL，却始终包含 HTL。PSC 的缺点包括：铅基钙钛矿材料的不环保性，基于有机 HTL 的 PSC 的可疑稳定性，以及基于金阳极的昂贵价格。

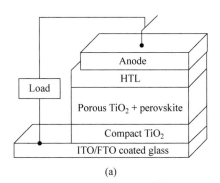
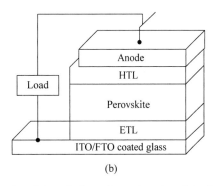

(a) (b)

Figure 7.10 Typical PSC structures, bulk heterojunction (a) and planar heterojunction (b) of PSCs

PSC also has inverted structure, where, the HTL is placed on the bottom glass anode, and the ETL and cathode are the most top layers in the device. Adopting inverted PSC structure brings forth the advantages of easier fabrication,

22. 如图 7.11(a)所示，单结太阳能电池只有在光子能量等于材料能隙（$h\nu = E_g$）时的这个特定吸收波长下，工作状态最优。对于光子能量高于光吸收材料带隙（$h\nu > E_g$）的入射光，只有一部分光子能量可以转换为电能，剩下的多余能量（$h\nu - E_g$）以热能损失的形式被浪费。另外，能量低于带隙（$h\nu < E_g$）的光子不能被吸收及转化为电能。由于太阳能电池在宽带太阳辐射下工作，光子的波长范围从紫外到红外（300 ~ 1700 nm），这种热能损失和亚带隙透明，导致了单结太阳能电池效率低下。

lower cost，higher throughput，larger material diversity，lower hysteresis effect in device operation. Please be noticed that in terms of functionality，the conventional OPV and the conventional PSC are mutually inverted structures，and vice versus.

7.2.4　Tandem Cells

As shown in Figure 7.11(a)，single junction solar cells can only have a maximum work condition when absorbing a certain wavelength of photons，the energy of which is equal to the bandgap（$h\nu = E_g$）. For incident light with a photon energy higher than the absorber bandgap（$h\nu > E_g$），only part of the photon energy can be converted to electrical energy and the remaining excess energy（$h\nu - E_g$）is wasted as thermalization loss. On the other hand，photons with energies lower than the bandgap（$h\nu < E_g$）cannot be absorbed and then converted into electricity. Since solar cells are operated under broadband solar irradiation with photons at wavelength range from ultraviolet to infrared（300~1700 nm），such thermalization loss and sub-bandgap transparency result in low efficiencies in single junction solar cells[22]. To overcome this obstacle of energy loss in solar cells，multijunction photovoltaic structures that comprise subcells with different bandgaps are developed，as shown in Figure 7.11(b) and Figure 7.11(c).

Stacking the solar cells in series（a 2-terminal structure，Figure 7.11(b)）will produce a large V_{OC}，and the active layers with different absorption regions in the tandem structure can allow the cell to absorb light over a wide wavelength range. By such，the tandem cell will possess open circuit voltage equal to the sum of those in the subcells，whereas its short circuit current is equal to that of subcell with smaller value. Apart from connecting the subcells in series，it has been demonstrated that connecting the cells in parallel（a 3-terminal structure，Figure 7.11(c)）can also result in improved efficiency. When the cells are connected in parallel，the short circuit current is ideally the sum of the current outputs from the two subcells，but the open circuit voltage will be，in the ideal case，equal to the open circuit voltage of the subcell with smaller value. Despite of the advantages, tandem solar cells suffer from difficulties in matching subcells. As carriers from the top and bottom subcells recombine at the interlayer, a good interlayer has to be chosen to allow efficient recombination and has to be transparent to reduce optical losses. Also, current matching between subcells has to be achieved in order to prevent charge accumulation on one of the subcells which deteriorates in efficiency[23]. During fabrication, it is also important that newly fabricated layers will not damage the layers below as more layers are fabricated. These factors present engineering challenges for tandem solar cell design.

23. 尽管有优点,叠层电池却苦于电池匹配的困难。因为上层和下层电池的载流子在中间层复合,选择出来的好的中间层,更允许有效的复合,且透明以减少光损失。另外,为了阻止一个子电池上能够破坏效率的电荷聚集,要达到子电池之间的电流匹配。

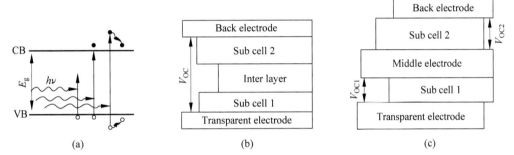

Figure 7.11　(a) Schematic illustration of electron-hole pair generation process in a material with bandgap of E_g, the energy of the excitation photons ($h\nu$) is smaller than, equal to, and greater than the material bandgap; and the tandem cells connected in (b) series and (c) parallel modes

7.3　Working Principles

7.3.1　Silicon Solar Cells

Before talking about the working principles for solar cells containing organic materials, basic operational mechanism is discussed using a silicon solar cell, Figure 7.12 shows its typical device structure. The base is a piece of p-type silicon (lightly doped with boron) with thickness of a fraction of a millimeter. A highly doped n-type silicon, with a thickness of a fraction of one micrometer was generated by doping with phosphorus of much higher concentration. Because of the built-in potential of the pn-junction, electrons migrate to the n-type region, generating electric power similar to an electrochemical battery.

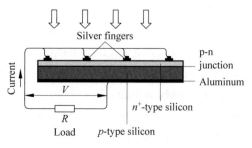

Figure 7. 12   A typical silicon solar cell

According to the theory of quantum transitions, radiation, as a stream of photons, interacts with a semiconductor in two ways (Figure 7.13). A photon with energy greater than the gap energy of the semiconductor can be absorbed and create an electron-hole pair. An electron-hole pair can recombine and emit a photon of energy roughly equal to the energy gap of the semiconductor. The probabilities of the two processes should be equal at equilibrium condition. This fact has a significant consequence to the efficiency of solar cells[24]. Because the potential energy of the electron-hole pair equals the value of the energy band gap, the best material for solar cells should have a band gap close to the center of the solar spectrum. Another factor that affects the efficiency of solar cells is the type of the energy gap. Depending on the relative positions of the top of the VB and the bottom of the CB in the wavevector space, the energy gap of a semiconductor can be direct or indirect. For semiconductors with a direct gap that the CB bottom and the VB top locate at the same wavevector, such as GaAs, $CuInSe_2$, and CdTe, a photon can directly excite an electron from the VB to the CB; the absorption coefficient is high, typically greater than 1×10^4 cm^{-1}. For semiconductors with an indirect gap, such as Ge and Si, the top of VB and the bottom of the CB are not aligned in the wavevector space, and the excitation must be mediated by a phonon, in other words, by lattice vibration. Therefore, the absorption coefficient is low, typically smaller than 1×10^3 cm^{-1}. A thicker active layer is required[25].

24. 能量大于半导体材料带隙能量的光子可以被吸收并产生电子-空穴对。电子-空穴对可以复合并发射能量大约等于半导体带隙的光子。在平衡情况下,两个过程的概率应该相等。这一事实对太阳能电池的效率产生重大影响。

25. 对于如 Ge 和 Si 这类具有间接带隙的半导体,其价带顶部和导带底部在波矢空间中不对齐,激发必须有声子的介入,换句话说,有晶格振动参与。因此,吸收系数低,通常小于 1×10^3 cm^{-1},需要更厚的活性层。

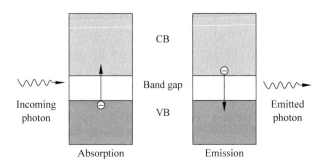

Figure 7.13   Interaction of radiation with semiconductors

Figure 7.14 shows a microprocess of how the electric power is generated by a Si-based solar cell. As shown in Figure 7.14(a), a photon generates an electron-hole pair in the p-type region. Because of the built-in electric field E_x, which points towards the p-type region, the electrons drift into the n-type region, whereas the holes stay in the p-type region. By connecting the two terminals in Figure 7.14(a) together, almost all the electrons generated by the photons can migrate to the n-type region, and complete the circuit[26]. The short-circuit current is the current of the electrons generated by sunlight.

26. 由于指向 p-型区域的内建电场 Ex 的缘故,电子向 n-型区域迁移,而空穴向 p-型区域迁移。将图 7.14(a) 中的 2 个端点连接在一起,几乎所有的光生电子都移到 n-型区域,形成闭合电路。

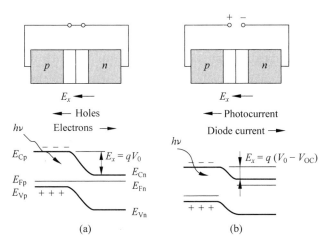

Figure 7.14   Working principles in a silicon solar cell,
(a) short circuit, (b) open circuit

As shown in Figure 7.14(b), if the two regions are not connected externally, the charges accumulated in the two regions generate a potential across the junction capacitance. The potential becomes a forward voltage for the diode. The

27. 如图 7.14(b)所示，如果这两个区域没有外部连接，则两个区域中累积的电荷将会在结电容上产生电势。这个电势成为二极管的正向电压。过渡区域厚度得以减小，产生了正向二极管电流。当二极管的正向电流等于光生电子的漂移电流时，就建立了一个平衡。两端电极上的电压，就是太阳能电池在光照下的开路电压。

thickness of the transition region is reduced，and a forward diode current is created. When the forward diode current equals the drift current of the electrons generated by the photons，an equilibrium is established. The voltage on the two terminals is the open-circuit voltage of the solar cell under illumination[27].

7.3.2　Organic Photovoltaics

Understanding the physics of OPVs is a must for their rapid development and successful implementation in energy production. Along with high efficiency，the stability and reliability in different environmental conditions are highly important issues that should be well understood for the success of this technology.

From the microscopic point of view，the band structure of organic semiconductors can be treated similarly as inorganic semiconductors. The VB is normally filled with electrons and CB is normally free of electrons. In organic semiconductors，i.e.，organic conjugated systems，HOMO is typically an occupied π orbital（bonding orbital）corresponding to VB，and LUMO is typically an unoccupied π^* orbital（antibonding orbital）corresponding to CB. The variations of HOMO and LUMO are accounted from the different hybridization states of the π-bonds which will result in different energy levels of an organic semiconductor. When an electron is excited from the HOMO to the LUMO of an organic semiconductor，the molecule itself is excited into a higher energy state，as opposed to the actual excitation of a free electron from the VB to the CB in inorganic semiconductors[28].

28. 当有机半导体的电子从 HOMO 被激发到 LUMO 时，分子本身被激发到更高的能量状态，而不是像在无机半导体中那样，从价带到导带产生自由电子的激发。

The carrier transport mechanism in organic semiconductors is also different from that of inorganic semiconductors. In organic semiconductors，thermally activated "hopping" of carriers occurs to overcome the energy barriers within the disordered conjugated π structure，thus allowing carrier transport within the organic semiconductor. This is highly different for charge transport in inorganic semiconductors，which can be described by movement of free carriers in the valence or conduction band. The hopping transport mechanism gives organic semiconductors a rather low mobility when compared to their inorganic counterparts. The low mobility, when compared to inorganic semiconductors, is a major disadvantage for organic semiconductors and consequently, different device structures were proposed to overcome this weakness.

Energy conversion by an OPV can be broken down into at least five important steps: ① exciton generation via light absorption, ② exciton diffusion to donor-acceptor interface, ③ exciton dissociation or charged carrier generation at donor-acceptor interface, ④ carrier transport to respective electrodes, and ⑤ carrier collection by the electrodes.

1. Exciton Generation via Light Absorption

In this first step of organic photovoltaic conversion, a basic requirement is that the optical gap of the material should be equal or close to the incident photon energy. In most amorphous organic materials, it is difficult to form electronic band structures due to the lack of both long-range and short-range molecular orders. The energy gap defaults to the difference between the frontier orbitals, i.e., the HOMO and the LUMO[29]. Upon the absorption of a photon, an electron in the organic semiconductor is excited from HOMO to LUMO. However, due to the low dielectric constant and localized electron and hole wavefunctions in organic semiconductors, strong Coulombic attraction exists between the electron-hole pair. The resulting bound electron-hole pair is called an exciton, with a binding energy of $0.1 \sim 1.4$ eV. Since an organic LUMO-HOMO excitation basically generates a tightly bound exciton instead of a free electron and a hole, the "optical energy gap" is therefore used instead of the conventional "electronic gap" that typically refers to the energy gap between the free holes at VB and the free electrons at CB in inorganic semiconductors. In organic materials, the relationship of "optical gap (E_{go})" versus "electronic gap (E_{ge})" may be expressed as $E_{ge} = E_{go} + E_B$, where E_B is exciton binding energy that represents a minimum energy needed to separate the electron from the hole in an exciton into a radical ion pair. E_{go} values are usually estimated directly from optical absorption band edge

29. 大多数非晶态有机材料,缺乏长程和短程的分子有序,很难形成电子的能带结构。能量带隙默认为前线轨道(即 HOMO 和 LUMO)之间的差异。

30. 有机材料中，"光学带隙（E_{go}）"与"电子带隙（E_{ge}）"的关系可以表示为 $E_{ge}=E_{go}+E_B$，其中 E_B 是激子的结合能，表示将激子中电子和的空穴分离成自由基离子对，所需的最小能量。E_{go} 的值通常可直接从光学吸收带边缘估计，E_{ge} 的绝对值可以通过电化学氧化还原分析来估计。

31. 有机半导体材料的一个主要问题是通常具有大的带隙和小的光子吸收范围，这导致光子在长波区域的吸收效率低。对于陆地应用，希望太阳能电池活性吸收材料的带隙范围为 1.3～2.0 eV。这可以通过加入一系列在太阳辐射范围内吸收光且具有不同带隙的有机给体-受体或有机染料来实现。

and the absolute E_{ge} values may be estimated by electrochemical redox analysis[30]. Absolute HOMO-LUMO levels may also be estimated from a "half" electrochemical analysis in combination with the optical absorption spectroscopy. If VB is defined as containing "free" holes, and CB is defined as containing "free" electrons, then for a donor-acceptor binary organic system, the self-organized or well aligned acceptor LUMO bands may then be called CB, and self-organized donor HOMO bands may then be called VB.

The absorption coefficient of organic materials is commonly high at $\sim 10^5$ cm^{-1}. Hence, although the thickness of the active layer of OPVs is limited by electrical conduction, a few hundreds of nanometers of the active layer are thick enough to absorb an adequate amount of light and show significant solar cell characteristics. For solar cell applications, solar radiation spans a wide range of wavelengths, with largest photon flux between 600 nm and 1000 nm (1.3～2.0 eV, on surface of the earth or AM 1.5) or 400 nm and 700 nm (1.8～3.0 eV, in space or AM 0). A main concern for organic materials is the commonly large band gap and small absorption range which lead to low absorption efficiency of photons in the long wavelength region. For terrestrial applications, it is desirable that the bandgaps of a solar cell span a range from 1.3 eV to 2.0 eV. This may be achieved by incorporating a series of different bandgap donor-acceptor or organic dyes that absorb light in that solar radiation range[31]. However, while the solar photon loss can be minimized in this manner, due to energy transfer processes where all high-energy excitons will eventually become lowest energy excitons, the open circuit voltage (V_{OC}) of the cell will also be reduced accordingly, as experimental studies have revealed a close correlation between the V_{OC} and the gap between the lowest LUMO level of acceptor and the highest HOMO level of donor. In reality, several widely used conjugated semiconducting polymers used in OPV studies have optical gaps higher than 2.0 eV. Widely used alkyloxy derivatized poly-p-phenylenevinylenes (PPV) has a typical optical gap of about 2.3 eV to 2.6 eV, well above the maximum solar photon flux range. This is why the photon absorption（or exciton generation）for PPV-based solar cells are far from getting optimized at AM 1.5. This "photon loss" problem is in fact very common in lots of reported organic photovoltaic materials and devices. However，one advantage of organic materials is the flexibility of its energy levels. They can be fine-tuned via molecular design and synthesis. Therefore，ample opportunity exists for improvement.

2. Exciton Diffusion

Once an organic exciton is photogenerated，it typically diffuses（e.g.，via intrachain or interchain energy transfer or "hopping"，including Förster energy transfer for a singlet exciton）to a remote site. At the same time，the exciton can decay either radiatively or non-radiatively to its ground state with typical lifetimes from picoseconds to nanoseconds. Alternatively，in condensed phases，some excitons may be trapped in defects or impurity sites. Both exciton decay and trapping would contribute to the "exciton loss"[32]. The average distance for an organic exciton travelling within its lifetime is called average exciton diffusion length（AEDL）. In non-crystalline and amorphous materials，the AEDL depends heavily on the spatial property（morphology）of the materials. For most conjugated organic materials，the AEDL is typically in the range of 5~70 nm. Since the desired first step of photovoltaic process is that each photogenerated exciton will be able to reach the donor-acceptor interface where charge separation can occur，one way to minimize the "exciton loss" would be to make a defect-free and donor-acceptor phase separated and ordered material. One example would be a donor-acceptor phase separated tertiary nanostructure（i.e.，donor domain，acceptor domain and interface）such that an exciton generated at any site of the material can reach a donor-acceptor interface in all directions within the AEDL. This can also be called a "bulk heterojunction" structure. In the alternative structure of bilayer solar cells，one limitation was that，if the donor or acceptor layer is thicker than the AEDL，excitons do not

32. 或者，在凝聚态中，一些激子可能被困在缺陷或杂质位点。激子衰变和俘获都会导致"激子损失"。

reach the interfacial region to separate before decay. This "exciton loss" is a serious problem. <u>On the other hand，if the photovoltaic active-layer is too thin or much shorter than the penetration depth of the light in the material，then "photon loss" due to poor light absorption would occur</u>[33]. This is also why "bulk heterojunction" type OPVs are attractive，as they not only minimize the exciton loss by increasing the donor-acceptor interface，but they can also offer enough thickness for effective photon harvesting.

3. Exciton Dissociation

After light absorption to form excitons and then their arrival at a donor-acceptor interface，the potential field at the interface due to the donor-acceptor frontier orbital energy level offsets，i. e.，ΔE as shown in Figure 7. 15，can then separate the exciton into a free electron at LUMO of acceptor (Step 3 in Figure 7. 15) and a free hole at HOMO of donor，provided that this field or energy offset is close to its optimal value or range as discussed in Section 9.

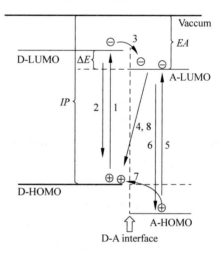

Figure 7. 15　Scheme of molecular orbitals and processes in an organic donor-acceptor binary system upon light incidence. Step 1 and 5：light absorption；Step 3 and 7：photoinduced charge separation；Step 2，4，6 and 8：recombination (notice that the difference between Step 4 and 8 is the origination of the excited electron：either optically pumped or by charge transfer from the donor LUMO)

This photoinduced charge separation process is also called "photon-doping", as it is a photoinduced (in contrast to chemical or thermal induced) redox reaction between the donor and the acceptor[34]. For a derivatized PPV donor and fullerene acceptor binary system, it has been experimentally observed that the photoinduced charge separation process at the PPV-fullerene interface was orders of magnitude faster than either the PPV exciton decay or the charge recombination. This means that optoelectronic quantum efficiency at this interface is near unity and a high-efficiency organic photovoltaic system is feasible.

4. Carrier Transport to Respective Electrodes

Once the carriers, either free electrons or holes, are generated, holes need to transport towards the high work function electrode, and electrons need to transport towards the low work function electrode. The driving forces for the carrier motion may include the field created by the work function difference between the two electrodes, and a "chemical potential" driving force that can be interpreted as a density potential driving force, i.e., particles tend to diffuse from a higher density domain to a lower density domain. In an organic donor-acceptor binary photovoltaic cell, the high-density electrons at the LUMO of acceptor nearby the donor-acceptor interface tend to diffuse to lower electron density region within the acceptor phase, and high-density holes at the HOMO of donor nearby the donor-acceptor interface tend to diffuse to the lower hole density region within the donor phase[35]. In the PHJ OPV, once an exciton was separated into a free electron at acceptor side and a free hole at donor side of the donor-acceptor interface, the electron will be "pushed" away from the interface toward the negative electrode by both the "chemical potential" and by the field formed from the two electrode work functions. The holes will be "pushed" toward the positive electrode by the same forces but in the opposite direction. With this chemical potential driving force, even if the two electrodes are the same, asymmetric photovoltage could still be achieved (i.e., the HOMO of donor would yield the positive and LUMO of acceptor would

34. 这种光诱导的电荷分离过程也称为"光子掺杂"，因为它是在给体和受体之间的由于光诱导（与化学或执作用对比鲜明）产生的氧化还原反应。

35. 载流子运动的驱动力可能包括两个电极之间的功函数差而产生的电场，以及"化学势"的驱动力。"化学势"驱动力可以解释为密度势能驱动力，即粒子倾向于从较高密度区域扩散到较低密度区域。在有机给体-受体的二元体系光伏电池中，在给体-受体界面附近的受体 LUMO 上的高密度电子倾向于扩散到受体中电子密度较低的区域，而在给体-受体界面附近的给体 HOMO 上的高密度空穴倾向于扩散到给体中空穴密度较低的区域。

36. 带隙中间的物质,无论是杂质和缺陷,还是故意掺杂的氧化还原物质,都可以通过为电子或空穴传输提供"跳跃"的位点来促进载流子扩散和电导率。然而,当电子-空穴在界面处分离后,由于 A-LUMO/D-HOMO 界面的势能降(图 7.15 中的步骤 4)和自由电子和空穴之间的库仑力,电子和空穴也可以重新复合。

37. 目前,载流子热"跳跃"和"隧穿"被认为是大多数报道的有机光伏系统的主要扩散和导电机制;因此,"载流子损耗"被认为是有机光伏材料和器件效率低下的另一个关键因素。

yield the negative electrodes). Mid-gap state species, either impurities and defects, or intentionally doped redox species, may also facilitate the carrier diffusion and conductivities by providing "hopping" sites for the electrons or holes. However, right after electron-hole is separated at the interface, they can also recombine due to both potential drop of A-LUMO/D-HOMO (Step 4 in Figure 7.15) and the Coulomb force between the free electron and hole[36]. Fortunately, the charge recombination rates in most cases are much slower than the charge separation rates (charge recombination rates are typically in nano- to milliseconds as compared to femto- and picoseconds charge separation rate), so there is an opportunity for the carriers to reach the electrodes before they recombine. Yet, in most currently reported OPVs, the diffusion of electrons and holes to their respective electrodes is not really fast due to poor morphology. If donor and acceptor phases are perfectly "bicontinuous" between the two electrodes, and that all LUMO and HOMO orbitals are nicely aligned and overlapped to each other in both donor and acceptor phases, like in a molecularly self-assembled thin films or crystals, then the carriers would be able to diffuse smoothly in "bands" towards their respective electrodes. Currently, carrier thermal "hopping" and "tunneling" are believed to be the dominant diffusion and conductivity mechanism for most reported organic photovoltaic systems; therefore, the "carrier loss" is believed to be another key factor for the low efficiency of organic photovoltaic materials and devices[37].

5. Carrier Collection by Electrodes

It has been proposed that when the LUMO level of acceptor matches the Fermi level of the low work function electrode, and the HOMO of donor matches the Fermi level of the high work function electrode, an ideal "Ohmic" contact would be established for efficient carrier collection at the electrodes. So far, there are no organic photovoltaic cells that have achieved this desired "Ohmic" alignment due to the availability and limitations of materials and electrodes involved. There were a number of studies, however, focusing on the open circuit voltage (V_{OC}) dependence on LUMO-HOMO level changes, electrode Fermi levels, and chemical potential gradients of the materials. The carrier collection mechanisms at electrodes are relatively less studied and are not well understood. It is believed that the carrier collection loss at the electrodes is also a critical contributing factor for the low efficiency of existing OPVs[38].

6. Energy Level Tilting under OPV Operations

In a single layer OPV, there is only one organic material sandwiched between two electrodes; and its electronic structure can be described by metal-insulator-metal model (MIM). Figure 7.16 describes the four situations for an OPV under MIM model. For an OPV device at open circuit condition (Figure 7.16(a)), the energy levels in the organic material are all flat, referring as flat levels. When incident light irradiates the device, the produced voltage is termed as open circuit voltage, which is not greater than the work function difference between the two electrodes and which is balanced with the built-in electric field in the device; the internal electric field is zero and there is no current in the device[39]. Figure 7.16(b) shows the situation of short circuit without external bias. Due to the difference between the two electrodes and the alignment of their Fermi levels in short circuit condition, built-in electric field is produced, which is often assumed a homogeneous distribution in the organic layer, leading to the energy level downward tilting toward the cathode. Considering the low carrier density in organic materials and the thinner thickness of organic films compared to inorganics, band bending in OPVs can be neglected. Under dark condition, there will be no current flow. After exciton formation by light incidence and charge-carrier production by the driving force of built-in electric field, electrons at LUMO will move to the lower levels (toward cathode) and holes at HOMO will move to higher levels (toward anode), and finally, some of carriers can reach electrodes and be collected. The device exhibits photovoltaic property[40]. If reverse voltages are applied (Figure 7.16(c)), due to the high energy barrier between the ITO and the LUMO level for electron injection and high barrier between

38. 载流子在电极处的收集机制的研究相对较少，并且尚未得到很好的了解。人们相信，电极处的载流子收集损耗也是导致现有 OPV 效率低下的关键因素。

39. 当入射光照射器件时，产生的电压称为开路电压，其值不大于两个电极材料之间的功函数差，并且与器件的内建电场相平衡；内部电场为零，器件中没有电流。

40. 在光入射形成激子并通过内建电场驱动力产生电荷载流子后，LUMO 处的电子将移动到较低的能级（朝向阴极），HOMO 处的空穴将移动到较高能级（朝向阳极），最后，一些载流子可以到达电极并被收集。该器件表现出光伏特性。

Al and the HOMO level for hole injection, carrier injection is difficult, leading to small dark current. Under illumination, the photogenerated carriers can drift under the strong electric field, leading to much larger currents compared to dark condition; hence the device can be used as photodetector or as solar cell to collect charges. In Figure 7.16(d), forward bias is applied, due to the small barrier between Al and LUMO level for electron injection and small barrier between ITO and HOMO level for hole injection, charges can be effectively injected into device, producing electrically pumped exciton. If the material fluoresces, the device behaves as an organic light-emitting diode (OLED).

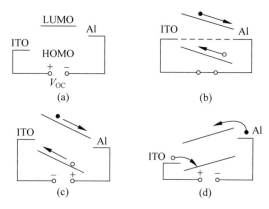

Figure 7. 16   MIM model of a single layer OPV at different conditions: (a) open circuit, (b) short circuit without external bias, (c) under reverse bias, (d) under forward bias

For BHJ and PHJ type OPV devices, the electronic structure in the devices and their energy-level tilting phenomena are similar to those in single layer OPVs. Figures 7.17(a) and (b) describe the flat levels under dark and open circuit condition, and the tilting of energy level under illumination and short circuit condition for PHJ device; those for BHJ device are described in Figures 7.17(c) and (d). One thing should be noticed is that the energy difference at the interface is determined by the intrinsic level-difference between the materials, and keep unchanged when the level tilting[41], e.g., from Figure 7.17(a) to Figure 7.17(b) and from Figure 7.17(c) to Figure 7.17(d), the values of E_{g1}, E_{g2}, δ_1, δ_2, δ_3 and δ_4 are unchanged.

41. 需要注意的一点是，界面处的能量差是由材料之间固有的能级差决定的，当能级倾斜时保持不变。

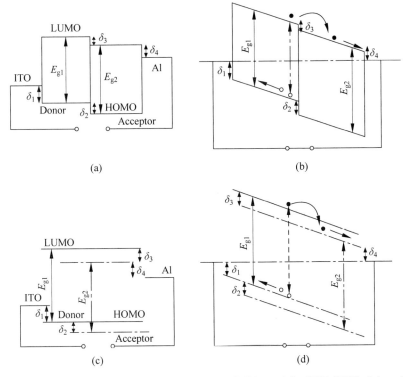

Figure 7.17   MIM models for PHJ OPVs (a) and (b), and for BHJ OPVs (c) and (d). (a) and (c) are under dark & open circuit; (b) and (d) are under illumination & short circuit

7. Energy Losses in Organic Solar Cells

Photon harvesting by OPVs encounters a number of optical and electrical losses and only the photogenerated charge carriers, which manage to survive and extract out of the device actually contribute to electricity[42]. To state these in details, we start from light absorption in an OPV device. Upon photon irradiation, only a fraction of incident light is absorbed because some of the incident photons are reflected back from the OPV and some are left from being absorbed (transmitted). The absorbed photons generate excitons that need to dissociate into free electrons and holes. For dissociation, excitons need to diffuse to the donor-acceptor interface where the LUMO level offset between donor and acceptor materials provides necessary force for dissociation. Due to short diffusion lengths, the excitons in the vicinity of the donor-acceptor interface dissociate efficiently, and those

42. OPV 中的光子利用会遇到许多光学和电学的损耗,只有光生电荷载流子在器件中存活下来并从电极中提取出来,才对产生电力有实际的贡献。

43. 由于有机材料中电荷载流子的迁移率低,即使在激子解离后,电荷载流子也会在异质结附近聚集并形成空间电荷,进一步影响器件的光伏性能。在给体-受体界面处分离的电荷载流子依然通过库仑力结合在一起并形成电荷转移状态(CTSs)。CTSs是极化的电子状态,是通过库仑力作用结合起来的负电荷和正电荷对,有时也称为激基复合物、界面电荷对、电荷转移激子和束缚的分子间自由基对。一些电荷转移态也可能以辐射或非辐射方式复合。

far from the interface decay without contributing to photocurrent. The problem of inefficient exciton dissociation was solved in BHJ structures where donor-acceptor materials are mixed together such that the donor and acceptor phase-widths are less than the exciton diffusion lengths. Exciton dissociation takes place by transfer of electrons to the LUMO of acceptor. The transfer of holes from donor to accepter is prevented by higher HOMO energy of the donor. Higher HOMO energy of the donor also helps in dissociation of excitons generated in the acceptor, that is, it can promote the acceptor holes transferring to the donor HOMOs (Step 7 in Figure 7.15). Exciton dissociation gives electrons in the acceptor phase and holes in the donor phase around the interface. Because of low charge carrier mobilities in organic materials, even after exciton dissociation the charge carriers accumulate near the heterojunction and formulate a space charge, which further controls the photovoltaic performance. The separated charge carriers at the donor-acceptor interface are still bound together by Coulombic force and form charge transfer states (CTSs). CTSs are electronic states of polarons, pairs of Coulombically bound negative and positive charges, and sometimes are also known as exciplex, interfacial charge pairs, charge transfer excitons, and bound intermolecular radial pairs. Some of the CTSs may also recombine radiatively or non-radiatively[43].

CTSs are a result of intermolecular overlap between donor and acceptor molecules. CTSs have direct effects on the photovoltaic performance by affecting the short-circuit current density (J_{SC}) and open-circuit voltage (V_{OC}). The maximum value of V_{OC} is determined by the energy of CTSs. However, it will be important to know whether the maximum photocurrent and maximum V_{OC} can be achieved simultaneously. Optimization of the PCE by tuning the properties of CTSs is a difficult task but should be carefully carried out. The creation of CTSs in OPVs is evident by the generation of photocurrent due to absorption of photons with energy well below the HOMO-LUMO gap of the active layer. Absorption and emission spectra from the OPVs are an important tool for the characterization of CTSs in OPVs. The results can infer the possibility of absorption and emission of photons having energy less than the band gap of the organic semiconductor，i. e.，the absorption and emission from CTSs. As the energy of CTSs determines the V_{OC}，it should be tuned by selecting material combinations such that it is lower than but as close as possible to the exciton energy. Also the losses due to triplet excitons decay should be avoided. It is quite obvious that the fundamental understanding of the processes involving CTSs is essential for optimization of the performance of OPVs.

The CTSs require an additional force to break apart and the necessary force is provided by an internal built-in electric field of the cell[44]. After dissociation of CTSs，the separated electrons and holes transport through respective channels (electrons in acceptor channels and holes in donor channels) to collect at the respective electrodes. The separated charge carriers may recombine or get trapped in localized states during transportation，which results in reduced photocurrent. The surviving charge carriers are extracted out of the cell via the drift-diffusion process and formulate a photocurrent. Charge transport usually happens via the hopping phenomenon from one localized state to another. As a whole，the extracted photocurrent does not solely depend upon photogeneration of excitons but also on the exciton diffusion，exciton dissociation，and charge transport properties of donor/acceptor materials[45]. The possible processes happening in an OPV from light incidence to collection of charge carriers are shown schematically in Figure 7.18.

8. Optimization of OPVs in the Spatial Domain

Here，a copolymer melt with structure of DBAB type (donor-bridge-acceptor-bridge) is taken as an example (Figure 7.19)，where D is a π-electron conjugated donor block，A is a conjugated acceptor block，and B is a nonconjugated and flexible bridge unit. This type of copolymers is well known to exhibit behavior similar to conventional amphiphilic systems such as lipid-water mixtures and surfactant solutions. The covalent bond connection between

44. CTS 需要额外的力来解离，必要的分解力由电池内部的内建电场提供。

45. 总之，最终提取的光电流不仅依赖于光生激子的产生，还取决于给体/受体材料的激子扩散、激子解离和电荷传输的特性。

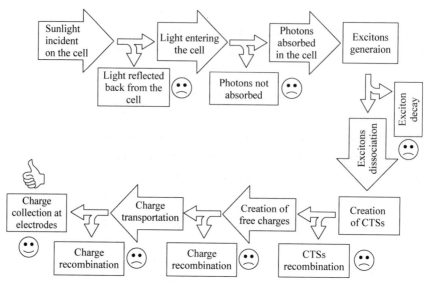

Figure 7.18　Scheme of photon harvesting by OPVs, with possible energy loss routes indicated by the unhappy faces

46. 众所周知,这种类型的共聚物表现出类似于传统双亲性体系的行为,例如脂质-水混合物和表面活性剂溶液。不同种类或不同嵌段之间的共价键连接,对材料可能的平衡状态施加了严格的约束;这导致了独特的超分子纳米结构,如薄片(LAM)、六角形(HEX)堆叠的圆柱或柱体、球体堆叠的体心立方体(BCC)晶格、六角形孔洞层(HPL)和形成至少两个双连续相:双连续的有序双金刚石相(OBDD)和螺旋相。

distinct or different blocks imposes severe constraints on possible equilibrium states; this results in unique supramolecular nano-domain structures such as lamellae (LAM), hexagonally (HEX) packed cylinders or columns, spheres packed on a body-centered cubic (BCC) lattice, hexagonally perforated layers (HPL), and at least two bicontinuous phases: the ordered bicontinuous double diamond phase (OBDD) and the gyroid phase[46].

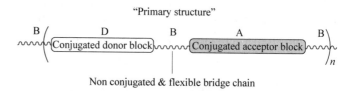

Figure 7.19　Scheme of a "primary structure" of -DBAB- type block copolymer

The morphology of block copolymers is affected by chemical composition, block size, temperature, processing, and other factors. For a triblock copolymer, a variety of even more complex and unique morphologies can be formed. Clearly, the block copolymer approach to photovoltaic function offers some intrinsic advantages over the bilayer or composite-blend systems.

As reported, in a donor-acceptor di-block copolymer system comprising MEH-PPV and polystyrene based C_{60} derivative, phase separation between the two blocks was indeed observed. However, the polystyrene-C_{60} acceptor block is not a conjugated chain system and the poor electron mobility or "carrier loss" problem in polystyrene phase still remains as an issue. On the other hand, it has been reported that when a conjugated donor block was linked directly to a conjugated acceptor block to form a direct p-n type conjugated di-block copolymer, while energy transfer from higher gap block to lower gap block was observed, no charge separated states (which is critical for photovoltaic functions) were detected.

To address several loss problems of organic photovoltaics discussed earlier, particularly the "exciton loss" and the "carrier loss" problems, optimization in spatial domain of the donor and acceptor materials has been investigated. Using this rationale, an organic photovoltaic device based on a -DBAB- type block copolymer and its potential "tertiary" supramolecular nanostructure were designed (Figure 7.20). In this structure, the HOMO level of the bridge unit is lower than both the donor and the acceptor HOMOs, and the bridge's LUMO level is higher than both the donor and acceptor LUMOs (Figure 7.20(a)). With this configuration, a wide bandgap energy barrier is formed between the donor and acceptor conjugated blocks on the polymer chain. This potential energy barrier separates the energy levels of the donor and acceptor blocks, retarding the electron-hole recombination encountered in the case of directly linked *p-n* type di-block conjugated copolymer system. At the same time, intramolecular or intermolecular electron transfer or charge separation can still proceed effectively through bridge σ-bonds or through space under photoexcitation. Additionally, the flexibility of the bridge unit would also enable the rigid donor and acceptor conjugated blocks more easily to self-assemble, phase separate, and become less susceptible to distortion of the conjugation[47]. Since both donor and acceptor blocks are π-electron conjugated chains, if they are self-assembled in planes perpendicular to the

47. 这个势垒将给体和受体嵌段的能级分开,延缓了在直接 *p-n* 连接的二嵌段共轭共聚物体系中遇到的电子-空穴复合。同时,在光激发下,分子内或分子间的电子转移或者电荷分离仍然能够有效地通过 σ 键桥或通过空间进行。此外,桥单元的柔性还将使刚性给体和受体共轭嵌段更容易自组装和相位分离,并且不易受到共轭扭曲变形的影响。

molecular plane like a π-π stacking morphology well-known in all π conjugated system (Figure 7.20(b)), good carrier transport in both donor and acceptor phases now become feasible.

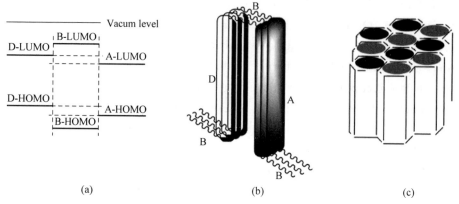

Figure 7.20　(a) Schematic diagram of intrachain energy levels in -DBAB type block copolymer，(b) "secondary structure" and (c) "tertiary structure" of -DBAB- type block copolymers

48. 这个"二级结构"风格,被确认可以极大地增强载流子迁移率,因为提高了 π 轨道重叠,这在有序碟型液晶相或者在衍生得到的自主装的区域规整型聚噻吩或者在模板上排列的聚对苯撑乙烯中得到了证明。最重要的,这个"二级结构"如图 7.20(b) 所示,实际上有利于水平方向上的激子解离和垂直方向上的电荷输运,这已经被实验观测到。

While the -DBAB- block copolymer backbone structure may be called "primary structure" (Figure 7.19), the conjugated chain with closely stacked and ordered morphology may therefore be called "secondary structure" (Figure 7.20(b)). This "secondary structure" style has been known to dramatically enhance carrier mobility due to improved π orbital overlap as demonstrated in ordered discotic type liquid crystalline phases, or in derivatized and self-assembled region-regular polythiophenes, or template aligned poly-p-phenylenevinylenes. Most importantly, this "secondary structure" as shown in Figure 7.20(b) is in fact favorable for the exciton diffusion in horizontal direction and charge transport in vertical direction, as has been experimentally observed[48]. Finally, through the adjustment of block size, block derivatization, and thin film processing protocols, a "tertiary structure" (Figure 7.20(c)) where a "bicontinuous" such as columnar (or "HEX") type of morphology of the donor and acceptor blocks is vertically aligned. By sandwiching such a layer of tertiary structure between two electrodes, OPV device with both electron and hole transport channels and effective exciton dissociation interface can be realized (Figure 7.21(a)). Even better, a thin donor layer may be inserted between ITO and active "HEX" layer, and a thin acceptor layer is inserted between metal electrode and active layer (Figure 7.21(b)). The terminal donor and acceptor layers would enable a desired asymmetry and favorable chemical potential gradient for asymmetric (selective) carrier diffusion and collection at respective electrodes[49]. Since the diameter of each donor or acceptor block column can be conveniently controlled via synthesis and processing to be within the organic AEDL of 5~70 nm, so that every photoinduced exciton will be in convenient to reach a donor-acceptor interface along the direction perpendicular to the columnar. At the same time, photogenerated charge carriers can diffuse more smoothly to their respective electrodes via a truly "bicontinuous" block copolymer columnar morphology.

49. 通过在两个电极嵌入夹层的三层结构,可以实现具有电子和空穴传输通道以及有效激子解离界面的 OPV 器件(图 7.21(a))。更好的是,可以在 ITO 和 "HEX" 活性层之间插入薄的给体层,在金属电极和活性层之间插入薄的受体层(图 7.21(b))。末端给体层和受体层将为各自电极上的不对称(选择性)载流子扩散和收集提供所需的不对称和有利的化学势梯度。

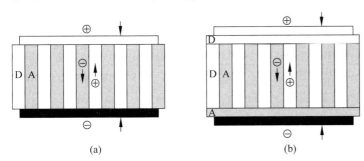

Figure 7.21　-DBAB- type block copolymer solar cells in the form of (a) columnar structure directly sandwiched between two electrode layers and (b) with terminal asymmetric active material layers

While the increased donor and acceptor interface area and phase morphology will dramatically minimize the exciton and carrier losses, it may also increase the carrier recombination at the same interfaces. However, by proper energy level manipulation via molecular engineering, the charge recombination rate can be reduced in comparison to the charge separation as will be discussed in the following. In many of the reported organic photovoltaic systems, the charge recombination typically occurs on the microseconds or slower timescale, which is in contrast to the ultrafast pico- or femtoseconds charge separation rate at the same interface. Therefore, the charge carrier recombination does not appear to be of a major concern for organic solar cell applications where the radiation is continuous. Additionally, with appropriate adjustment of donor and acceptor block sizes and their substituents, energy levels, or with attachment of better photon energy-matched sensitizing dyes on the polymer backbone, it is expected that the photon loss, the exciton loss, and the carrier loss (including charge recombination) issues can all be addressed and optimized simultaneously in one such block copolymer photovoltaic device.

9. Optimization of OPVs in the Energy-Time Domain

To address the optimal energy levels in a paired donor-acceptor organic light harvesting system, first, energy gaps in both donor and acceptor should be fine-tuned via molecular engineering to match the photon energy, as both can absorb photon and incur charge separation at donor-acceptor interface. Furthermore, the difference between the acceptor LUMO and donor HOMO is linearly related to the value of open circuit voltage (V_{OC}) for an OPV device, as shown in Figure 7.22(a). To drive the charge separation, a critical requirement is the enough magnitude of energy offset between the donor and the acceptor[50]. A current widely cited view is that the frontier orbital energy offset between the donor and the acceptor should be at least over the exciton binding energy E_B (i.e., the minimum energy needed to overcome the electric Coulomb forces and separate the tightly bound and neutral exciton into a separate or "free" electron and hole). However, if the energy offset is too large, despite of the resulted small V_{OC} (middle of Figure 7.22(b)), Marcus "inverted" region (i.e., large reorganization energy for charge transfer, see Chapter 3) may slow down charge separation, which is not desirable for light harvesting functions. On the other hand, when the LUMO energy offset is too small (right of Figure 7.22(c)), charge separation appears to become less efficient, or even no photovoltaic effect can be observed. In reality, there are exceptional cases. For example, in many donor-bridge-acceptor systems

50. 针对一对有机给体-受体组成的光收集系统的最优能级问题,首先,因为给体/受体两者都可以吸收光子并在给体-受体界面处产生电荷分离,应该通过分子工程对给体和受体的能量带隙进行微调,以匹配光子能量。此外,受体 LUMO 和给体 HOMO 之间的差异与 OPV 器件的开路电压(V_{OC})呈线性关系,如图 7.22(a)所示。为了驱动电荷分离,一个关键的要求是给体和受体之间有足够的能级差。

with positive energy offset electron transfer from donor to acceptor via a higher energy level bridge unit still occurs effectively. Therefore, an analysis of optimal donor-acceptor energy offsets is necessary in order to achieve efficient charge separation and large V_{OC}; and particular attentions should be given to exciton decay, charge separation, and recombination processes in both donor and acceptor.

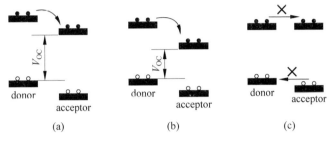

Figure 7.22   Schemes of the driving force for exciton dissociation and its relationship to V_{OC} of an OPV

(a) value of V_{OC}; (b) loss of voltage; (c) no charge transfer

In an "ideal" organic donor-acceptor binary solar cell, both donor and acceptor should harvest photons and contribute to photovoltaic effects. For organic solar cells, the charge separated state is the desired starting point. However, charge separation (Step 3 and 7 in Figure 7.15) is also competing with exciton decay (Step 2 and 6 in Figure 7.15)[51]. The ratio of rate constants between charge separation and exciton decay can therefore be defined as exciton quenching parameter (EQP, mathematically represented as Y_{EQ}). The parameter Y_{EQ} reflects the efficiency of exciton to charge conversion. It was experimentally observed that the charge separation could be orders of magnitude faster than the exciton decay in a MEH-PPV/fullerene donor-acceptor binary pair. Second, charge separation (Steps 3 and 7 in Figure 7.15) is also competing with charge recombination (Steps 4 and 8 in Figure 7.15). The ratio of charge separation rate constant over charge recombination rate constant may therefore be defined as recombination quenching parameter (RQP, mathematically represented as Y_{RQ}). For any light harvesting applications, such as solar cell applications, it is desirable that both Y_{EQ} and Y_{RQ} parameters are large.

51. 对于有机太阳能电池,电荷分离态是期望的起点。但是,电荷分离(图 7.15 中的第 3 和 7 步)也和激子衰减(图 7.15 中的第 2 和 6 步)相竞争。

有机光电子双语教程

52. 对于任何光收集的应用,例如太阳能电池应用,希望参数 Y_{EQ} 和 Y_{RQ} 都很大。具体而言,在电子转移的动态状态下,基于马库斯理论,研究发现,存在一个最佳的给体-受体能级差,使得激子-电荷转换最有效(或 EQP 达到最大值),以及存在第二个最佳能级差,使得电荷复合与电荷分离相比,相对较慢(或 RQP 变得最大)。如果最大 RQP 与最大 EQP 相距太远,这时的最佳 RQP 条件无关紧要,因为此时最大 RQP 条件下的电荷分离可能太慢。

53. 在开发具有高效有机光收集系统(包括有机光伏电池、光电探测器或任何人造的光-电合成器和转换器)时,期望的能量差、期望的电荷分离重组能以及不期望的电荷复合重组能,对于分子结构和能级的微调,都至关重要。

Specifically，in electron transfer dynamic regime and based on Marcus theory，it has been found that，there exists an optimal donor-acceptor energy offset where exciton-charge conversion is most efficient（or EQP reaches its maximum），and a second optimal energy offset where charge recombination is relatively slow compared to charge separation（or RQP becomes largest）. If the maximum RQP is too far away from maximum EQP，then this optimum RQP is insignificant as the charge separation at this maximum RQP might be too slow[52]. The molecules should be designed and developed such that the maximum RQP is close to or coincides with maximum EQP. There also exists a third energy offset where charge recombination becomes fastest. The molecules should be designed and developed such that this maximum charge recombination is far away from maximum EQP. For a donor-acceptor binary photovoltaic system，there exists a fourth optimal donor-acceptor energy offset，where the EQP product of both donor and acceptor become largest，so that both donor and acceptor can effectively contribute to photoinduced charge separation. This final optimal donor-acceptor energy offset is related to the exciton binding energy， the optical excitation energy gaps， and the reorganization energies of the charge separation of both donors and acceptors. While there exists a desired donor（or acceptor）charge separation reorganization energy where Y_{EQ} has maximum value，there also exists an undesired charge recombination reorganization energy value where Y_{RQ} becomes minimum. The desired energy offset，the desired charge separation reorganization energy，and the undesired charge recombination reorganization energy values are critically important for molecular structure and energy level fine-tuning in developing high efficiency organic light harvesting systems，including organic photovoltaic cells，photodetectors，or any artificial photo-charge synthesizers and converters[53].

7.3.3　DSSCs and PSCs

1. DSSCs

A DSSC operates on the principle that dye molecules are

excited upon absorption of sunlight，and electrons are injected into the CB of the TiO$_2$ semiconductor in an *n*-DSSC，or holes are injected into VB of a photocathode in *p*-DSSC. The electrons are transported through the TiO$_2$ layer，or holes are transported through the photocathode，and finally the photon-generated charges are collected out to an external circuit，where useful work is performed on a load；the electrons or holes then return to the electrochemical cell at the counter electrode. In n-DSSCs，electrons from counter electrode reduce triiodide to iodide，the iodide is responsible for regenerating dyes back to its electronic ground state by donating electrons to the cation of dye，which then simultaneously converts iodide back to triiodide. Similarly，in p-DSSCs，holes from counter electrode are transferred to electrolyte redoxes that are responsible for the generation of neutral dyes via anion-dye oxidation[54]. Figure 7.23 is the working principle for n-DSSCs，with positive processes labeled in grey background；and Equation（7-2）to Equation（7-6）describe these processes.

54. 在 n 型染料敏化太阳能电池中，对电极上的电子把 I$_3^-$ 还原成 I$^-$，I$^-$ 可通过将电子传给染料阳离子，将染料再生回到基态，同时 I$^-$ 转换回 I$_3^-$。相似地，在 p 型染料敏化太阳能电池，对电极处的空穴被传输到电解质中的氧化还原对，以便通过对染料阴离子的氧化来再生染料。

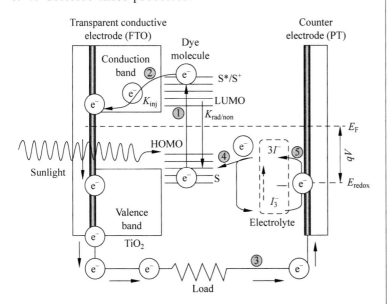

Figure 7.23　Working principle of n-DSSCs，labeled with positive processes for photon to electricity conversion

Now using TiO$_2$ photoanode based n-DSSC as an example，we explain the detail microprocesses. Upon absorption of photons，dye molecules are excited from the

55. 一旦电子被注入宽带隙半导体的 CB，例如纳米结构的 TiO_2 薄膜，染料分子（光敏剂）就会被氧化（式(7-3)）。注入的电子在 TiO_2 纳米颗粒之间传输，最后提取到负载，其中完成的功以电能的形式表示（式(7-4)）。含有 I^-/I_3^- 氧化还原离子对的电解质作为氧化还原介质，应用于 TiO_2 光电极和含有 Pt 涂层的对电极之间。

HOMO to the LUMO states. This process is represented by Equation (7-2). Once an electron is injected into the CB of the wide bandgap semiconductor, such as nanostructured TiO_2 film, the dye molecule (photosensitizer) becomes oxidized (Equation (7-3)). The injected electron is transported between the TiO_2 nanoparticles and then extracted to a load where the done work is delivered as an electrical energy (Equation (7-4)). Electrolytes containing I^-/I_3^- redox ions are used as redox mediators between the TiO_2 photoelectrode and the Pt coated counter electrode[55]. Therefore, the oxidized dye molecules (photosensitizer) are regenerated by receiving electrons from the I^- ion that get oxidized to I_3^- (Tri-iodide ions). This process is represented by Equation (7-5). The I_3^- substitutes the internally donated electron with that from the external load and reduced back to I^- ion at counter electrode, (Equation (7-6)). The movement of electrons in the CB of the wide bandgap nanostructured semiconductor is accompanied by the diffusion of charge-compensating cations in the electrolyte layer close to the nanoparticle surface. Therefore, generation of electric power in DSSC causes no permanent chemical change or transformation. This whole process is repeated many times to convert solar illumination into electrical energy. It is possible for this turnover number (number of regeneration cycles achievable by a single dye molecule) to reach $>10^8$, which translates to a remarkable lifetime of $15 \sim 20$ years for a stable DSSC based on a typical turnover frequency of ~ 0.1 Hz.

$$S^0 + \text{photon} \rightarrow S^* \tag{7-2}$$

$$S^* + TiO_2 \rightarrow e^-_{(TiO_2)} + S^+ \tag{7-3}$$

$$e^-_{(TiO_2)} + C.E. \rightarrow TiO_2 + e^-_{(C.E.)} + \text{electrical energy} \tag{7-4}$$

$$S^+ + \frac{3}{2}I^- \rightarrow S + \frac{1}{2}I_3^- \tag{7-5}$$

$$\frac{1}{2}I_3^- + e^-_{(C.E.)} \rightarrow \frac{3}{2}I^- + C.E. \tag{7-6}$$

An important and unique feature of the DSSC, unlike most other solar cell technologies, is the physical separation of charge generation, which occurs at the semiconductor-dye interface，from charge transport，which occurs within the semiconductor and electrolyte[56]. This is unlike inorganic or organic photovoltaics，where the materials used for generating charges are the same ones used to transport them，which makes it harder to optimize. For DSSCs，the greater design freedom allows the dye to be designed to maximize absorption across the solar spectrum，while the semiconductor and electrolyte can be separately designed to enhance charge transport properties. The more interesting feature in DSSC structure is the synergetic effect created by the unique combination of DSSC components. For example，although TiO_2 itself is not an electrical conductor，it is able to function effectively as an electron transporting mediate in DSSCs. Likewise，although ruthenium dyes can easily degrade when their solution is exposed to sunlight，they remain remarkably stable when adsorbed onto the TiO_2 in DSSCs，with a lifetime of over 15 years under outdoor irradiation based on extrapolation from accelerated device testing.

The overall performance of the DSSC is dependent on an intricate balance of the complex interactions between the various cell materials and components. Particularly critical is the dynamics occurring at the dye-sensitized electrode and electrolyte interface，where dye excitation，electron injection，and dye regeneration occur，as shown in Figure 7.24 for a typical DSSC using a ruthenium dye and an iodide-triiodide redox electrolyte. It is also at this interface where recombination losses that lead to major cell inefficiencies take place. The injected electrons at the interface can recombine with either the oxidized sensitizer or the oxidized redox couple（triiodide）. In addition，the excited dye at the interface can itself relax back to the ground state. Fortunately，the timescales for these recombination events are sufficiently slower than the pathways that direct electrons out to the external circuit. The well-timed and concerted effort by all the cell components acting together is what makes the DSSC design so remarkable and enables the respectable efficiencies that are observed.

56. 与大多数其他太阳能电池技术不同，DSSC的一个重要而独有的特征是，半导体-染料界面处的电荷产生，与半导体和电解质内的电荷传输，二者之间的物理分离。

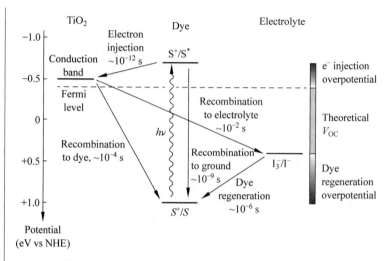

Figure 7.24　Energetics and dynamics of an n-DSSC

2. PSCs

Perovskite solar cells（PSCs）are a kind of solar cells based on multiple thin films. In a PSC, the incident lights are absorbed in the perovskite semiconductor to generate electrons and holes, which then transport through carrier transport layers and are collected at the respective electrodes. The work principle of perovskite is similar to OPV, excepting for the exciton dissociation process, because Wannier exciton is formed in perovskite material, whereas the excitons formed in OPVs are mainly Frenkel type. The binding energy of Wannier exciton is much low than that of Frenkel exciton; therefore, in PSCs, excitons can be dissociated easily by thermal energy at room temperature, and hence no extra driving force is needed[57].

7.4　Device Characterization

7.4.1　J-V Curves

The performance of solar cells is primarily characterized by the photon-response, i.e., the plot of photocurrent density vs voltage as shown in Figure 7.25(a). Usually, the bright J-V curve is measured under standard illumination conditions using the AM 1.5G solar spectrum（Air Mass 1.5 Global; integrated irradiance of 1000 W・m^{-2} or 100 mW・

57. 在钙钛矿太阳能电池中，入射光被钙钛矿半导体吸收后产生的电子和空穴，在载流子传输层传输，并在相应的电极上收集。除了激子解离过程，钙钛矿的工作原理与OPV相似，因为在钙钛矿材料中形成的是万尼尔激子，而在OPV中主要形成的主要是弗仑克尔激子。万尼尔激子的结合能远低于弗仑克尔激子。因此，在钙钛矿太阳能电池中，激子可以在室温下很容易地被热能解离，不需要额外的解离驱动力。

cm^{-2}). To judge the electrical property of the device, sometimes, dark current without illumination should also be recorded. Figure 7.25(b) is an equivalent circuit for a simplified OPV device, based on which, the J-V behavior in the circuit can be expressed as

$$J = J_0 \left[\exp \frac{q(V - JR_sA)}{nk_BT} - 1 \right] + \frac{V - JR_sA}{R_{sh}A} - J_{ph} \quad (7\text{-}7)$$

where J and V are current density and voltage, respectively, k_B is Boltzmann constant, T is absolute temperature, q is elementary charge, A is device area, n is ideal factor, J_0 is the reverse saturated current density, J_{ph} is saturated photocurrent; R_s and R_{sh} are series and shunt resistance. Equation (7-7) includes three parts connected by symbol of plus or minus, representing currents from the diode, the resistor and the photocurrent. When $V = 0$, the current from Equation (7-7) is the theoretical short circuit current (J_{SC}). Note that large J_{SC} requires small R_s and large R_{sh}[58].

58. 式(7-7)包括三部分,用减号或者加号连接,分别表示来自二极管、电阻和光电流的电流。当 $V = 0$ 时,式(7-7)中的电流为理论短路流(J_{SC})。请注意,大的 J_{SC} 电流需要小的 R_s 和大的 R_{sh}。

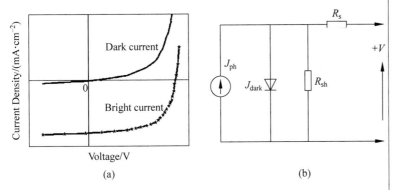

(a)   (b)

Figure 7.25 Steady-state response of current density versus voltage (a) and equivalent circuit of a solar cell (b)

Figure 7.26 describes the device parameters of open circuit voltage (V_{OC}), short circuit current (J_{SC}) and fill factor (FF). V_{OC} refers to the maximum photovoltage that can be produced in the device under illumination; it also means the condition when the applied reverse voltage cancels with the photovoltage under short circuit condition. In OPV cells, the V_{OC} is mainly determined by the energy gap between acceptor LUMO and donor HOMO. In n-DSSC cells, V_{OC} physically reflects the potential difference between the Fermi level of the photoanode and the Nernst potential of

59. 在 OPV 电池中，V_{OC} 主要由受体 LUMO 和给体 HOMO 之间的能隙决定。在 n-DSSC 电池中，V_{OC} 在物理上反映了光阳极费米能级与电解质中氧化还原离子对的能斯特电位之间的势能差。含有电子传输层（ETL）的 PSC 电池，V_{OC} 受控于 ETL 的 CB 与空穴传输层（HTL）的 VB 之间的能量差，而不含 ETL 的器件，V_{OC} 的决定性因素是钙钛矿 CB 与 HTL 的 VB 之间的能量差。

the redox mediator in the electrolyte. In PSCs with electron transport layer（ETL），V_{OC} is dominated by the energy difference between the CB of ETL and VB of hole transport layer（HTL）；while in devices without ETL，the determining factor for V_{OC} is the energy difference between perovskite CB and HTL VB[59]. J_{SC} refers to the current density under short circuit without external bias. Physically，J_{SC} expresses the total extractable current from the system，and will be higher if the active material can absorb over a broader range of solar spectrum. From the bright current curve，parameters of V_{OC} and J_{SC} can be obtained directly，which correspond to the intercept of the x- and y- axis，respectively.

The parameter FF is the ratio of the maximum power density（P_{max} in Figure 7.26）achieved by the device to its theoretical maximum power density（product of J_{SC} and V_{OC}）as shown in Equation（7-8）：

$$FF = \frac{P_{max}}{J_{SC} \times V_{OC}} \tag{7-8}$$

60. FF 图示的是 J-V 曲线形状的矩形情况；物理上，FF 表示所制造的太阳能电池的质量，其值随着分流电阻的增加而增加，随着串联电阻和过电位的增大而降低。

FF graphically represents how rectangular the shape of the J-V curve；physically，FF expresses the quality of the fabricated solar cell；it increases with the shunt resistance and decreases with series resistance and over potentials[60].

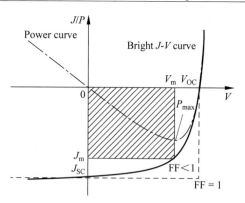

Figure 7.26　Description for J_{SC}, V_{OC} and FF of a solar cell

7.4.2　Device Efficiencies

The most important evaluation parameter for solar cell is device efficiency，which includes power conversion efficiency（PCE or η_p），external quantum efficiency（η_{EQE}）or termed as incident photon-to-current conversion efficiency (IPCE), and internal quantum efficiency (η_{IQE}) or termed as absorbed photon conversion efficiency (APCE). Among these efficiencies, PCE is a more practical parameter for device performance evaluation, the larger the value, the more promising the solar cell; while the other two are generally used for device mechanism investigation[61]. PCE can be calculated using:

$$PCE = \eta_p = \frac{J_{SC}V_{OC}FF}{P_{in}} \times 100\% \tag{7-9}$$

where J_{SC} is the short-circuit current density (mA $\cdot$ cm^{-2}), V_{OC} is the open circuit voltage (V), P_{in} is the power intensity of the incident light (mW $\cdot$ cm^{-2}), and FF is the fill factor. EQE (IPCE) is defined as the percentage of the number of charge carriers collected in the external circuit to the number of incident photons at a particular excitation wavelength:

$$IPCE(\lambda) = \eta_{EQE}(\lambda) = \frac{J_{SC}(\lambda)}{e\phi(\lambda)} \times 100\%$$

$$\approx \frac{J_{SC}(\lambda)}{\lambda P_{in}(\lambda)} \times 100\% \times 1240 \ (W \cdot nm \cdot A^{-1})$$

$$\tag{7-10}$$

where e is elementary charge, $\phi(\lambda)$ is the photon flux (cm^{-2} $\cdot$ s^{-1}) and λ is the wavelength of interest (nm). Two typical IPCE curves plotted as a function of λ are displayed in Figure 7.27. Internal quantum efficiency, IQE (APCE), is the fraction of photoelectron from the absorbed photons, i.e., the absorbed photon conversion efficiency. IQE is different from EQE by an absorption factor at each wavelength. Considering the process of light energy to electrical energy conversion, there are photon absorption, exciton dissociation, charge transport and charge collection, which can be expressed as η_A, η_{ED}, η_{CT}, and η_{CC}, respectively. We have

$$\eta_{EQE} = \eta_A \eta_{ED} \eta_{CT} \eta_{CC} \tag{7-11}$$

$$\eta_{IQE} = \eta_{ED} \eta_{CT} \eta_{CC} \tag{7-12}$$

$$\eta_{IQE}(\lambda) = \frac{\eta_{EQE}(\lambda)}{\eta_A(\lambda)} = \frac{\eta_{EQE}(\lambda)}{1 - 10^{-A(\lambda)}} \tag{7-13}$$

61. 太阳能电池最重要的评价参数是器件效率,包括功率转换效率(PCE 或 η_p)、外量子效率(η_{EQE})或称为入射的光子电流转换效率(IPCE),以及内量子效率(η_{IQE})或称为吸收光子转换效率(APCE)。在这些效率中,PCE 是器件性能评价的实用参数,其值越大,太阳能电池技术越有前景;而其他两个通常用于器件的机理研究。

where, A is the absorbance for photons at wavelength of λ (nm).

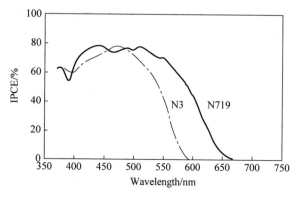

Figure 7. 27 Incident photon-to-current conversion efficiency (IPCE) of two ruthenium sensitizer dyes (N3 and N719) based DSSCs

7. 4. 3 Theoretical J_{sc}

Based on the photon-response curve of IPCE ($\eta_{EQE}(\lambda)$) of a solar cell, solar irradiation spectrum ($\phi(\lambda)$) and material absorption edge (λ_{Eg}), the theoretical $J_{SC}(\lambda)$ can be determined via Equation (7-14). This is simply calculated by evaluating the integral over the spectral range of the response of the device under test.

$$J_{sc} = \int_0^\lambda \eta_{EQE}(\lambda)\phi(\lambda)d\lambda \qquad (7\text{-}14)$$

where $\phi(\lambda)$ is the number of incident photons per wavelength, obtaining from the solar spectrum. Assuming that η_{EQE} is 100%, based on Equation (7-14), the theoretical maximum attainable photocurrent of solar cell, $J_{SC}^{max}(\lambda)$, can be obtained by integrating from zero to the cutoff wavelength of light absorption[62]. Figure 7. 28 demonstrates the theoretical $J_{SC}(\lambda)$ (Figure 7.28(a)) for a given IPCE and the theoretical maximum photocurrent attainable in a solar cell (Figure 7. 28(b)) with a given optical gap under AM 1. 5G irradiation (100 mW · cm^{-2}).

It can be seen, in many solar cell devices, that the short circuit current calculated from IPCE measurement is different to J_{SC} from J - V measurement, with either lower or higher

62. 假设 η_{EQE} 为 100%，基于式(7-14)，从零到光吸收截止波长进行积分，获得太阳能电池的最大理论光电流 J_{SC}。

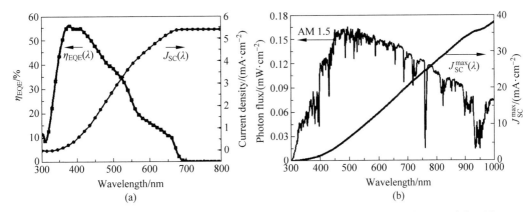

Figure 7.28 (a) An IPCE curve ($\eta_{EQE}(\lambda)$) of a solar cell and the calculated theoretical J_{SC} (J_{SC} at 800 nm); (b) Dependence of theoretical maximum $J_{SC}^{max}(\lambda)$ on their absorption edges under AM 1.5G irradiation

values derived from IPCE. The factors for the J_{SC} calculated from IPCE lower than that of from J-V measurement are: ①Barrier for the photocurrent which is large under low light intensity or monochromatic illumination but lowered by photodoping of the buffer at AM 1.5 illumination; and ②large number of micro shunts. If the cell is irradiated only on a limited area (as in IPCE measurement), the non-illuminated part acts as a shunting load. Seen from the active solar cell, this load is in parallel with the input resistance of the IPCE current amplifier. Although the total shunt resistance of the cell (measured as a macroscopic quantity by J-V analysis) is high, the local shunt resistance seen from the active solar cell part may be small. Thus, the current is drained throughout the shunting load. Reducing the cell area to the area of illumination may increase the IPCE. On the other hand, the reason for lower J_{SC} from J-V curve than that from IPCE curve often lies in the fact that irradiation under AM 1.5 generally has stronger light intensity compared to IPCE measurement. Intensive light incidence produces more charge carriers hence more space charges in the device, forming larger energy barriers for charge transport/collection and hence leading to decreased photocurrent density. Therefore, the J_{SC} value from J-V curve under AM 1.5 illumination is lower than that calculated from IPCE measurement under weak irradiation[63].

63. 基于 IPCE 的 J_{SC} 小于基于 J-V 测量的，因素是：①在低光子强度或者单色光情况下，光电流的势垒较大，而在 AM 1.5 辐照下由于光掺杂，这个势垒降低了；②存在大量的微并联电阻。如果光照仅仅发生在有限的面积时（IPCE 测量就是这样），没有被照射的部分将作为并联负载。从工作的电池来看，这个负载与 IPCE 电流放大器的输入电阻是平行的。虽然电池的总并联电阻较高（基于 J-V 宏观量的分析），工作电池的局部并联电阻可能较小。因此，电流通过并联负载被漏掉。将器件的面积降低至辐照面积，可能增加 IPCE。另外，基于 J-V 曲线的 J_{SC} 小于基于 IPCE 曲线的原因，通常是由于这样一个事实：AM 1.5 的光入射通常比 IPCE 测量时的光入射强度大，强光产生了更多的电荷载流子，因此器件的空间电荷较多，形成了较大的电荷传输/收集势垒，导致了降低的光电流密度。因此，基于 AM 1.5 辐照 J-V 曲线的 J_{SC} 比基于较弱辐照 IPCE 测量的 J_{SC} 小。

本章小结

1. 内容概要：本章讨论了含有有机材料的太阳能电池器件,包括有机太阳能电池、染料敏化太阳能电池和钙钛矿太阳能电池。首先对太阳辐射和太阳能电池的发展进行了简单介绍,然后对这三种太阳能电池的器件结构及工作原理给予了详细介绍。最后,讨论了太阳能器件的性能表征参数,对太阳能电池的电流电压曲线的特征给予了理论描述。

2. 基本概念：太阳辐射、光伏、有机太阳能电池、染料敏化太阳能电池、钙钛矿太阳能电池、活性层、本体异质结、平面异质结、激子解离、电荷转移、给体、受体、填充因子、短路电流、开路电压、功率转换效率、理论短路电流。

3. 主要公式：

(1) 黑体辐射：
$$\frac{\mathrm{d}J_E}{\mathrm{d}E_p} = \frac{2\Omega}{h^3 c^2} \frac{(h\nu)^3}{\exp\left(\dfrac{h\nu}{k_B T}\right) - 1}$$

(2) 太阳能电池的电路电流：
$$J = J_0 \left[\exp\frac{q(V - JR_S A)}{nk_B T} - 1\right] + \frac{V - JR_S A}{R_{sh} A} - J_{ph}$$

(3) 填充因子公式：
$$FF = \frac{P_{max}}{J_{SC} \times V_{OC}}$$

(4) 功率转换效率公式：
$$PCE = \eta_p = \frac{J_{SC} V_{OC} FF}{P_{in}} \times 100\%$$

(5) 入射光子转换效率(外量子效率)：
$$IPCE(\lambda) = \eta_{EQE}(\lambda) = \frac{J_{SC}(\lambda)}{e\phi(\lambda)} \times 100\% \approx \frac{J_{SC}(\lambda)}{\lambda P_{in}(\lambda)} \times 100\% \times 1240 \ (W \cdot nm \cdot A^{-1})$$

(6) 外量子效率、内量子效率及材料吸光度之间的关系：
$$\eta_{IQE}(\lambda) = \frac{\eta_{EQE}(\lambda)}{\eta_A(\lambda)} = \frac{\eta_{EQE}(\lambda)}{1 - 10^{-A(\lambda)}}$$

Glossary

Active layer	活性层
Black body radiation	黑体辐射
Bulk heterojunction	本体异质结
Carrier transport	载流子传输
Charge collection/extraction	电荷收集/提取
Charge transfer	电荷转移
Charge transport	电荷传输
Collection	收集
Donor	给体
Dye sensitized solar cell	染料敏化太阳能电池

Effective conjugation	有效共轭
Energy	能量
Energy resource	能源
Environment friendly	环境友好的
Exciton diffusion	激子扩散
Exciton dissociation	激子解离
Exciton generation	激子产生
Exhaust	用尽,耗尽,排出
Fill factor（FF）	填充因子
Inorganic solar cell	无机太阳能电池
Light radiation	光辐射
Living environment	生存环境
Molecular D-A structure	给体-受体（D-A）的分子结构
Open circuit voltage（V_{OC}）	开路电压
Organic solar cell	有机太阳能电池
Output power	输出功率
Performance characterization	性能表征
Photon absorption	光吸收
Photovoltaic	光伏
Planar heterojunction	平面异质结
Power conversion efficiency（PEC）	能量转换效率
Recombination	复合
Resource	资源
Short circuit current（J_{SC}）	短路电流
Solar radiation	太阳辐射
Tandem structure	串联结构,叠层结构
Visible light	可见光

Problems

1. True or false questions

(1) The AM 1.5 solar radiation is perpendicular to the earth surface.

(2) In terms of organic and inorganic solar cells，OPV and DSSC belong to all-organic devices，while PSC belongs to hybrid device.

(3) Among the three solar cells containing organic materials（OPV，DSSC and PSC），OPV generally exhibits the highest power conversion efficiency.

(4) In the active layer of OPV device with high performance，there are generally two organic materials.

(5) In OPVs the optical pumped excitons can be dissociated by thermal activation.

（6）In OPVs, the main driving force for exciton dissociation is the interface between donor and acceptor.

（7）In OPVs, the open circuit voltage is determined by the work function difference between anode and cathode.

（8）In DSSCs, the optical pumped excitons can be dissociated by thermal activation.

（9）Excitons in PSCs are mainly Frenkel type.

（10）Among the three solar cells containing organic materials, OPV generally exhibits the highest operation stability.

2. Please (1)tell the reasons for developing solar cell technology, and (2)discuss the development of organic-containing solar cells in terms of: (a) what are they, (b) the milestones, (c) the overall performance.

3. Please explain the structure of OPV device in detail.

4. Please explain the structure of DSSC device in detail.

5. Please explain the structure of PSC device in detail.

6. Please point out the main optoelectrical conversion processes in an OPV device.

7. Please discuss the working principle of a DSSC device.

8. Please explain the mechanism of a PSC device.

9. Please draw I-V curves for an OPV device both under dark and irradiation conditions. Indicate parameters of open circuit voltage, short circuit current, and fill factor in the curve. Give the equation for PCE (power conversion efficiency).

CHAPTER 8   ORGANIC LIGHT-EMITTING DIODES

After brief historical introduction，this chapter gives a comprehensive study on the mechanism for organic electroluminescence，especially in the aspects of carrier injection，carrier recombination to produce excitons and their space distribution. In the meantime，device structure and corresponding functionality of each component are explained. Then，characterization parameters for organic electroluminescence devices are discussed in deep，giving concise concepts for various important parameters in terms of device electrical and optical properties，as well as device efficiency. Finally，mainly focusing on drive-voltage，effects on device lifetime are stated.

8.1   Introduction

Electroluminescence（EL）refers to the light that emits from an activated media excited by an applied electric field. EL represents，in fact，the direct conversion of electrical energy into light energy by various means of electrical excitation of the optical modes of a condensed matter，without recourse of any intermediate energy form，such as heat[1]. Organic EL is also called organic light emitting diode （OLED），in which organic materials are used as the active materials. Phenomena of organic EL were discovered in 1960s based on crystal antheracene（Figure 8.1(a)）. Around 1963，both groups of M. Pope and R.E. Viso made a model organic EL device comprising a $10 \sim 20$ μm thick anthracene crystal as the active material，and silver paste on both sides of anthracene as electrodes（Figure 8.1(b)）. When the applied voltage is higher than 400 V，blue light can be observed. In 1966，M. Helfrich and W.G. Schneider reported the blue EL from thick anthracene crystal（$1 \sim 5$ mm）using liquid electrodes，blue EL was observed when bias of $50 \sim 1000$ V was applied（Figure 8.1(c)）.

1. 实际上，EL 表示利用各种电学手段激发凝聚态物质的光学模式后，无需任何中间的能量形式（如热能），将电能直接转换为光能的方式。

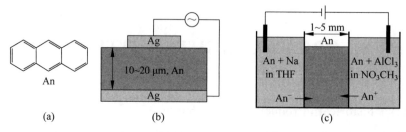

Figure 8.1　The discovery of organic EL in 1960s. （a）Anthracene molecule，（b）model organic EL device comprising anthracene single crystal and silver paste electrodes，（c）thick anthracene crystal in liquid electrodes for blue EL

2. 由于驱动电压很高——这是蒽晶体片必须制备成厚膜的缘故，以及蒽单晶的机械脆弱性，在随后的几年中，对有机固体电致发光应用的研究并不是很深入。

Owing to the high driving voltages - made necessary by the large layer-thickness of the anthracene crystal platelets，and due to the mechanical fragility of the single crystals，research on the applications of electroluminescence in organic solids was not very intensive in the following years[2]. This changed drastically after C. W. Tang in 1987 developed a thin double-layer device with evaporated organic thin films （Figure 8.2）. With this heterostructure，a promising device performance with voltages smaller than 10 V，maximum brightness of 1000 nit （1 nit = 1 cd · m^{-2}），and external quantum efficiency （EQE） up to 1% was achieved. In such a heterostructure OLED under operation，holes and electrons are injected from electrodes （usually inorganic） into organic layers，where they are then transported to the emissive layer （EML）. The opposite charges form an exciton，which decays to electronic ground state while emitting a photon[3]. There are three striking novelties in Tang cell：①very thin organic layers of 50 nm or less to enable low driving voltages，②separated hole and electron conducting layers to provide efficient injection and recombination by permitting optimization of electron/hole injection and transport，and simultaneously，moving the recombination zone far away from electrode to prevent exciton quenching by electrodes，and ③the introduction of different electrodes for anode and cathode，e. g.，ITO for anode and low work-function metal for cathode，to improve electron injection. Shortly after the introduction of heterostructure-based thin film OLEDs，two-component emitting-layer consisting of emitter molecules doped into an appropriate host matrix was demonstrated，

3. 这种异质结 OLED 工作时，空穴和电子从电极（通常是无机材料）注入有机层，然后在各自的有机层中被输送到发光层（EML）。相反的电荷复合形成激子，激子衰减到电子基态，伴随着光子发射。

leading to a great increase in device efficiency，which is mainly accounted from the decrease of self-quenching of the emitting dopants[4].

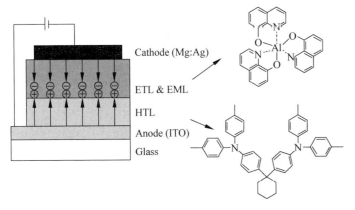

Figure 8.2　The double layer heterostructure OLED reported by C. W. Tang in 1987

In 1990，R. H. Friend firstly reported a polymer OLED，which employed polymer PPV as emission layer and EQE of 0.05% was observed. The white OLED（WOLED）was first developed by J. Kido in 1994，demonstrating the potential of OLED for lighting applications. In the late 1990s，phosphorescent materials were introduced into OLED（PhOLED）as dopant emitters by S. R. Forrest and M. E. Thompson. PhOLEDs are based on the utilization of triplet excitons that statistically take up 75% of the total excitons by injection method. Currently，the main strategy to achieve high-efficiency phosphorescence is to incorporate a heavy metal atom into an organic molecule，so that spin-orbit coupling can occur，promoting both intersystem crossing between singlet and triplet states and radiative decay of the triplet exciton. Very recent mechanism-based breakthrough in OLED fields is the demonstration of thermal activated delayed fluorescence（TADF）in the year of 2015 by C. Adachi. TADF-based OLEDs employ fluorescent emitters with partial radiative components originated from the electrical pumping of the triplet states，which transfer to the emissive singlet states by reverse intersystem crossing（RISC）with the help of thermal energy. By such，both singlet and triplet electrically pumped excitons can be converted to light

4. 在引入异质结薄膜 OLED 后不久,发光分子掺杂到适当主体基质中的双组分发光层的研究被报道出来。主要是由于发光分子自淬灭的减少,这种结构大大提高了器件发光效率。

5. 基于 TADF 的 OLED 采用的是荧光发光物质，但是这些荧光分子的部分发光起源于电激发的三线态，这些三线态在热辅助下，通过反向系间窜跃（RISC）转化为发光单线态。这样，电激发的单线态和三线态激子都可以转换为光发射，因此 100% 的理论内量子效率（IQE）是可以预期的。

6. 此外，OLED 材料和制造技术的最新进展已经实现了大尺寸显示器、专业照明甚至普通照明的应用。这些新颖的光源不仅提供了能源效率的前景，而且由于新颖的外形特征，还可以为房间照明设计提供全新的方法，以及应用在其他照明技术无法达到的透明和柔性器件中。

7. 对于基于 OLED 的电视、照明和其他大面积器件的应用，所有这些必须进一步减少，以便充分发挥其潜力，特别是在需要低成本的照明市场中，这些产品必须同时实现消费者可接受的寿命、色彩质量、均匀性、亮度和效率。

emission，hence 100% theoretical internal quantum efficiency (IQE) is expected[5].

During the development of OLEDs，the materials and technology have advanced rapidly. In just over 30 years，quantum efficiency （or the fraction of charge that is converted into light） has increased more than 20-fold，approaching the theoretical limit for IQE. Besides promising EL efficiency，other advantages of OLEDs include self-emission，large viewing angle，flexible，low driving voltage and low cost. Since discovery，OLEDs have evolved from a scientific curiosity to a commercially viable technology incorporated into hand-held devices for bright，vibrant displays that offer beautiful colors and unmatched viewing angles. Furthermore，recent progress in OLED materials and manufacturing technologies has realized large-scale displays，specialty lighting，and even general lighting. These novel light sources，not only offer promise of energy efficiency，but also，may enable entirely new approaches to room lighting design due to novel form factors；and for the transparent and flexible devices that are not easily accessible with other lighting technologies[6].

Despite these advances，the challenges lie in scaling down all OLED manufacturing process steps，such as cost of substrate manufacturing，organic thin film deposition，and electrode deposition. All these must be further decreased for OLED-based televisions，lighting，and other large-area applications in order for OLEDs to achieve their full potential，especially in low-cost markets such as lighting，products，which must simultaneously achieve lifetime，color quality，uniformity，brightness，and efficiency acceptable to a consumer[7]. To save energy and money for consumers，efficiency demands are also very challenging. Fortunately，resolution，pixel pitch，and ambient contrast requirements are relaxed as compared to displays. However，the biggest challenge to OLEDs for lighting from the perspective of device materials is combining high efficiency with longevity. The search continues for stable and efficient materials and device architectures，and for tackling increasing important challenges that have come to the fore in manufacturing technology: substrate and electrode processing, light extraction, and cost.

8.2 Working Principles

8.2.1 Types of Electroluminescence

Electroluminescence (EL) refers to the process of light radiation via the interaction between active components and electric field; if the active components are organic materials, then it is termed as organic EL (OEL). The classification of ELs can be carried out based on the following three aspects: ①pumping source (direct or alternative electric field), ②excitation method (injection or noninjection), ③radiative decay scheme (fluorescence or phosphorescence)[8].

As shown in Figure 8.3, both direct and alternative electric fields can serve as driving source for EL process; but the former is only applied to injection type device, while devices under the latter scheme can be injection and noninjection types. EL devices driven by direct electric-field comprise active components of crystal, microcrystal or thin film;

8. EL 的划分可从三个角度考虑：①激发源类型（直流或者交流电场）；②产生激发态的方式（载流子注入或者非载流子注入）；③光辐射跃迁的机制（荧光发射或者磷光发射）。

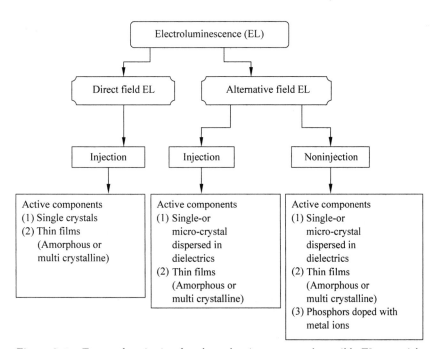

Figure 8.3 Types of excitation for electroluminescence and possible EL materials

the active components in alternative electric field driven EL devices can be microcrystal dispersed in dielectric material, thin film, and phosphors doped with metal ions. For EL devices based on films, there are: ① thin film EL (TFEL), ② inorganic light emitting diode (LED), and ③ metal-insulator-metal (MIM) type EL.

TFEL based EL belongs to noninjection ELs (Figure 8.4 (a)), with basic structure of the emitting layer sandwiched by high dielectric-constant insulator and electrode in sequence. Normally, the emitting materials are doped as dispersion in the emitting layer; and electrons in the insulators are accelerated by high alternating electric field, which then collide with emitting material to endow excitation, followed by the radiative decay of electrons in the emitting material. Drawbacks of TFEL include high driving voltage and low efficiency. Additionally, the preparation of TFEL often needs high-temperature methods, such as vacuum deposition, sputtering and chemical vapor deposition (CVD); the emitting layer needs to be annealed under high temperature of 500 ℃ to 600 ℃, requiring high temperature substrate. LED is inorganic p-n junction based EL of injection type (Figure 8.4(b)). Upon applying electric field, carriers inside the semiconductor, as well as those injected from electrodes, can be driven to p-n junction, where emission can be produced via band to band recombination[9]. Generally, the preparation of inorganic LED device requires making multiple layers by the technique of molecular beam epitaxial deposition with highly pure materials, hence the cost is high. In the case of organic ELs, they are injection based EL with MIM-like structure (Figure 8.4(c)), in which there are no p-n junction and free carriers. Almost all the positive and negative charges are injected from electrodes, and then moving oppositely, and partially recombine with each other to form excitons, then emission can be produced via radiative decay.

Comparing between LED and organic EL, there are mainly three differences: ①In LED, stable p-n junction can be obtained via doping; but no stable and reproducible p- and n- type organic semiconductors are available to form stable p-n junction, without the influence of chemical reaction and diffusion. ②Free carriers exist in LED before exerting bias;

9. LED 是基于无机半导体 p-n 结的注入型 EL,如图 8.4(b)所示。加载电场,半导体内的载流子以及从电极注入的载流子,可以被驱动到 p-n 结,在此通过带到带的复合产生发光。

while in organic EL，carriers are injected from electrodes，there are no free carriers before applying voltage．③In LED，carriers transport via band，which is fast and mobile，and the recombination of electron and hole at p-n junction leads to band to band emission；In organic EL，due to the disordered films，carriers transport via hopping，which tends to be localized and polarized，the recombination of electron and hole results in relatively localized exiction and the emission is exciton type[10]．Considering the injection of positive and negative carriers and the recombination of electron and hole for emission，organic EL is also called organic light emitting diode（OLED）．

10．有机 EL 与 LED 相比,主要有三点区别：①LED 可通过掺杂形成稳定的的 p-n 结,而有机材料不可能得到稳定的、重复性很好的 p-型和 n-型半导体材料来形成稳定的 p-n 结,而不受化学反应和扩散的影响；② 无机 LED 中,在加电场之前就存在自由载流子,而有机 EL 中,载流子是通过电极注入的,在没有外加电场时,是不存在自由载流子的；③ 在 LED 中,载流子以较快、流动性较奻的能带模式输运,电子与空穴在 p-n 结处的复合产生能带到能带之间的光辐射；在有机 EL 中,由于薄膜的无序性,载流子以跃进方式输运,倾向于定域化和极化,电子和空穴的复合产生相对定域化的激子,光辐射是激子型的。

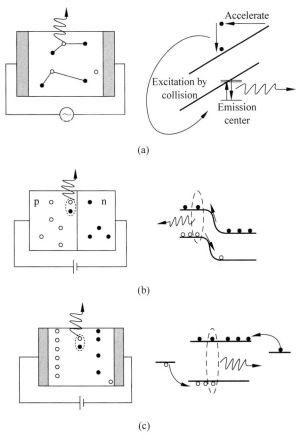

Figure 8. 4　Emission schemes for thin-film based EL devices：（a）Intrinsic EL under high electric field（TFEL），（b）EL based on p-n junction of semiconductors（LED），（c）electrode/insulator/electrode（MIM）type EL via charge injection，which includes single or multiple layer wide bandgap inorganic or organic materials

11. 由于有机半导体本身通常没有本征载流子,因此 OLED 器件中,参与发光过程的载流子需要通过电极注入:空穴从阳极(通常是透明的导电氧化物薄膜,如 ITO)注入到空穴传输层(HTL)的最高占用分子轨道(HOMO),电子从阴极(通常为低功函数的金属,如 Mg、Ca、LiF/Al)注入到电子传输层(ETL)的最低空置分子轨道(LUMO)。

8.2.2 Overview of OLED Working Principle

Figure 8.5 depicts the working principles of an OLED. As organic semiconductors are normally no intrinsic carriers available in themself, the carriers, which are involved in the light emitting processes in an OLED device, are injected through electrodes: holes are injected from anode (normally transparent conducting oxide film such as ITO) to the highest occupied molecular orbital (HOMO) of hole transporting layer (HTL), electrons are injected from cathode (normally low work function metal such as Mg, Ca, LiF/Al) to the lowest unoccupied molecular orbital (LUMO) of electron transporting layer (ETL)[11]. Under the applied voltages, the injected holes and electrons are driven by the electric field towards each other in opposite directions and then recombine inside of organic layers to form excitons (the preferred location is the emitting layer (EML)), finally the excitons decay to the electronic ground state with emission light.

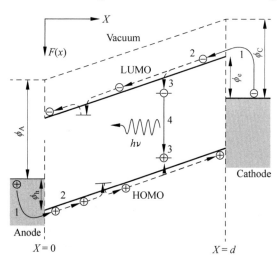

Figure 8.5 Micro processes in OLED operation

8.2.3 Injection and Transport Barriers at Interfaces

1. Energy Barriers in an OLED

During OLED operation, electrons and holes are injected from the cathode and the anode, respectively, and then transport oppositely through multiple organic layers respectively to the anode and cathode, if no recombination occurs. Along their path, electrons and holes traverse several interfaces due to their movements from one material-layer to another. The word interface here refers to the contact area between two different solid materials. At each interface, there is the possibility of an energy barrier that the electron (or hole) must overcome in order to enter the next material. Energy barriers increase the driving voltages of OLEDs, and also result in charge-accumulation regions that increase the probability of non-radiative recombination[12].

As shown in Figure 8.6, the magnitude of interfacial energy barriers depends on the relative electronic positions of the two adjacent materials. For organic-electrode interface, the hole injection barrier refers to the energy difference between anode Fermi level and the HOMO of the contacting material (Δ_h in Figure 8.6), and the electron injection barrier refers to the energy difference between cathode Fermi level and the LUMO of the contacting material (Δ_e in Figure 8.6). For an organic-organic interface, the energy barrier refers to HOMO difference (hole barrier, Δ_{HOMO} in Figure 8.6) and LUMO difference (electron barrier, Δ_{LUMO} in Figure 8.6) of the two organic materials;. Under operation, after hopping from the cathode's Fermi level into the LUMO of the electron-transporting organic layer (labeled A in Figure 8.6), the electron travels through the bulk of an organic semiconductor by hopping from the LUMO of one molecule to that of a neighboring molecule. Please notice that even moving within the same material layer from one molecule to another, it is not entirely barrier-less because molecules polarize when they are negatively charged, resulting in a polaron binding energy[13]. When an electron approaches an interface to a different organic material (labeled B in Figure 8.6), having a different LUMO level, it will come across an energy barrier (Δ_{LUMO}); the electron can either hop into the LUMO of B or recombine with holes in the HOMO of material B. Electrons and holes recombine either radiatively (i.e., emitting light) or nonradiatively (i.e., generating heat). In the meantime, in order for holes to be available in the HOMO of material B, they must be provided from the

12. 界面在这里指的是两个不同固体材料的接触区域。在每个界面上，都可能存在一个能量势垒，电子（或空穴）必须克服这个势垒才能进入下一个有机材料。能量势垒增加了 OLED 的驱动电压，还导致电荷积累区域，从而增加了非辐射复合的可能性。

13. 请注意，即使（电子）在同一材料层内从一个分子移动到另一个分子，也不是完全没有势垒的，因为分子带负电时会极化，从而产生极化子结合能。

14. 同时,为了使材料 B 的 HOMO 中有可以利用的空穴,必须从阳极提供。空穴从阳极的费米能级跳入空穴传输材料的 HOMO 级(图 8.6 中标记为 B)。

anode. Holes hop from the Fermi level of the anode into the HOMO level of the hole-transporting material (labeled B in Figure 8.6)[14].

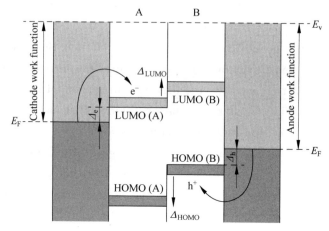

Figure 8. 6 Schematic energy-level diagram of a simple, two-layer OLED. The arrows indicate the movement of electrons (e^-) and holes (h^+). E_v represents the vacuum level

2. Contact Barriers

15. 特别地,在 OLED 器件中存在两个重要的界面,涉及有机半导体和电极之间的接触,它们被称为电接触,并且已被证明电接触界面的问题更多。

High-efficiency OLEDs rely heavily on good-quality interfaces, such as good physical contact and small energy barriers. Particularly, in OLED devices there are two important interfaces involving the contact between organic semiconductor and the electrode, which are termed as electrical contacts and have proven to be much more problematic[15]. Electric contact has been discussed in detail in Chapter 4 of this textbook.

In OLED devices, the contacts between the electrode and the organic active-component are generally considered as Ohmic contact, which means that carrier density at the contacts are far more higher than the bulk of organic materials. In the situation of intrinsic carrier is higher than the injection carrier under low driving voltages, and the situation of capacitance-determined carrier density at higher electric field, all Ohmic contacts are the same. This means that the quantities of the injected electrons and holes are the same. Determined by injection barrier and the applied electric field, carrier injection from electrodes can be thermal injection and quantum tunneling (Details can be referred in Chapter 4).

Decreasing injection barrier facilitates charge injection and hence reduces turn-on voltage of the device. The first strategy to decrease injection barrier is the selection of energy-matching electrode in contact. The minimum injection barriers are achieved when the frontier-orbitals (i. e., HOMO and LUMO levels) in organic semiconductor are aligned with the Fermi levels of the respective electrodes. At the cathode interface, the LUMO of the electron-transporting material should be aligned with the cathode Fermi level, and at the anode interface, the HOMO level of the hole-transporting material should be aligned with the anode Fermi level. This is what is referred to as energy-level matching[16]. In order to mitigate the charge-injection energy barriers in an OLED, it is often necessary to modify the electrode surface by incorporating a buffer layer of some sort between the electrode and the organic layer. Buffer layers are generally very thin (ca. $1 \sim 10$ nm), and are used to provide better energy-level alignment across an interface.

Although the charge injection barrier is typically determined by the energy difference between the Fermi level of the electrode and the contact organic material, in the real case, things are different. Due to the redistribution of carriers caused by many factors, such as interfacial charge transfer, electron cloud reorganization, interfacial chemical reaction, etc, there always exists interfacial dipole layer at contacts, which will influence the injection barrier. The interfacial dipole layer changes the potential field, leading to the shift of energy levels in organic materials[17]. As shown in Figure 8. 7, if the negative charge locates at the electrode side, and the positive charge at the side of organic film, there will be a Δ down-shift in organic film compared to the electrode Fermi level. As a result, the electron injection barrier (ϕ_{Be}) decreases from $\phi_M - EA$ to $\phi_M - EA - \Delta$, and the hole injection barrier (ϕ_{Bh}) increases from $IP - \phi_M$ to $IP - \phi_M + \Delta$.

16. 在阴极界面处,电子传输材料的 LUMO 能级应该与阴极费米能级对齐,在阳极界面处,空穴传输材料的 HOMO 能级应该与阳极费米能级对齐。这就是所谓的能级匹配。

17. 由于多种因素导致的电荷重新分布,如界面间电荷转移、电子云重新分布、界面化学反应等,在接触处总是存在界面偶极层,它将影响电荷注入势垒。界面偶极层改变了势能场,使有机材料的能级发生移动。

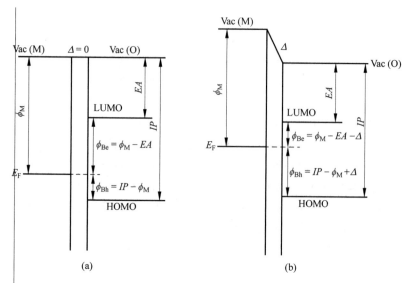

Figure 8.7　Schematic views for energy levels at electric contact，without dipole layer（a），with dipole layer（b）

3. Anode Materials and ITO-Anode Modification

To inject hole from anode into an OLED，the basic requirement is the alignment of anode Fermi level to the HOMO of the hole transport layer（HTL）that contact to the electrode. Good anode materials possess the following features：① high conductivity，② good chemical and morphological stability，③high work function，④transparency in visible-light range if the anode side is the direction for light emission[18]. High work function metals such as gold and silver，transparent conducting oxides such as indium tin oxide（ITO），carbon black and conducting polymers can be used as anode for OLEDs. These materials possess work functions in the range of 4.5 eV to 5.0 eV，which can effectively inject holes in to organic materials with HOMO levels lying in 5.0 eV to 5.5 eV.

ITO is the most utilized anode material in OLEDs，its work function lies at 4.0 eV to 4.5 eV. The advantages of ITO are good stability，high transparency（$E_g = 3.5$ eV to 4.3 eV，transmittance up to 90% at visible light range），and low resistivity（about 2 $\Omega \cdot$ cm to 4×10^{-4} $\Omega \cdot$ cm）. The electrical conduction of ITO relies on the oxygen vacancy and the n-type doping of tin into indium oxide during

18. 为了从阳极注入空穴到 OLED 器件，基本要求是阳极费米能级和与之接触的空穴传输层（HTL）的 HOMO 相匹配。良好的阳极材料有以下特点：①高电导率；②优良的化学及形态稳定性；③高功函数；④如果阳极一侧是光输出方向，则阳极材料要求在整个可见光范围内是透光的。

preparation. One oxygen vacancy offers two excessive electrons in In_2O_3, while the substitution of Sn with four valence electrons to In with three valence electrons can produce one excessive electron, both of which can produce n-doping effect. ITO film is often prepared by vacuum sputtering, chemical vapour deposition and spray pyrolysis, etc, which are all high temperature technique. Low temperature preparation of ITO film includes pulsed ablation, ion beam assisted deposition and negative sputtering ion beam technology. In order to improve hole injection efficiency with ITO anode, modification have been done by several methods:

1) Oxygen Plasma Treatment

Treating ITO surface with oxygen plasma can not only clean the surface and improve the work function of ITO hence reducing hole injection barrier from ITO to organic thin film, but also improve the wetting property of ITO surface hence improving the quality of organic film deposited on it[19]. Figure 8.8 schematically shows the set up for oxygen plasma. In the procedure, oxygen gas is introduced into the inductance or capacitance coupled radio frequency energy field. Oxygen plasma is produced via ionization. Typically, the capacitance coupled oxygen plasma is simple and easy for large area, and is widely used in OLED preparation.

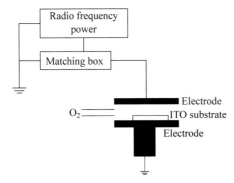

Figure 8. 8　Set up for oxygen plasma treatment to ITO substrate based on capacitance coupling

It is found that oxygen plasma treatment to ITO can increase ITO surface work function due to the increase of oxygen and the decrease of Sn concentration at ITO surface,

19. 采用氧等离子体处理 ITO 表面,不仅可以清洁 ITO 表面而且可以提高 ITO 的功函数,从而减小从 ITO 向有机薄膜的空穴注入势垒,还可以提高 ITO 表面的浸润性能,改善有机材料在 ITO 薄膜上的成膜性能。

20. 研究发现 ITO 氧等离子体处理，可以提高 ITO 表面的功函数。其原因是，ITO 表面氧原子含量的提高和锡原子的降低导致 ITO 表面电子的减少，因而功函数增大。因为氧等离子体处理仅仅对 ITO 表面有效，不影响内部，所以 ITO 内部的载流子浓度不变，其导电性也不变。

21. 如图 8.9 所示，当 ITO 表面吸附酸的单分子层时，质子化的表面将吸附带负电的阴离子，形成偶电层，使得 ITO 功函数增大。类似地，当 ITO 表面吸附碱单分子层时，形成的偶电层使得 ITO 功函数减小。

which lead to the decrease of electrons in ITO surface hence increasing work function. Because the plasma treatment only affects ITO surface，but not the bulk，the carrier density of the bulk does not change，hence ITO conductivity also does not change[20]. Oxygen plasma treatment is the most often used method to improve the work function of ITO because it is convenient and can improve device driving voltage significantly，the stability and lifetime of OLEDs can also be improved.

2) CF_3H Treatment

Similar to the process of oxygen plasma treatment，trifluoromethane (CF_3H) plasma treatment on ITO surface is another effective way to modify ITO surface. Introducing CF_3H plasma on ITO surface induces polymerization reaction，that is，the fluorine atoms in CF_3H can be bound to ITO surface to form CF_x buffer layer with high ionization energy and low resistivity，which can effectively increase the work function of ITO surface. CF_3H plasma treatment，on the one hand，prevents In atoms diffusing to organic layer，on the other hand，improves the hole injection efficiency due to the increase of ITO work function and prevent the degradation of ITO surface. It has been found that the OLED devices fabricated on low-frequency CF_3H plasma treated ITO exhibit low driving voltage and long life time.

3) Acid and Base Monolayer

The surface of ITO can be modified by acid or base adsorption，with which the work function of ITO can be greatly changed，up to ± 0.7 eV. As shown in Figure 8.9，when acid monolayer is adsorbed on ITO surface，the protonated surface will attract negative cations，leading to the formation of dipole layer which increases the work function of ITO. Similarly，the adsorption of base monolayer on ITO surface leads to dipole layer with opposite direction，hence ITO work function decreases[21]. The most effective acid and base are H_3PO_4 and $N(C_4H_9)_4OH$ (Figure 8.9(c) and (d)). Although the influence of acid or base on the work function of ITO is significant，it is not stable. For example，when NPB (4, 4'-bis (N-(1-naphthyl)-N-phenyl-amino) biphenyl) is deposited on acid treated ITO substrate，due to the interaction between nitrogen in NPB and hydrogen in acid, the dipole layer is destroyed and the ITO work function will decrease.

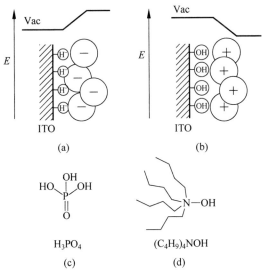

Figure 8. 9　Charges of electron potential in ITO surface by acid (a) and base (b) adsorption, molecular structure for phosphoric acid (H_3PO_4) (c) and tetrabutylammonium hydroxide ($N(C_4H_9)_4OH$) (d)

4. Al-Cathode Modification

In OLED devices, the most commonly used cathode is Al, which possesses moderate activity; and once oxidized, only a layer of compact Al_2O_3 is formed on the Al surface, preventing further oxidation of Al. However, the work function of Al is 4.3 eV, not matching with most of the LUMOs in organic materials. Therefore, Al is generally paired with LiF when serving as cathode in OLEDs[22]. The working principles of LiF modification between the electron transport layer Alq_3 (tris (8-hydroxyquinolinato) aluminum (Ⅲ)) and Al cathode are intensively studied and are summarized as following:

Investigation by ultraviolet photoemission spectroscopy (UPS) has been demonstrated that inserting LiF between Alq_3 and Al leads to the downward shift of Alq_3 energy levels as well as its vacuum level. As shown in Figure 8. 10 (a), without LiF, due to the dipole layer potential (1.0 eV)

22. OLED 器件中,最普遍使用的阴极是 Al,它拥有适中的活性,当被氧化时,仅仅在 Al 表面形成一层致密的 Al_2O_3,防止了 Al 的进一步氧化。然而,Al 的功函数是 4.3 eV,与大多数有机材料的 LUMO 不匹配。因此,Al 一般和 LiF 配对使用作为 OLED 的阴极。

formed between Al and Alq_3, electron injection barrier from Al to Alq_3 reduces from 1.5 eV to 0.5 eV. When LiF is inserted (Figure 8.10 (b)), interface dipole potential increases to 1.6 eV. Therefore, electron injection barrier decreases to -0.1 eV, leading to energetically favorable electron injection.

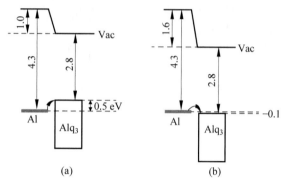

Figure 8.10   Electron injection barrier from Al cathode without (a) and with (b) LiF insertion

Combining X-ray photoemission spectroscopy analysis and theoretical calculation, it has been concluded that when LiF is inserted between Alq_3 and Al, LiF will be decomposed into Li with concomitant formation of AlF_3. Then, the Alq_3 radical cation is produced via reaction between Li and Alq_3. These reactions can be expressed as following:

(a) $3LiF \longrightarrow 3Li + 3/2F_2$ $\Delta G_f = 421.5$ kcal $\cdot$ mole^{-1}

(b) $Al + 3/2F_2 \longrightarrow AlF_3$ $\Delta G_f = -340.6$ kcal $\cdot$ mole^{-1}

(a) + (b) $3LiF + Al \longrightarrow AlF_3 + 3Li$

$\Delta G_f = 80.9$ kcal $\cdot$ mole^{-1}

(c) $3LiF + Al + 3Alq_3 \longrightarrow AlF_3 + 3Li^+Alq_3{}^-$

$\Delta G_f \approx 0$ kcal $\cdot$ mole^{-1}

$\Delta G_f \approx 0$ kcal $\cdot$ mole^{-1} in reaction (c) means that when LiF, Al and Alq_3 are present together, the formation of Alq_3 cation radical is possible. The produced Alq_3 cation radical can greatly improve electron injection and transport[23].

8.2.4   Generation of Excitons by Charge Recombination

After charge injection from electrode, electrons and holes move oppositely under electric field, and part of them

23. 反应(c)中 $\Delta G_f \approx$ 0 kcal $\cdot$ mole^{-1} 意味着当 LiF、Al、和 Alq_3 同时存在时，形成 Alq_3 阴离子自由基是可能的。生成的 Alq_3 阴离子自由基极大地增强了电子注入和传输。

can recombine to produce excitons. The charge recombination process can be defined as a fusion of a positive (e. g., hole) and a negative (e. g., electron) charge carriers into an electrically neutral entity, though the positive and negative charge centers on it do not necessarily coincide. The radiative decay of such an entity, or following its evolution the process of producing light from successive excited states, is called recombination radiation[24].

Generally, if the oppositely charged carriers are generated independently far away of each other (e. g., injected from electrodes) volume-controlled recombination (VR) takes place. The carriers are statistically independent of each other, the recombination process is kinetically bimolecular (Step 1 in Figure 8. 11). It naturally proceeds through a Coulombically correlated electron-hole pair (e - h) leading to various emitting states in the ultimate recombination step (mutual carrier capture) as shown in Step 2 of Figure 8. 11[25]. As a result, the overall recombination probability becomes a product of the probability of the VR formation of a charge pair (Equation (8-1)) and the capture probability for exciton formation (Equation (8-2)), the formula is shown in Equation (8-3):

1. $R_{eh} < r_{VR}$: Coulombically correlated interaction (VR);
$R_{eh} > r_{VR}$: dissociation of e-h correlation in VR process

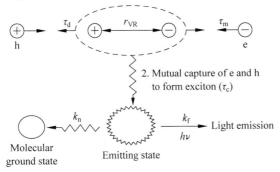

2. Mutual capture of e and h to form exciton (τ_c)

Figure 8. 11　Recombination of oppositely charged, statistically independent carriers (e, h) can lead to the creation of an excited state through a Coulombically correlated charge pair. Note, VR = volume-controlled recombination, R_{eh} = distance between the injected electron and hole, r_{VR} = the critical distance for VR process

24. 电荷复合过程,可以定义为一个正的(例如空穴)和一个负的(例如电子)载流子聚合成一个电中性的实体,尽管这个实体上的正负电荷中心不一定重合。这种实体的辐射衰变,或其后续演化的激发态产生的发光过程,称为复合辐射。

25. 自然地,它通过库仑作用相关联的电子-空穴对(e-h)来进行,导致在最终复合步骤(载流子互相捕获)中形成各种发光状态,如图8.11的步骤2所示。

$$P_R^{(1)} = \left(1 + \frac{\tau_m}{\tau_t}\right)^{-1} \tag{8-1}$$

$$P_R^{(2)} = \left(1 + \frac{\tau_c}{\tau_d}\right)^{-1} \tag{8-2}$$

$$P_R = P_R^{(1)} P_R^{(2)} = \left(1 + \frac{\tau_m}{\tau_t}\right)^{-1} \left(1 + \frac{\tau_c}{\tau_d}\right)^{-1} \tag{8-3}$$

where τ_t is the lifetime of the injected free carriers; τ_m and τ_d are the mean time for VR formation of charge pairs and their dissociation time, respectively; and τ_c is the mutual capture time of a VR-related charge pair, P_R is the probability for recombination, the superscript (1) and (2) refer to the process of 1 and 2 described in Figure 8.11.

1. Langevin and Thomson Recombination

There are two limiting cases of the VR recombination: ①the Langevin-like, and ②the Thomson-like recombination. The classic treatment of carrier recombination can be related to the notion of the recombination time[26]. The recombination time (τ_{rec}), in an OLED, represents a combination of the carrier motion time (τ_m), i.e., the time to get the carriers within capture radius (it is often assumed to be the Coulombic radius $r_{VR} = \dfrac{e^2}{4\pi\varepsilon_0 \varepsilon k_B T}$), and the elementary capture time (τ_c) for the ultimate recombination event (actual annihilation of charge carriers), $\tau_{rec}^{-1} = \tau_m^{-1} + \tau_c^{-1}$. Following the traditional description for recombination processes in ionized gases, a Langevin-like and Thomson-like recombination can be defined if $\tau_c \ll \tau_m$ and $\tau_c \gg \tau_m$, respectively. In solid-state physics, these two cases have been distinguished from each other by a comparison of the mean free path for optical phonon emission, λ, with a critical capture radius r_c[27]. One has Thomson recombination if $\lambda \ll r_{VR}$ and Langevin recombination if $\lambda \gg r_{VR}$. Two subcases should be considered when $\lambda = (D\tau_0)^{\frac{1}{2}}$, and $\lambda = \nu_{th}\tau_0$. In the former, the momentum (p) exchange cross-section is larger than the energy (E)-loss cross-section. In the latter, the reverse is true[28]. Here, D is the diffusion coefficient of a carrier, τ_0 is

26. 载流子复合的经典处理，可以与复合时间的观念相关。

27. 在固体物理中，这两种情况可以通过比较光声子发射的平均自由程 λ 和临界捕获半径 r_C，相互区分。

28. 前者中，动量（p）交换截面大于能量（E）损失截面；后者中，情况相反。

the lifetime of the p and E charge carrier states, respectively, and ν_{th} is the thermal velocity of the carrier. Due to the low carrier mobility (μ) in organic solids, one would expect to deal with the first subcase with $\mu < 1 \text{ cm}^2 \cdot v^{-1} \cdot s^{-1}$ and the mean free path for elastic scattering $< 10 \text{ Å}$. This value of λ is clearly much lower than $r_{VR} \cong 150 \text{ Å}$ (assuming the dielectric constant of organics $\varepsilon = 4$), strongly suggesting a Langevin-like model to be appropriate to describe the recombination process in organics[29]. Its signature is a field and temperature-independent ratio,

$$\gamma_{eh}/\mu_m = e/\varepsilon_0 \varepsilon = \text{const} \qquad (8\text{-}4)$$

where γ_{eh} is the bimolecular (second order) recombination rate constant, μ_m the sum of the carrier mobilities, and ε is the dielectric constant.

The kinetic description of bimolecular reactions in condensed media leads to the time (t)-dependent rate constant

$$\gamma(t) = 4\pi D R_{eh} \left[1 + R_{eh}/(\pi D t)^{\frac{1}{2}}\right] \qquad (8\text{-}5)$$

Equation (8-5), which seemed to be consistent with experimental data on time evolution of many chemical reactions in liquids, is not adequate to describe the reaction kinetics in disordered solids. In disordered solids, the carrier motion is only partially diffusion-controlled, carrier hopping across a manifold of statistic distribution in energy and space molecular sites must be defined and taken into account[30]. The detail is out of the scope of this textbook.

The long-time balance between recombination and drift of carriers as expressed by the γ/μ ratio has been analyzed using a Monte Carlo simulation technique and shown to be independent of disorder. Consequently, the Langevin formalism would be expected to obey recombination in disordered molecular systems as well. However, the time evolution of γ is of crucial importance if the ultimate recombination event proceeds on the time scale comparable with that of carrier pair dissociation ($\tau_c/\tau_d \approx 1$). The recombination rate constant becomes, then, capture, rather than diffusion, controlled, so that Thomson-like model would be more adequate than Langevin formalism for the

29. λ 值显然远低于 $r_C \cong 150 \text{ Å}$（假设有机物的介电常数 $\varepsilon = 4$），强烈地建议朗格文模型更适合描述有机物中的复合过程。

30. 在无序固体中，载流子的运动仅仅部分地受控于扩散，载流子在多种多样统计分布的能量范围内和空间分子位置上的跳跃，必须得到定义和考虑。

31. 然而,如果最终载流子复合过程与载流子对解离过程在时间尺度上相当($\tau_c / \tau_d \approx 1$),则 γ 随时间的演变至关重要。然后,复合速率常数成为捕获控制而不是扩散控制,因此对于复合过程的描述,汤姆森的模型将比朗格文公式更充分。

32. 当代表二线态粒子的电子和空穴在有机固体中复合时,形成单线态或三线态的激发态。如果两个载流子都是自由的,则由于自旋的统计分布,产生的三线态是单线态的三倍。

33. 单线态和三线态可以根据材料结构特征相互转换。例如,如果满足自旋—轨道耦合的条件,则从单线态到三线态的向下系间穿越(ISC)可以有效地发生;如果单线态和三线态的能量分裂很小,则在热能辅助下,可能会反向系间窜跃(RISC)。

description of the recombination process[31]. Detail discussion is given in Section 4.5.

2. Multiplicity of Excitons in Recombination Process

The multiplicity of an electronic state is defined by its spin quantum number (s) as $2s + 1$. The most often occurring singlet, doublet and triplet states are defined by their spin quantum numbers 0, 1/2, and 1, respectively. When an electron and a hole, representing doublet species, recombine in an organic solid, an excited state of either singlet or triplet character is formed. If both carriers are free, three times more triplets than singlets are produced due to spin statistics[32]. This means that theoretically, the creation probability of singlets is $P_S = \dfrac{1}{4}$, and triplets is $P_T = \dfrac{3}{4}$ under electrical pumping. Theoretically, this holds as long as the conduction gap (E_g) or the electron-hole energy gap, E_{eh}, is larger than or comparable with excited singlet energy (E_S).

However, the real system is more complicate. Figure 8.12 schematically shows the involved energy levels in an OLED. S_1, T_1 and E_T represent the excited singlet, triplet and trap states (in some cases, the trap state may be the electron conduction state). Single and triplet states can mutually transfer depending on material structural features. For example, if the condition for spin-orbital coupling is satisfied, inter system crossing (ISC) from singlet to downward triplet can efficiently occur; if the energy splitting of singlet and triplet is small, with the assistant from thermal energy, reverse inter system crossing (RISC) may occur[33]. And depending on relative energy of the singlet and triplet excitons, triplet-triplet annihilation (TTA) to form singlet may occur. Furthermore, if there are a large amounts of trap states, one of the two recombining carriers (either electron or hole) is trapped, the spin statistics of doublet species, here charge carriers, break down. In an extreme case, when the electron-hole energy is much lower than the excited singlet energy, for example, due to deep trapping of electrons, only triplets are energetically feasible, the recombination yield of singlet production drops down to 12.5%, meaning that two trapped singlet excitons are annihilated to form one free excited singlet exciton[34].

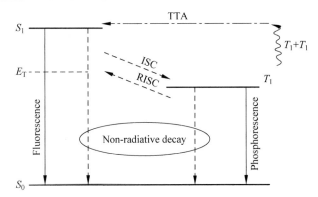

Figure 8.12　Schematic diagram of the mutual conversion between the excited singlet and triplet states, possibly occurred in an OLED

Since singlets fluoresce and triplets phosphoresce, the efficiency limit for fluorescent and phosphorescent OLED is respectively depended on their formation. As discussed above, the origination of singlet and triplet exciton is very complicate in OLEDs, which greatly influences device efficiency. Theoretically if all the excited states decay via radiative way, device efficiency reaches the up-limit. Taking internal quantum efficiency (IQE), i.e., the fraction of radiative decay from injected carriers, as evaluation parameter, several cases can be described: ①in fluorescence OLEDs, only carrier injection dominates the excited states formation, IQE $\leqslant$ 25%; ② if with special molecular structure, there exists RISC process, the 75% electrically pumped triplets can convert to singlet, IQE of the fluorescent OLED can be 100%, and the emission is the combination of prompt fluorescence and delayed fluorescence; ③ if there exists TTA process to produce singlet states, possibly due to the nearly doubled energy of the singlet exciton compared to triplet exciton, maximum IQE of the fluorescent OLED is expected to be $25\% + \frac{1}{2} \times 75\% = 62.5\%$, i.e., two triplets can produce one singlet; ④If the energy of electron-hole pair is smaller than the singlet excited state and higher than the triplet state, singlet excitons are in a large extent trapped in

34. 进一步，如果存在大量的陷阱态，两个复合载流子的其中之一（或者电子或者空穴）是被陷阱捕获的，则自旋二线态（这里是电荷载流子）的统计会被打破。在极端情况下，当电子-空穴能量远低于激发的单线态能量时，例如，由于电子被俘获在深陷阱能级，只有产生三线态在能量上是可行的，可能的发光是纯的磷光或者延迟荧光。这时复合产生的单线态可降至12.5%，意味着2个被捕获单线态激子通过湮灭形成了自由的单线态激子。

35. 如上讨论，OLED器件中单线态和三线态激子的来源非常复杂，这也极大地影响器件效率。理论上，如果所有的激子都辐射衰减，器件效率达到上限。用内量子效率(IQE)，即器件发生辐射跃迁部分占注入载流子数的比例来讨论，有几种情况：①在荧光 OLED 中，注入形成的激发态占主导，IQE≤25%。②如果在某种特殊分子结构下，存在反系间窜跃，电激发的 75%三线态可以转变为单线态，这时荧光 OLED 的IQE 有望达到 100%，这时的发光是即时荧光和延迟荧光的组合。③如果材料中存在T-T 聚变产生单线态的过程，这可能是单线态激子的能量几乎是三线态激子能量的 2倍，荧光器件的最大 IQE＝25%＋1/2×75%＝62.5%，即 2 个三线态变为 1 个单线态。④如果电子空穴的能量低于单线态激发态，而高于三线态，那么单线态激子将在很大程度上被捕获在传导电子或者空穴的能态上，荧光器件IQE≤12.5% 并且是延迟荧光，意味着 2 个被捕获的单线态通过湮灭产生一个自由的激发单线态，用来发光。⑤如果是磷光 OLEDs，在某些情况下，旋转耦合比较强（如在含有重金属的配合物），导致高效率的系间窜跃 ISC 过程，最高 IQE 是 100%。因为除了注入时 75% 的三线态，注入产生的 25% 单线态激子，可以通过 ISC 转变为三线态。

36. 无论是哪种激发状态，通过电激发形成的总激子在理论上包括 25% 的单线态和 75% 的三线态，这是由于对密度比例为 1∶3 的单线态和三线态的电子轨道的随机占据。

37. 主体发光器件的结构简单，发光中心的数量大于客体发光器件的。

the electron or hole conduction state, the IQE of the fluorescent OLED $\leqslant 12.5$ and is a kind of delayed fluorescence, meaning that two trapped singlet excitons are annihilated to form one free excited singlet exciton for emission; ⑤ If the OLED emitting via phosphorescence (PhOLED), in some cases, strong spin-orbital coupling exists (e.g., in complexes comprising heavy metals), which leads to efficient ISC process, the maximum IQE can be 100%. This because, besides the electrically pumped 75% triplet states, the 25% electrically-pumped singlet excited states can convert to triplet states by ISC process[35].

8.2.5 Radiative Decay of Excitons

After recombination of electrons and holes in the OLED device, excited species are produced (Figure 8.13(a)), which can be in the forms of: ① electrically excited molecules, ②exciplexes or ③excimers. No matter what kinds of excited species they are, the total excitons formed by electrical pumping theoretically comprise 25% of single states and 75% of triplet states due to the random occupation of electronic orbitals with density ratio of 1 : 3 for singlet and triplet states[36]. These excitons, then, give birth to EL by radiative decay to ground states in the emitting materials. Because OLEDs can be thought as a flat emission light, the EL distribution on the surface of a substrate conforms to Lambertian light characteristics (Figure 8.13(b)), that is, suppose the intensity of light perpendicular to the emitting surface is I_0, the light intensity (I) with angle of θ deviating from the normal line satisfies: $I = I_0 \cos\theta$.

In an OLED, the light emission can originate from host or dopant material. Figure 8.14 shows the host and dopant luminescence devices with three-layer structure. The structure of a host luminescence device is simple and the number of luminescence center is more than that of dopant device[37]. However, some organic emitters, which have strong luminescence in dilute solution, show week luminescence in concentrated solution or in solid state due to concentration quenching; so doping this kind of emitters in other host material benefits to obtain high EQE.

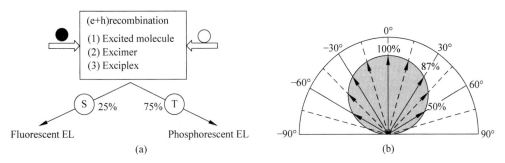

Figure 8.13　Emission species in OLEDs (a)，distribution of OLED emission (b)

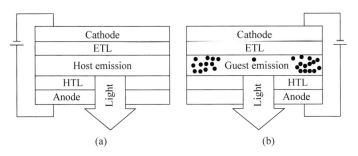

Figure 8.14　Structure of OLEDs based on host (a) and dopant (b) emitters

The use of doped luminescence can also increase the flexibility of device design. Because the carrier injection potential barrier between the luminescent layer and the adjacent two organic layers，and the carrier transport properties of the luminescent layer can be adjusted by selecting the host material without considering the dopant. The luminescence properties of devices can be controlled by choosing appropriate dopant materials[38]. The requirement of doped OLEDs on the molecular design of functional materials is also relaxed，because the electrical characteristics and optical characteristics can be considered separately：the electrical properties are controlled by the host material of the light-emitting layer，while the requirement of the light-emitting performance of the device can be realized mainly by doped emitters. In this case，the energy transfer between host and guest should be considered. Generally，there are three mechanisms for energy transfer：① cascade energy transfer which is negligible in thin films such as OLEDs；② Förster and ③Dexter energy transfers（Details are discussed in Chapter 3.

38. 使用客体掺杂发光，还可以增加器件设计的灵活性。因为发光层和相邻两个有机层之间的载流子注入势垒以及发光层载流子的传输特性，可以通过发光层主体材料调控，而不用考虑掺杂材料。器件的发光性能，可以通过选择适当的掺杂分子来调控。

8.3 Device Structure

8.3.1 Single Layer OLED

The simplest architecture of OLED devices is showed in Figure 8.15. It is a single organic layer device. The hole-injecting electrode is typically a high work function material and is usually designed to be transparent so that the light can be transmitted out. Typically, indium-tin-oxide (ITO) on a transparent substrate (glass or plastic) is used because of its high conductance, transparence and a fairly high work function[39]. The electron-injecting electrode is typically a low work function metal. Generally, organic compounds tend to have low electron affinity, and so the low work function metals tend to have much better electron injection. However, the high reactivity of the low work function metal restricts their application[40]. The problem was solved by using metal alloys such as Mg:Ag or Li:Al where the composite has a low work function and the relatively low reactivity.

39. 典型地，使用的是在透明衬底（玻璃或塑料）上的铟锡氧化物（ITO），因为它具有高的导电性和透明度，并且具有相当高的功函数。

40. 通常，有机化合物往往具有低的电子亲和能，因此低功函数金属对于有机化合物往往具有更好的电子注入。然而，低功函数金属的高反应活性，限制了它们的应用。

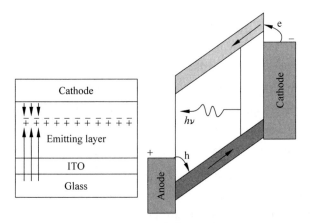

Figure 8.15　Single layer OLEDs

One disadvantage of single layer devices is that, unless the energy levels of the cathode and anode are extremely well matched to the relative energy levels of the organic compound, the injected electron and hole currents are not balanced. The dominant carrier can then cross the entire structure without meeting the carrier of the opposite sign to recombine. In addition, most organic thin films support preferentially the transport of either electron or hole. Thus, the use of these unipolar films favors the localization of the recombination zone either near the cathode or the anode, depending on whether the hole or the electron is the more mobile species[41]. As a result, the vicinity of recombination zone to an electrode will be subjected to the effect of emission quenching. Under these situations, a lot of energy will be wasted, leading to the low efficiency in the conversion of electrical energy to optical energy.

The single organic layer does not always mean that only one kind of organic molecules are in the layer. Mixing several organic functional materials together, such as carrier injection/transport materials and organic emitter by co-evaporation or solution casting, forms a functional single organic layer, and as long as the adverse processes (non-radiative energy transfer, the formation of charge transferring state and carriers trapping etc.) can be effectively avoided, a single layer device with good performance can be obtained[42].

8.3.2　Double-Layer OLED

Imaginably, better carrier balance can be achieved by using two organic layers, one of which is matched to the anode and is responsible for the hole transport (i.e., HTL), while the other is optimized for the electron injection and transport (i.e., ETL). Each sign of carriers is blocked at the interface between the two organic layers and tends to stay there until the counterpart is found (Figure 8.16). In this configuration, since excitons are formed far from the electrode, the quenching effect will be reduced. C. W. Tang at Kodak proposed this two-layer structure and demonstrated, for the first time, the bright OLED devices with high efficiency. In another work, they extended the two-layer structure with a modified emission zone by doping fluorescent dye, which enhanced the efficiency and tuned the color from green to red.

41. 此外,大多数有机薄膜优先电子或者空穴一种载流子的传输。因此,取决于空穴还是电子的传输性更强,应用这些单极性薄膜,复合区倾向于定位在阴极或阳极附近。

42. 将儿种有机功能材料混合在一起,如把载流子注入/传输材料以及有机发光材料通过共蒸发或溶液成膜,形成一个功能性的有机层,只要可以有效地避免不利过程(非辐射能量转移,形成电荷转移状态和载流子捕获等),性能良好的单层器件是可以获得的。

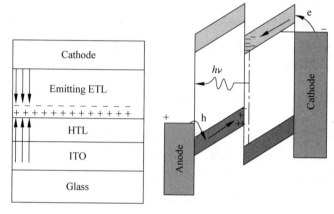

Figure 8.16　Double layer OLEDs

The advantages of double layer device lie in three aspects：①the double layer structure releases the difficulty in the choice of anode and cathode，because they only needs to individually match with HOMO of HTL or LUMO of ETL，respectively，which can simultaneously reduce the carrier injection barrier at anode and cathode interface[43]. Thus，low driven voltage and more balanced carrier injection can be easily achieved；② the double-layer structure relaxes the requirements of electric properties for HTL and ETL，because the holes and electrons move in different organic molecules；and ③ the double-layer structure renders carrier recombination zone to locate near the interface of HTL and ETL，which is far away from electrodes and minimizes the excitons quenching by electrodes，hence partially accounted for the high device efficiency.

8.3.3　Triple-Layer OLED

Another improvement in OLED devices is the introduction of a third layer specifically chosen for its high luminescent efficiency （Figure 8.17（a））（emitter layer - EML）[44]. The feature of the EML is that the EML has a relative low LUMO than that of HTL and a relative high HOMO than that of ETL. At this condition，the electrons can be blocked inside the EML near the HTL and the holes can be also blocked inside the EML near ETL. The possibility of electron and hole recombination can be improved. On the other hand，because three organic function layers perform their own functions，materials can be separately optimized for the hole transport，electron transport and/or luminescence[45].

J. Kido modified the triple-layered structure into what he termed as "confinement layer" (Figure 8.17(b)). He inserted a function layer between HTL and ETL into the OLED device. The confinement layer has the highest LUMO than both of ETL and HTL and the lowest HOMO than both of ETL and HTL. The confinement layer can partially block the electron and hole in ETL and HTL respectively[46]. By changing the thickness of this layer，it is possible to emit light either from the hole transporting layer or the electron transporting layer. If the "confinement layer" is properly constructed，emission from both layers can be observed simultaneously. This mixed emission allows the generation of white light.

45. 另一方面，由于三个有机功能层分别承担自己的功能，因此可以针对空穴传输、电子传输和/或发光分别优化材料。

46. 约束层的 LUMO 高于 ETL 和 HTL 的，而 HOMO 则低于 ETL 和 HTL 的 HOMO。约束层可以分别地，部分阻挡 ETL 和 HTL 中的电子和空穴。

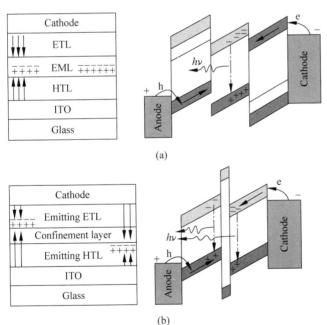

Figure 8.17　Triple layer OLEDs，the middle layer is EML (a) and the middle layer is a confinement layer (b)

8.3.4　Multi-Layer OLED

In order to further improve the OLED efficiency，particularly in phosphorescent OLED (PhOLED)，the charge carriers and excitons shall be manipulated respectively. More functional organic layers, such as hole injection layer and exciton block layer etc., can be introduced into OLED devices as shown in Figure 8.18. Figure 8.18(a) is a typical PhOLED structure. In this kind of device, the phosphorescent emitters are doped in a wide bandgap hole transport materials. In order to achieve a highly efficient phosphorescent emission, a hole block layer (HBL) with a HOMO level lower than that of host materials of EML shall insert between EML and ETL. This block layer can block the further movement of holes toward the ETL. Meanwhile, the triplet excitons usually have a long diffusion distant because of their longer lifetime. <u>The triplet excitons can also diffuse into HBL and ETL. To prevent this diffusion, the triplet energy of HBL shall be larger than the triplet energy of phosphorescent emitter. This can make sure the charge recombination in EML and to confine triplet exciton in EML</u>[47]. Furthermore, an electron block layer (EBL) can also be added between the EML and HTL to block electron moving

47. 三线态激子也可以扩散到 HBL 和 ETL 层中。为了阻止这种扩散, HBL 的三线态能量应大于磷光发光体的三线态能量。这可以确保电荷在 EML 中的复合, 并可以约束三线态激子在 EML 中。

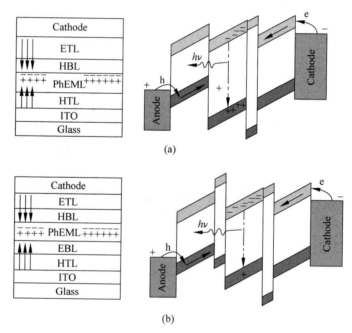

Figure 8.18　Multiple-layer OLEDs: (a) with hole/exciton blocking layer, (b) with both hole/exciton and electron/exciton blocking layers

into HTL and the confinement of triplet excitons in EML as shown in Figure 8.18(b). Same as HBL, the triplet energy of EBL shall be larger than the triplet energy of emitter.

8.3.5 White OLED

White light is important for the illumination application. The structure of white OLED is complicated because more emission colors are needed to achieve white emission. Generally, there are two kinds of methods to obtain white color. One is the mixing of red, green and blue colors to achieve white light emission. The other is the combination of blue color and its complementary color of yellow. Figure 8.19 shows three structures to build white OLED. The first approach puts all the required color emitters (red, green and blue or blue and yellow) in one EML as shown in Figure 8.19 (a). <u>Here, the EML can be two/three emitters mixing together or one emitter with the white emission spectrum (including dimer or exciplex emission etc.)</u>[48]. The second white OLED structure uses multi emission layers as shown in Figure 8.19(b). In this structure, each EML can emit one or two colors, the white emission is achieved by the mixing of each color light emitting from different EML. The third architecture of white OLED utilizes down-conversion mechanism. A photoluminescence layer (usually emitting yellow or orange color) is applied on the top of glass substrate

48. 在这里，EML 可以是两种或三种发光体混合在一起，也可以是一个具有白色发射光谱的发光体（包括二聚体或激基复合物的发光等）。

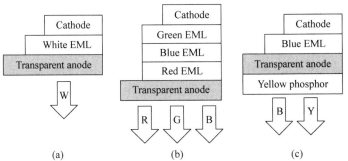

(a)   (b)   (c)

Figure 8.19 WOLEDs: (a) A single host layer doped with dopant emitters of different luminous colors, (b) two or three layers with different luminous colors, (c) a single layer emission and a photon excited luminous layer with complementary color on the transparent electrode

of a blue OLED as shown in Figure 8.19(c). A part of blue light emitting from blue OLED can excite the emitter in photoluminescence layer. The final white light output of the device will include blue light directly from blue OLED and yellow/orange light from photoluminescence layer[49].

49. 蓝色 OLED 发出的部分蓝光,可以激发光致发光层中的发光体。最终输出的白光将包括直接来自蓝色 OLED 的蓝光和来自光致发光层的黄色/橙色光。

8.4 Device Characterization

In OLEDs, electricity is directly converted into light, which means that the optical properties of OLEDs are tightly correlated to their electrical characteristics. Therefore, the evaluation of the overall light output and its relation to the driving current are of fundamental interest, and the understanding and tailoring of OLED emission color also fall in central of device physics and practical application problems[50]. This section provides an overview of various characterization parameters.

50. 在 OLED 中,电能直接转化为光能,这意味着 OLED 的光学性质与电学性质密切相关。因此,对整体光输出及其与驱动电流关系的评估是(研究的)基本意义所在,另外理解和调控 OLED 器件发光颜色,也是器件物理和实际应用问题的中心。

8.4.1 *J-U-B* Curve and EL Spectrum

After preparation of an OLED device, the first thing for performance evaluation is to measure current-voltage-brightness (J-U-B) curve, along with the collection of EL spectrum at different driving voltage, where, J is current density, generally with unit of mA · cm^{-2}; U is driving voltage with unit of voltage (V); B is the brightness of the output light, with unit of cd · m^{-2}. An example is shown in Figure 8.20. Based on J-U-B curve and EL spectra, main parameters for device evaluation can be obtained. Before talking about them, some typical features in the J-U-B curve will be stated first.

1. Terminology for Light Brightness

Brightness is also called luminance, which is an attribute of visual perception in which a source appears to be radiating (radiator) or reflecting (reflector) light. It is not necessarily proportional to light intensity. The standard international (SI) unit of luminance is cd · m^{-2}, also written as nits.

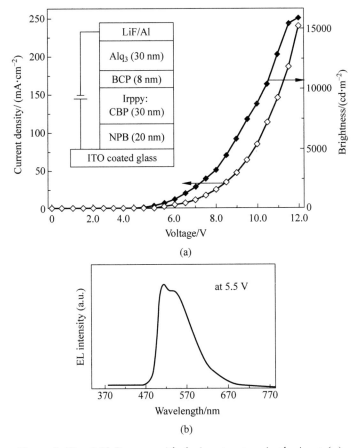

Figure 8.20 *J-U-B* curve with device structure in the inset (a) and EL spectrum at 5.5 V (b)

Another terminology for brightness is luminous flux or luminous power, which is the measure of the perceived power of light. It differs from radiant flux (i.e., the measure of the total power of electromagnetic radiation such as infrared, ultraviolet, and visible light) in that luminous flux is adjusted to reflect the varying sensitivity of the human eye to different wavelengths of light[51]. The SI unit of luminous flux is the lumen (lm).

2. Nonlinear *J-U* of OLEDs

The non-linearity is recurrent feature of current-voltage characteristics of all operating OLEDs, independent of the number and configuration of organic layers. It is associated with the fact that the driving current is due to injection of charge at the electrodes: holes at the anode and electrons at the cathode (and not being a result of the bulk generated

51. 另一个亮度的术语是光通量或光功率,指的是光被感知的功率。它与辐射通量(即电磁辐射的总功率的度量,诸如红外线、紫外线和可见光等)的不同之处在于,光通量根据人眼对不同波长光的不同灵敏度进行了调整。

52. 非线性，是所有OLED工作时电流-电压特性的反复特性，与有机层的数量和有机层结构无关。它与这样一个事实相关，即驱动电流缘于电极上电荷的注入：阳极注入空穴、阴极注入电子（而不是本征载流子的结果）。

carriers)[52]. Double logarithmic plots of the current vs. applied voltage allow to distinguish the power law behavior of the current. Figure 8.21 shows an example for a double-layer phosphorescent OLED based on the metal organic phosphor Irppy. The J - F curves are well reproduced from run to run except the lowest field region，where，the difference in the work functions of the electrodes becomes comparable with the applied field. Three general regimes can be distinguished as indicated in the figure：（A）leakage or diffusion-limited current，（B）space charge limited current with an exponential distribution of traps，and（C）space charge limited current with filled traps.

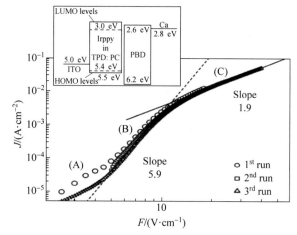

Figure 8. 21　The current-field characteristics of a double-layer PhOLED based on Irppy. The energy levels of OLED materials are given in the inset

3. Light Output from OLED

Generally，there are two limiting cases leading to simplified interrelations between external light output and injection currents. ① Injection-controlled EL（ICEL）：The light output for OLEDs operating in the ICEL mode cannot be related directly to the driving current which must not be identified with the recombination current that is not known from electrical measurements[53]. For the ICEL mode，the EL is a "side" effect of the current flow and its intensity is proportional to the product of the electron and hole injection currents；② Volume-controlled（space charge limited）EL

53. 在 ICEL 模式下工作的 OLED 的光输出，不与驱动电流直接相关，不能直接从电学测量中不可获得的复合电流中确定。

(VCEL): In this case, the measured current is simply the recombination current. The light output remains directly proportional to the driving current. Typically, these two cases occur at low and high driving voltages, respectively.

8.4.2 Characterization Parameters

Except stability parameters, other parameters for OLED performance can be derived from the J - U - B curve of an OLED device, plus the EL spectra at different driving voltages.

1. Turn-on and Driving Voltages

Turn-on voltage refers to the voltage when the brightness of OLED device is 1 cd · m^{-2}. Small turn-on voltage is desired for OLED device because it indicates a good Ohmic contact between electrodes and organic functional layers[54]. In this case, carriers can be easily injected into organic layer without overcoming large energy barrier. The balanced holes and electrons injection can be easily achieved. Driving voltage is the voltage that a device can normally operate, generally at 20 mA · cm^{-2}. Sometimes, the voltages, under which the device brightness is 100, 1000 and 10000 cd · m^{-2} can also be thought as driving voltages for the evaluation of display, lighting and phosphoresce OLED, respectively. Figure 8.22 gives an example.

54. 开启电压是指 OLED 器件亮度为 1 cd · m^{-2} 时的驱动电压。OLED 器件需要较低的开启电压,因为低的开启电压表明电极和有机功能层之间具有良好的欧姆接触。

2. Device Efficiency

There are several expressions for OLED efficiencies. Current efficiency (CE or η_C) is brightness (luminance) per current density (Equation (8-6)); its SI unit is cd · A^{-1}. Luminous efficiency (LE), also termed as power efficiency (η_p), refers to the ratio of luminous power and input electric power (Equation (8-7)), so its unit is lm · W^{-1}.

$$\eta_c = \frac{B \cdot (S \times 10^{-4})}{(J \cdot S) \times 10^{-3}} = \frac{B}{J} \times \frac{1}{10} \tag{8-6}$$

$$\eta_p = \frac{\pi \cdot B \cdot (S \times 10^{-4})}{U \cdot [(J \cdot S) \times 10^{-3}]} = \frac{\pi}{U} \times \frac{B}{J} \times \frac{1}{10} \tag{8-7}$$

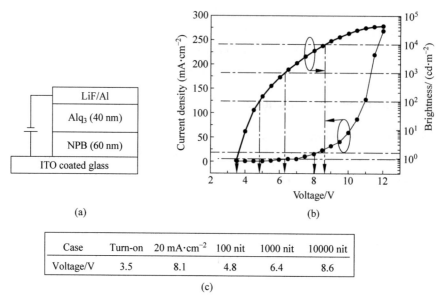

Case	Turn-on	20 mA·cm⁻²	100 nit	1000 nit	10000 nit
Voltage/V	3.5	8.1	4.8	6.4	8.6

(c)

Figure 8.22　Device structure (a), J-U-B curves (b) and the determination of turn-on and driving voltages based on the J-U-B curves (c)

55. 第三种器件效率是量子效率,包括内量子效率(IQE)和外量子效率(EQE)。IQE 是指光辐射光子数目与注入载流子数目的比率;而 EQE 是指器件输出的光子与注入载流子的比率。需要记住的重点是,EML 中所产生的光,通常只有一定的比例可以在 OLED 的表面获得。在平面 OLED 中,由于在各个功能层中折射率的不匹配,大部分的发射光在玻璃基板、ITO 和有机材料层中通过光波导模式而损失。

where B is brightness (cd · m⁻²), S is emission area (cm²), J is current density (mA · cm⁻²), U is driving voltage (V).

The third expression for device efficiency is quantum efficiency, including internal quantum efficiency (IQE) and external quantum efficiency (EQE). IQE refers to the ratio of the radiative photons to the injected carriers; while EQE is the ratio of the output photons to the injected carriers. It is important to remember that, generally, only a certain fraction of the light generated in the EML is available for the face detection from an OLED. In a planar OLED, a large fraction of the emitted light is lost to wave guiding modes in the glass, ITO, and organic layers due to refractive index mismatching[55]. As shown in Figure 8.23, a photon, which is produced by a radiative decay process of an excited state in an organic layer of an OLED, needs to pass through the multiple organic layers as well as the ITO glass before coming out of the device. Therefore, it suffers many loss paths within the device, such as reflection, absorption and refraction, etc., hence leading to significant smaller EQE value compared to the IQE value.

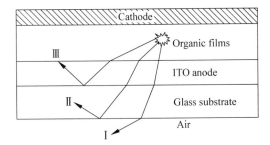

Figure 8.23　The fate of a photon by radiative decay of
an excited state in OLEDs

The expressions for IQE of fluorescence and phosphorescence OLEDs are shown in Equations (8-8) and (8-9), respectively; and the EQE of an OLED is shown in Equation (8-10)

$$\mathrm{IQE}_{\mathrm{Fl}} = \eta_i \eta_r (\chi_s + \chi_{\mathrm{RISC}} + \chi_{\mathrm{TT}}) \phi_{\mathrm{PL}} \qquad (8\text{-}8)$$

$$\mathrm{IQE}_{\mathrm{Ph}} = \eta_i \eta_r ((1 - \chi_s) + \eta_{\mathrm{ISC}} \times \chi_s) \phi_{\mathrm{PL}} \qquad (8\text{-}9)$$

$$\mathrm{EQE} = \eta_i \eta_r \eta_e \chi \Phi_{\mathrm{PL}} \qquad (8\text{-}10)$$

where η_i, η_r and η_e are injection, recombination and out-coupling efficiencies, respectively; χ_s, χ_{RISC} and χ_{TT} are the portion of singlet exciton by injection, from reversed ISC, and from TT annealing, respectively; η_{ISC} is the efficiency of inter system crossing, ϕ_{PL} is the photoluminescent quantum yield, and χ is the portion of singlet/triplet exciton.

At a fixed voltage of U, different efficiencies have the following relationship:

$$\mathrm{EQE}\,(U) = \eta_e \cdot \mathrm{IQE}(U) \qquad (8\text{-}11)$$

$$\eta_p\,(U) = \frac{\pi}{U(V)} \cdot \eta_c\,(U) \qquad (8\text{-}12)$$

$$\eta_p\,(U) = \mathrm{EQE}(U) \cdot \frac{\bar{E}(\mathrm{eV})}{U(V)} \qquad (8\text{-}13)$$

$$\eta_c\,(U) = \frac{1}{\pi}\mathrm{EQE}(U) \cdot \bar{E}(\mathrm{eV}) \qquad (8\text{-}14)$$

8.4.3　Chromaticity Coordinates and Color Saturation

Chromaticity is an objective specification of the quality of a color regardless of its luminance. Chromaticity consists of two independent parameters, specified as hue and saturation; the former tells the appearance of the color (red, green, red, or else), the latter describes the purity of the concerned color[56]. Chromaticity of colors is often shown

56. 色度是不考虑亮度的颜色质量的客观指标。色度由两个独立的参数组成,指定为色调和饱和度;前者描述了颜色的外观(红色、绿色、红色或其他),后者描述了相关颜色的纯度。

57. 马蹄形截面的中心被称为白点(0.33,0.33),表示标准白色或饱和白色。中间圆圈包围的区域是白光区域。在蓝色一侧,曲线上的 A 点(0.08,0.13)是饱和蓝光。画一条从 A 点与中心白点 W 的连接直线,AW 线之间的其他蓝光点,如 B 点,是不饱和蓝光。B 的饱和度等于 WB 和 WA 的比率(WB/WA)。在马蹄形内部,两个不同色点(C 和 D)之间的点(E)的颜色可以通过混合一定比例的 C 色和 D 色来获得。

based on a standard chrominance curve formulated by Commission International de l'Eclairage (CIE), as shown in Figure 8.24. Its profile is a horseshoe-like curve. Points on the horseshoe-curve represent a saturated light. The color (hue) of the light changes from blue at the left side to green on the top of curve and finally to red at the right side. The center of horseshoe-shaped section calls white point at (0.33, 0.33), respecting standard white or saturated white. The area surrounded by the middle circle is the area of white light. At blue side, point A (0.08, 0.13) on the curve is a saturated blue light. Drawing a line from A connected with the center white point W, other blue light points between line AW, such as point B, is unsaturated blue light. Saturation of B equals to the ratio of WB and WA (WB/WA). Inside the horseshoe shape, the color of the point (E) between two different color points (C and D) can be obtained by mixing a certain proportion of C color and D color[57]. Top right in Figure 8.24 illustrates this method, that is, to get color at point E, the mixing ratio of C and D components should be $a : b$.

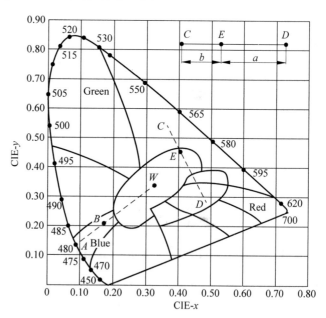

Figure 8.24   CIE chrominance curve

In terms of display application, the national television system committee (NTSC) standard CIE coordinates of red, green and blue are（0.67，0.33），（0.21，0.71）and（0.14，0.08），respectively. Another color parameter for display is color saturation index，which indicates the color quality of a display. Red，green and blue（RGB）are the three basic colors of a display；and the triangle area with the RGB points as vertex is termed as color gamut. Color saturation index is the ratio of color gamut of a display and that of a NTSC RGB，as shown in Figure 8.25. The more approaches to unity the color saturation index，the better the color quality of a display.

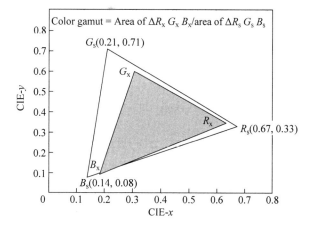

Figure 8.25　Calculation of color saturation for a display

For WOLED，except for CIE color coordination，color temperature and color rendering index（CRI）are other characterization parameters to be evaluated. Color temperature refers to the temperature at which a blackbody emits radiant energy competent to evoke a color is the same as that evoked by radiant energy from a given source[58]. For example，the color of an incandescent lamp is the same to the color of a black body with temperature of 2527 ℃，then，its color temperature is 2527 + 273 K = 2800 K. Low color temperature（2700 K to 3000 K）generally means warm white，while cold white often possesses high color temperature around 6000 K. CRI is a quantitative measure of the ability of a light source to faithfully reproduce the colors of various objects in comparison with an ideal or natural light source. Theoretically，CRI can be calculated by comparison between the spectra of a standard light source and the

58. 色温是指当一个黑体发出的辐射能量引起的颜色与一个给定的光源发出的颜色一样时,黑体的温度。

emitting WOLED. Experimentally, 14 samples with standard color are irradiated with the emitting WOLED and a standard light source to collect the color difference, ΔE_i, then the CRI at a fixed color can be calculated as,

$$\text{CRI}_i = 100 - 4.6\Delta E_i \qquad (8\text{-}15)$$

For a general evaluation of CRI with a light source, only 8 standard samples are selected, and the CRI of the light source can be obtained by,

$$\text{CRI} = \frac{1}{8}\sum_{i=1}^{8}\text{CRI}_i \qquad (8\text{-}16)$$

The color of an object to the human eye is strongly affected by the CRI of the light source. The smaller the CRI, the worse the reproducing of the color-appearance of an object by the tested light source[59].

8.4.4　Device Lifetime

Generally speaking, the lifetime of an OLED depends not only on the elementary device itself, but also on the operation and/or storage conditions, such as, for example, applied voltage, temperature, and humidity[60]. The intended application influences useable lifetimes as well: in some cases, loss of half of emission intensity may be tolerable; in others, the maximum allowed loss may be as small as 3%. Changes in emission color, drive voltage, and consumed power exemplify other important characteristics that may limit device lifetime. Nonetheless, OLED lifetimes are most commonly defined in terms of time it takes for the emission intensity to decay to a certain percentage of initial value while the emission-inducing current is unchanged. As there is no commonly accepted standard with respect to the initial value of emission intensity for the lifetime determination, values of initial current density or emission intensity are normally provided along with the lifetimes[61]. Figure 8.26 gives an example, showing a device with initial brightness of 100 nit and the time that the brightness decreases to half of the initial value (e.g., lifetime = 610 h).

59. 人眼对物体颜色的感知受到光源 CRI 的强烈影响。CRI 越小,被测光源对物体颜色外观的再现就越差。

60. 一般来说,OLED 的寿命不仅取决于器件基本结构本身,还取决于器件工作和/或存储条件,例如施加的电压、温度和湿度。

61. 尽管如此,OLED 寿命最通常的定义是,在驱动发光的电流保持不变的条件下,器件发光强度衰减到初始值的一定百分比所需的时间。由于没有普遍接受的、用于寿命测定的、发光初始强度的标准,因此驱动电流密度或初始发光强度值通常与寿命一起提供。

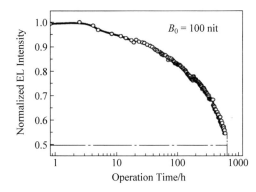

Figure 8.26　A device lifetime experiment

Under constant current density condition, the luminance decays of OLED conform to the following formula:

$$L(t)/L_0 = \exp\left[-(t/\tau)^{\beta}\right] \tag{8-17}$$

where L_0 and $L(t)$ are initial luminance and the luminance at time t after continue operation, respectively. τ and β are constants relative to material stability and device structure and can be fitted from measured data using least square method. Considering that the time for the brightness decreasing to half of the initial value ($t_{1/2}$) as lifetime, based on Equation (8-17), we have

$$t_{1/2} = \frac{C}{(L_0)^n} \tag{8-18}$$

where C and n are constants. Equation (8-18) means that device lifetime is negatively related to initial brightness L_0. Accordingly, one can carry stability experiment on accelerated mode, i.e., operating the device at a high initial brightness to get a lifetime at a short time. Several couples of (L_0, $t_{1/2}$) can be used to obtain value C and n in Equation (8-18), then, the lifetime at any initial brightness can be obtained using Equation (8-18)[62].

A large body of research toward developing long-lived and efficient OLEDs has yielded many important empirical observations about OLED degradation. Aside from their obvious value for designing practical devices, the commonly observed empirical characteristics of OLED degradation are also useful to validate mechanistic hypotheses. From the beginning of OLED device development, the luminance decays were reported to decelerate gradually during

62. 相应地,稳定性试验可以在加速模式下进行,即器件在较高的初始亮度下获得较短的寿命。数组(L_0,$t_{1/2}$)可以用来确定式(8-18)中的 C 和 n。然后任何初始亮度下的寿命都可以通过式(8-18)获得。

63. 从 OLED 器件研发开始就有报道,器件的亮度衰减速率随着工作时间增加而逐渐减小,明显遵循一些非线性函数。类似于一阶化学反应动力学方程中的指数函数,通常不适合描述 OLED 亮度的衰减。其他简单的动力学方程也无法充分描述典型 OLED 的衰减。

64. 衰减速率分别是电流密度和温度的函数,衰减程度是通过器件的总电荷量的函数。另外,电压似乎不会直接引起亮度损失。

operation, following some distinctly nonlinear function. The exponential function similar to that in first-order chemical kinetics equation does not generally fit the OLED decay. Other simple kinetic equations are also unable to describe typical OLED decays adequately[63].

A remarkably general and conceptually important property of luminance decay is its dependence on current rather than on the voltage. Just as the electroluminescence in OLEDs is determined by current, its decay appears to be current driven too. The rate of degradation is a function of current density and temperature, respectively, the extent of degradation is a function of total charge passed through the device. Voltage, on the other hand, does not appear to induce luminance loss directly[64]. It is easy to demonstrate that, for a wide range of applied voltages (or, more pertinently, internal electric fields), there is no luminance decay in the absence of forward current. This does not mean that the applied voltage has no effect on the rate of luminance degradation but rather that it is more appropriate to consider the voltage effect as a perturbation of degradation induced by current.

Initially, the specific form of relationship between luminance decay rate and current density was determined to be approximately linear, which led to the classifying of the degradation process as "Columbic", that is, determined by the total injected charge. Because the luminance and current density are also related in an approximately linear manner, the dependence of decay rate on current may be considered as the dependence on luminance. Later, it had become clear that there are substantial and systematic deviations from the linear relationships; usually the luminance decay rates increase with current densities faster than expected. In some cases, the super linearity was so great that the decay rates scaled approximately proportionally to luminance (or current density) squared. For practical purposes of extrapolating device lifetime, the usual lifetime-luminance relationship is usually amended by exponentiation luminance with an additional variable parameter, which is usually called acceleration factor or coefficient. Because of the difficulties associated with measuring lifetimes when current densities are widely varied (low current densities lead to impractically long durations of experiment, whereas high current densities are limited by various destructive phenomena owing to electric field and temperature), it is currently unclear how general is the relationship between lifetime and exponentiated luminance[65]. Note that Equation (8-18) gives an analogous expression for the exponential-dependence of lifetime on luminance.

The observations of linear and super linear relationships between the luminance decay rate and current density or luminance provide important clues about degradation mechanism. They suggest that the OLED degradation is driven by the excited states generated by recombining charges. The observed linearity is indicative of the first-order kinetics with respect to concentration of the excited states, and the super linear relationships suggest some involvement of higher-order kinetics as observed in bimolecular reactions. For comparison, the same observations suggest that the OLED degradation mechanism driven by the reactions of charge carriers without recombination is less likely because their concentrations are not generally expected to scale linearly with current density.

Another generality with respect to OLED degradation is a strong dependence on temperature. As a rule, the operation-induced luminance decay rate monotonically increases with temperature until temperature is kept below a certain, material-dependent limit. Under such conditions, significant degradation occurs only when device operates, although minor temperature-induced changes can occur even in the absence of current. Exceeding the temperature limit typically results in marked and irreversible changes in device characteristics or even a total failure (defined as complete loss of electroluminescence)[66]. This type of degradation occurs readily even when device is not operated. The device- and material-specific temperature limit can usually be traced to glass transition temperature of some material in OLED. A typical OLED material remains a rigid solid in amorphous state when it is within an intended operational range of

65. 为了实际应用的器件寿命外推,一般把寿命-亮度关系用一个带有可变参数的亮度指数函数来修正,通常称为加速度因子或系数。当电流密度变化很大时,寿命测量存在困难(低电流密度导致不切实际的实验持续,而高电流密度受到电场和温度引起的各种破坏性现象的限制),因此目前尚不清楚,寿命与亮度指数函数之间的关系有多普遍。

66. 超过温度极限,通常会导致器件特性发生明显且不可逆转的变化,甚至完全失效(定义为电致发光的完全消失)。

temperatures. Mobility of molecules of the size of a typical OLED material is essentially nonexistent in that state. Exceeding the glass transition temperature （which， particularly in thin layers，may be somewhat modified by adjacent materials or dopants） results in the material becoming a supercooled liquid and molecular mobility becoming significant factor. Many different physical phenomena such as mixing，phase separation，dewetting，and crystallization become possible in liquid state. Electrical short and electrode delamination frequently occur when temperature limit is exceeded，especially when forward or reverse voltage is applied[67].

67. 许多不同的物理现象，如混合、相分离、去浸润和结晶，在液态下成为可能。当超过温度极限时，特别是在当施加正向或反向电压的情况下，经常发生电短路和电极剥离。

Besides material aspects （such as thermal and electrochemical stabilities），other OLED lifetime related issues include：① encapsulation to prevent H_2O and O_2 for dark spots，② unbalanced carrier in device （heat-production and exciton-quenching by excessive carrier），③ large injection barrier reduced device instability，④ doping in the emitting-layer improves device stability，⑤ decreasing driving current improves device stability.

本章小结

1. 内容概要：

本章在简单介绍了有机电致发光的发展历史后，对有机电致发光的机理，特别是载流子的注入、载流子复合产生激子，以及激子在有机材料中的空间分布进行了详细阐述。同时对有机发光器件的结构的作用和功能给予了相应的讲解。接着，对有机发光器件的表征参数做了深入的介绍，对器件的电学、光学和效率等重要参数概念给予了明确的定义。最后，阐述了各种物理因素对器件寿命的影响，主要讲述了驱动电流与器件寿命的关系。

2. 基本概念： 电致发光、有机发光二极管、OLED、WOLED、阴极、阳极、载流子注入、载流子传输、载流子复合、注入势垒、接触势垒、空穴传输层、电子传输层、阻挡层、主体发光、掺杂发光、EL 光谱、亮度、流明、光取出、开启电压、驱动电压、电流效率、功率效率、外量子效率、内量子效率、载流子复合、辐射衰变、CIE 色坐标、色域、色彩饱和度、色温、显色指数、器件寿命。

3. 主要公式：

（1）载流子 VR 复合概率：$P_R^{(1)} = \left(1 + \dfrac{\tau_m}{\tau_t}\right)^{-1}$

（2）载流子形成激子的捕获概率：$P_{\mathrm{R}}^{(2)} = \left(1 + \dfrac{\tau_{\mathrm{c}}}{\tau_{\mathrm{d}}}\right)^{-1}$

（3）总载流子复合概率：$P_{\mathrm{R}} = P_{\mathrm{R}}^{(1)} P_{\mathrm{R}}^{(2)} = \left(1 + \dfrac{\tau_{\mathrm{m}}}{\tau_{\mathrm{t}}}\right)^{-1} \left(1 + \dfrac{\tau_{\mathrm{c}}}{\tau_{\mathrm{d}}}\right)^{-1}$

（4）Langeiv-like 复合速率常数：$\gamma_{\mathrm{eh}} = e\mu_{\mathrm{m}}/\varepsilon_0 \varepsilon = const \times \mu_{\mathrm{m}}$

（5）时间依赖的 Langeiv-like 复合速率常数：$\gamma(t) = 4\pi D R_{\mathrm{eh}} \left[1 + R_{\mathrm{eh}}/(\pi D t)^{\frac{1}{2}}\right]$

（6）电流效率：$\eta_{\mathrm{c}} = \dfrac{B \cdot (S \times 10^{-4})}{(J \cdot S) \times 10^{-3}} = \dfrac{B}{J} \times \dfrac{1}{10}$

（7）功率效率：$\eta_{\mathrm{p}} = \dfrac{\pi \cdot B \cdot (S \times 10^{-4})}{U \cdot \left[(J \cdot S) \times 10^{-3}\right]} = \dfrac{\pi}{U} \times \dfrac{B}{J} \times \dfrac{1}{10}$

（8）荧光内量子效率：$\mathrm{IQE}_{\mathrm{Fl}} = \eta_{\mathrm{i}} \eta_{\mathrm{r}} (\chi_{\mathrm{s}} + \chi_{\mathrm{RISC}} + \chi_{\mathrm{TT}}) \phi_{\mathrm{PL}}$

（9）磷光内量子效率：$\mathrm{IQE}_{\mathrm{Ph}} = \eta_{\mathrm{i}} \eta_{\mathrm{r}} \left[(1 - \chi_{\mathrm{s}}) + \eta_{\mathrm{ISC}} \times \chi_{\mathrm{s}}\right] \phi_{\mathrm{PL}}$

（10）外量子效率：$\mathrm{EQE} = \eta_{\mathrm{i}} \eta_{\mathrm{r}} \eta_{\mathrm{e}} \chi \Phi_{\mathrm{PL}}$

（11）内量子效率与外量子效率的关系：$\mathrm{EQE}(U) = \eta_{\mathrm{e}} \cdot \mathrm{IQE}(U)$

（12）功率效率与电流效率的关系：$\eta_{\mathrm{p}}(U) = \dfrac{\pi}{U(V)} \cdot \eta_{\mathrm{c}}(U)$

（13）功率效率与外量子效率的关系：$\eta_{\mathrm{p}}(U) = \mathrm{EQE}(U) \cdot \dfrac{\bar{E}(\mathrm{eV})}{U(V)}$

（14）电流效率与外量子效率的关系：$\eta_{\mathrm{c}}(U) = \dfrac{1}{\pi} \mathrm{EQE}(U) \cdot \bar{E}(\mathrm{eV})$

（15）指定颜色的显色指数：$\mathrm{CRI}_i = 100 - 4.6\Delta E_i$

（16）光源的显色指数：$\mathrm{CRI} = \dfrac{1}{8} \sum\limits_{i=1}^{8} \mathrm{CRI}_i$

（17）器件寿命：$L(t)/L_0 = \exp\left[-(t/\tau)^{\beta}\right]$

（18）器件寿命与初始亮度的关系：$t_{1/2} = \dfrac{C}{(L_0)^n}$

Glossary

Aggregation emission	聚集发光
Carrier transport	载流子传输
Cathode	阴极
Charge transport	电荷传输
CIE color (Commission International de I'Eclairage)	国际照明委员会色坐标
Color gamut	色域
Color rendering index	显色指数
Color saturation index	色彩饱和度
Color temperature	色温
Current efficiency	电流效率

Driving voltage	驱动电压
Electroluminescence	电致发光
Emission	发射
Excimer emission	基激二聚物发射
Exciton diffusion	激子扩散
Exciton generation	激子产生
External quantum efficiency	外量子效率
Flexible	柔性的
Glass substrate	玻璃基底
High brightness	高亮度
High monochromatism	高单色性
Injection	注射,射入
Interference	干涉
Internal quantum efficiency	内量子效率
Intramolecular charge transfer material	分子内电荷转移化合物
Lifetime	寿命
Light emitting diode	发光二极管
Light radiation	光辐射
Luminous efficiency	功率效率(流明效率)
Mobility	载流子迁移率
Optical transition	光跃迁
Output wavelength	输出波段
Recombination	复合
Stability	稳定性
Turn on voltage	开启电压
Ultraviolet light	紫外光
Visible light	可见光
WOLED（white organic light emitting device）	白光有机发光器件

Problems

1. True or false questions：

(1) OLED，i. e.，the organic light-emitting diode，relates to the emission of organic material under light illumination.

(2) In OLEDs，the exciton in vicinity to cathode can be easily destroyed by cathode metal.

(3) In OLEDs，the excitons are thoroughly assigned to singlet multiplicity.

(4) The emitting species in OLEDs exclude the type of exciplex.

(5) The main energy transfer scheme in an OLED with dopant emitter is cascade type.

(6) In an OLED device, the fluorescence emission via the way of reverse inter system conversion from triplet to singlet is absolutely forbidden.

(7) Generally speaking, the surface distribution of an OLED emission satisfies Lambertian light distribution.

(8) Typically, the performance of single layer OLED is worse than that of double layer OLED.

(9) In double layer OLEDs, the basic energy level alignment is both the matching of anode with LUMO of the hole transport layer and the matching of cathodc with HOMO of electron transport layer.

(10) In white OLEDs with downward energy-conversion scheme, cascade energy transfer is one of the dominate energy transfer mechanism.

2. Please state the discovery of organic electroluminescence; and then discuss the emerging of double layer OLED device, pointing out its novelties.

3. Pease point out the main conversion processes of electrical energy to optical energy in an OLED device, and please explain the formation of excitons in OLED devices.

4. Please theoretically deduce the maximum internal quantum efficiency (IQE) of an OLED device when the emitting mechanism is ① fluorescence, ② phosphorescence.

5. What is TADF, please schematically draw a figure for this process, and then explain the performance of OLED based on TADF.

6. Please enumerate characterization parameters for OLED performance, and give explanation for each.

7. An emitter exhibits the PL spectra in solution and solid state as the following figure:

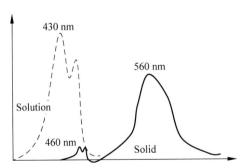

(1) Please explain the origination of the PL difference in solution and solid state.

(2) Based on an OLED device with structure of ITO/HTL/EML/ETL/LiF/Al, where HTL is hole transport layer, EML is emission layer, ETL is electron transport layer, please answer the following question:

a) If the EML contains 1% of the above emitter, and it is the only emitter in the device, generally speaking, what is the electroluminescence (EL) emission peak of this device?

b) If the EML contains 100% of the above emitter, what is the most possible emission peak(s)?

c) If with 1% of the above emitter doping, the EL spectra is as following, please state possible reasons for this EL spectrum.

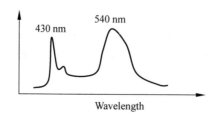

8. Please discuss the device structure of OLED devices according to the number of organic layers, and then point out the emitting zone of each device.

9. Please explain and compare different device efficiencies of an OLED device.

CHAPTER 9   ORGANIC SENSORS

After background introduction to sensor，this chapter discusses the main theoretical models for sensing，and various sensing schemes based on organic semiconductors，including those based on analyte properties of chemical conductometricity，potcntiometricity，fluorescence，and thermo-/iono-/bio-chromatism. Finally，organic photodetector is discussed in detail.

9.1   Concept of Sensor

Detection or identification of specific composition，external stimuli（pressure，temperature，touch and light），chemical activity，concentration，and the presence of specific element/or ion is one of the most sought after requirements in a variety of areas such as medicine，environment，industry，offices and homes. The most common and conventional methods developed for detection of specific analytes are based on high-performance liquid chromatography and gas chromatography. However，these techniques are time consuming and expensive. Therefore，a simple method is required for enabling rapid detection of analytes efficiently and effectively at low cost，and sensor is thought to be such a method. A sensor is a device that transforms information such as chemical composition，chemical activity，concentration and stress into detectable physical signals（such as current，mass，acoustic，optic）that could be analyzed using the available methods[1]. An ideal sensor should be portable，affordable，easy to fabricate and should respond to the particular analyte quickly and selectively，independent of ambient conditions. Such ideal sensors are，however，yet to be realized although several successful sensors are commercially available. Limiting features for a typical sensor are as follows：selectivity，sensitivity，detection limit，response and recovery time and packaging size[2]. Therefore，there is a constant demand for developing sensors，by using

1. 传感器是这样一种装置：可以把物质的化学组分、化学活性、浓度或者应力等信息转换成可探测并可被现有方法分析的物理信号（例如电流、质量、声和光）。

2. 典型传感器的限制性特征有选择性、灵敏度、探测极限、检测限、反应和恢复时间以及器件尺寸。

3. 传感器可大致分为这样的类别，诸如生物传感器、化学传感器和物理传感器。根据应用对象，传感器可以分为气体传感器、离子传感器、湿度传感器、热传感器、光传感器和压力传感器等。根据传感原理，传感器可分为电阻（电导）式、电位式、电容式、场效应、色度、荧光和谐振等类别。

new sensing materials, which may result in an improvement in selectivity and sensitivity of the device.

Sensors can be classified broadly into categories such as biological, chemical, and physical sensors. According to the application targets, sensors can be divided into gas sensor, ion sensor, humidity sensor, thermo-sensor, photo-sensor and pressure sensor, etc. Depending on the transducing principle, sensors are divided into resistive (conductive), potentiometric、capacitive, field effect, colorimetric, fluorescent, and resonating sensors, etc[3]. The operating principle of a sensor could be any of the combination of the parameters that are presented in Figure 9.1. Generally speaking, the metal-oxide or conducting polymers (chemiresistors) use the resistive transducing principle; the field-effect transducer, which is normally associated with the field-effect transistors, shows the change in threshold voltage or conductivity of the channel depending upon the exposure and type of analytes; capacitive transducers normally exhibit variations in dielectric constant; the fluorescence/colorimetric principle demonstrates the modification in the optical spectrum with the presence of analytes; and the resonating transducing principle displays change in the resonating frequency of the resonator.

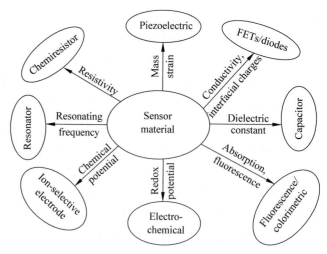

Figure 9.1　Most commonly used mechanism in sensors and the arrow indicates the change in physical parameter upon interaction with analyte

Organic sensors generally refer to sensors based on active components of organic materials, such as organic semiconductors. The sensors based on organic semiconductors are more inherently suitable for chemical, biological, and gas sensing applications compared to their inorganic counterparts. The reasons behind the fact are that inorganic sensing devices have limited selectivity, moderate sensitivity, higher operating temperature, and lack of flexibility for some specific sensing applications such as strain or pressure sensors; while those based on organic semiconductors (small molecular or conjugated polymers (CPs)) possess advantages of good selectivity, quick response, low cost, light weight, easy processing, tunable electronic and optic properties, large area fabrication and especially potential for fabricating flexible or wearable sensors[4]. Their lower operating temperature, portability, disposability, and multipara-metric sensing characteristics are also very attractive. Furthermore, the weak van der Waals force between organic molecules makes them reproducible upon the application of an appropriate cleaning substance. On the other hand, there are several inherent drawbacks of organic semiconductors, such as lower carrier mobility and deprived environmental stability, which make them less suitable as direct sensing elements.

9.2 Mechanism in Organic Sensors

9.2.1 Interaction Models

The interaction of the analyte with organic semiconductor depends on many parameters such as film morphology, structure, work function, ionization potential, and electron affinity. Therefore, many factors have to be optimized for the design of organic semiconductors to be used as sensing material. These include high surface-to-volume ratio for enhanced response, chemical functionality to improve selectivity, crystallinity or structural ordering to have efficient charge transfer and fast adsorption/desorption kinetics for quick response/recovery, etc[5].

In order to sense molecule, the gas molecules shall be

4. 与无机传感器相比，基于有机半导体的传感器，其本质上更适合于化学、生物和气体传感应用。这个事实背后的原因是，无机传感器的选择性有限，只有中等的灵敏度，需要比较高的工作温度，在诸如应变和压力等特殊的传感应用方面缺乏柔韧性；而基于有机半导体材料(小分子或共轭聚合物)的传感器具有良好的选择性、快速的反应、低成本、质量轻、容易制备、可调控的光电性质、可大面积制备，特别是在制备柔性或者可穿戴传感器方面具有潜力。

5. 这些包括高的表面-体积比来增强响应，通过化学功能团修饰来提高选择性，晶体或结构有序来产生高效的电荷转移和快速的吸附/脱附动力学以实现快速响应/恢复等。

interactive with the surface of sensing material. Generally, this means that the molecules are hold at the specific sites on the surface of sensing materials. Thus, the model to describe the molecular adsorption is essential for understanding the sensing mechanisms.

1. Langmuir Model

The Langmuir model describes the relationship between the available sites in organic semiconductors which can interact with a number of adsorbed molecules and is useful to describe the adsorption of gas on monolayers. In this model, it is assumed that: ① molecules are adsorbed at specific sites on the surface of sensing materials, and the total coverage is limited to a single layer and ② there is no interaction between adsorbed molecules/atoms on different sites and the enthalpy of desorption is independent of the number of sites and is the same for all the sites[6].

Based on these assumptions, the fundamental kinetic equation for the Langmuir isotherm is given by the following equation:

$$\frac{P}{\theta} = \frac{1}{\chi} + P \tag{9-1}$$

where $\chi = \left(\frac{\tau_a}{n_0}\right) \times \frac{I}{P}$, with τ_a, n_0, I and P being the surface lifetime of gas molecules after absorption, total number of sites available for adsorption, impingement rate, the pressure of the adsorbate, respectively, and θ is the number of occupied sites as a fraction of total sites. This model is used to describe many conducting polymer-based sensors and is also used in conjunction with other models.

2. Brunauer, Emmett and Teller (BET) Model

The Langmuir model does not hold good if there is any interaction between adsorbed species. In this case, adsorbed species are not confined to single mono-layer as a result of which, the heat of adsorption of an adsorbate molecule adsorbing on organic semiconductors will differ significantly and accordingly, can be described by the following equation:

6. 在这个模型中假设了：①分子是吸附在传感材料表面的特殊位点，而且总的覆盖度局限于一个单层；②在不同吸附位点的吸附分子之间没有相互作用，脱附焓与位点数量无关，并且对于所有位点都是相同的。

$$V = \frac{Y V_{max} P / P_V}{\left(1 - \dfrac{P}{P_V}\right)\left[1 + \dfrac{(Y-1)P}{P_V}\right]} \qquad (9\text{-}2)$$

where V is the adsorbed gas volume, V_{max} denotes the volume needed to form a monolayer on the whole surface, P_V denotes the vapor pressure of the adsorbed gas, Y is a constant number and $Y \approx \exp\left[(\Delta H_{des} - \Delta H_{vap})/(RT)\right]$, where H_{des} is the heat of desorption, H_{vap} is the heat of vaporization, R is the ideal gas constant and T is the temperature. This model has also been used widely to describe the mechanism in organic semiconductors.

3. Freundlich Model

This adsorption model is an empirical function that expresses the relationship between the magnitude of adsorption and the pressure at a constant temperature as follows:

$$\frac{\chi}{m} = K P^{\frac{1}{n}} \qquad (9\text{-}3)$$

where m and χ are masses of adsorbed vapor and adsorbate at an equilibrium pressure P, respectively. n and K are the constants that are dependent on the adsorbate and vapor intrinsic nature at a particular temperature. This model was used to explain the sensitivity of ultra-thin films of conducting polymer for NO_2. Other than the above models, modified models have also been proposed such as diffusion limited and reaction-diffusion model, which will not be discussed here.

9.2.2 Conductometric Sensors - Chemiresistors

In a chemiresistor, which consists of one or several pairs of electrodes, as depicted in Figure 9.2(a), the change of electrical resistance of sensing material is measured upon interaction with the analyte and a simple ohmmeter may be used to collect the data. Organic semiconductors, though intrinsically affected by many, still unsolved, problems, such as low mobility and poor stability, have a very unique property that makes them interesting for chemosensing: they can be chemically tailored in order to obtain specific properties[7]. This aspect, together with the low cost of

7. 在化敏电阻器中,如图 9.2(a)所示,有一对或者多对电极,敏感材料电阻的变化,在与被分析物作用时得到检测,简单的欧姆表可以用于收集数据。有机半导体虽然存在许多内在的影响因素,如迁移率低和稳定性差,却拥有很多独特性质,使它们的化学传感令人感兴趣:为了获得特定的性质,可以对它们进行化学裁剪。

deposition and fabrication techniques, makes them really attractive for future electronic concepts.

In general, the conductivity of organic semiconductors ranges from semiconductive to insulating regions in the pristine state, which can be altered by controlled addition of chemical species in small non-stoichiometric quantity termed as dopants. Doping generally refers to the oxidation/reduction of the electronic system of organic semiconductors. Upon exposure of analytes to organic semiconductors, the change in conductance/resistance could arise due to either change in doping levels or conformational changes. This change can happen in solid form or in solution and can be achieved chemically or electrochemically[8]. Since this redox process is reversible, sensitive transducers can be developed using this parameter. Other than redox processes, physiological parameters, such as pressure, temperature and moisture, also lead to the change in the resistance of organic semiconductors. Besides, the piezoresistive property of organic materials was also used to transduce mechanical strain into an impedance/conductivity change.

8. 当有机半导体暴露于分析物时，电导/电阻的变化可以由于掺杂水平或构象的变化而产生。这种变化可以以固体或溶液形式发生，可以通过化学或电化学方式实现。

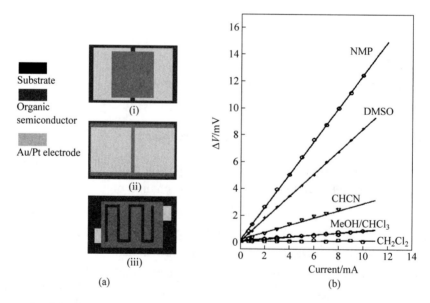

Figure 9.2　(a) Chemiresistor - planar (i and ii) or interdigitated (iii), (b) voltage change before and after exposed to the saturated vapors of organic solvent

Sensors based on organic film sandwiched between electrodes are the simplest configuration of the conductivity based sensors, and such an experimental result is shown in Figure 9.2(b). The voltage change before and after exposed an organic sensor to the saturated vapors of organic solvent are present in the figure. The sensor is made of a derivative of polyamine and the analyte is saturated vapors of organic solvent. It can be seen that: ① for a fixed current, voltage increased after exposed to organic solvent, inferring decrease of conductivity of the polyamine; ② NMP is the most sensitive, and CH_2Cl_2 is the least sensitive.

9.2.3 Sensors based on OFET Devices

An organic field-effect sensor is, first of all, a transistor, i.e., an active device, able to produce an output that is the amplified copy of the input. Alternatively, the device can behave as a bistable component, able to switch between the on and off states depending on the input. In a field-effect device, the input signal is a voltage, applied through a capacitive structure to the device channel. An organic field effect transistor (OFET) is comprised of a multilayered structure where the gate capacitor is formed by a metal, an insulator, and a thin organic semiconductor layer as shown in Figure 9.3(a)(ⅰ) and (ⅱ). On the semiconductor side, two metal contacts, source and drain, are used for extracting a current that depends both on the drain-source voltage and on the gate-source voltage. To obtain detection for physical parameters, again the external stimulus must reversibly affect one of the different layers of the device and result in variation of one or more of the electronic parameters (mobility, threshold voltage, etc.) that can be extracted from the output curves. A thin organic semiconductor layer is needed because this device does not work in inversion mode, i.e., the carriers that accumulate in the channel are the same that normally flow in the bulk in the off state. Therefore, if the semiconductor is too thick, the off-current is too high and the switching ability of the device is compromised[9]. For the same reason, this structure is used only for low-mobility semiconductors, as high mobility semiconductors would give

9. 因为该器件不能工作在反转模式下,有机半导体层必须是薄的,即在沟道中累积的载流子与通常在关状态下在内部流动的载流子相同。因此,如果半导体太厚,则关闭状态下电流太高,器件的开关能力被迫让步。

10. 为了实现 OFET 传感器，有必要在组成 OFET 器件的其中一层中，实现对器件外部刺激的特定灵敏度和选择性。如果这是化学刺激，通常需要对器件的某一层进行特定的化学功能化。

rise，even in thin film devices，too high OFFT-currents. To obtain a sensor from an OFET，it is necessary to achieve，in one of the layers that form the device，a specific sensitivity/selectivity toward an external stimulus applied to the device. If this is a chemical stimulus，often a specific chemical functionalization of one of the device layers is required[10]. Figure 9.3(b) shows the detection of ssDNA and dsDNA solution by exposing them to OFET devices，the output current of OFET exhibits difference at the same gate voltage compared to that without exposing；and dsDNA exhibits more sensitivity compared to ssDNA.

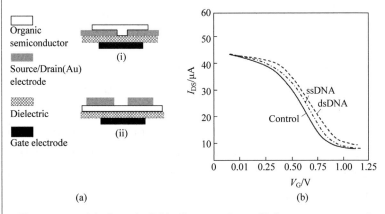

Figure 9.3　(a) Organic field-effect transistor (i) bottom-contact and (ii) top contact，(b) *I-V* curves of an OFET device before and after exposed to ssDNA and dsDNA with concentration of $5\mu M$

9.2.4　Potentiometric Sensors - Ion-Selective/ Electrochemical Sensors

In electrochemical sensors，changes，either in current or potential，are observed upon interaction with the analyte. The interaction of organic semiconductors with the analyte can give rise to a current，which is governed by Faraday's law. The transducers based on electrochemical methods can easily be fabricated using microelectrodes. Because of this reason，this method offers several advantages for detection of ions or biological molecules as compared to macroscopic electrodes because ① they provide higher spatial resolution due to small geometric area（i.e. selectivity）and ② have faster electron transfer and mass transport rate and so have quick response. The excellent redox activity of conjugated polymers has been utilized in fabricating many bio- and chemical sensors to detect ions as well as gases.

Some membrane sensors based on the potential method can be used to detect various ions，which are known as ion sensors. Figure 9.4 is a schematic diagram of ion sensor based on organic materials. The ion sensor is coated with an organic ion sensitive membrane，which can selectively transport ions into the membrane.

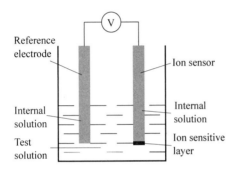

Figure 9.4　Structure of potentiometric sensor for ion sensing

The chemical potential of organic semiconductors may alter in the presence of an analyte，leading to a change in the open-circuit potential or work function. This change is proportional to the logarithmic of concentration of the analyte and is given by the Nernst equation as follows：

$$E = E_0 + 2.303 \left(\frac{RT}{zF} \right) \lg a_{H^+} \qquad (9\text{-}4)$$

where E and E_0 are the half-cell and standard half-cell reduction potential，respectively，with R，T，F，z and a_{H^+} being the universal gas constant，temperature，Faraday constant，number of electrons transferred in the cell reaction，and proton activity，respectively. The sensitivity of ion-selective electrodes（ISEs）is therefore limited to 59 mV for a tenfold sample activity change in case of a monovalent ion. Based on the change in the chemical potential，ion sensors having extremely high selectivity and detection limits down to nM（10^{-9} M）levels were developed for in situ sensing of ions and biological substances in appropriate aqueous media and physiological backgrounds[11].

11. 因此对于一价离子，为了使样品活度产生 10 倍的变化，离子选择性电极（ISE）的灵敏度局限于 59 mV。基于化学势的变化，实现了具有极高的离子选择性和低至 nM（10^{-9} M）检测限的离子传感器，这可用于在适当的水介质和生理环境中对离子和生物物质的原位检测。

12. 图 9.6 展示了一个聚合物在给定电压下,对不同离子的响应。可以看到,①少量 Li$^+$ 产生电流的急剧变化,②Na$^+$ 对电流的影响居中,③K$^+$ 的响应最慢。因此,这个聚合物对 Li$^+$ 最敏感,对 K$^+$ 最不敏感。

Figure 9.5 exhibits an experimental result from a electrochemical analysis, with Li$^+$ samples of different concentration. As can be seen, increasing Li$^+$ concentration, the oxidation potential position moves to higher values, making the material to be more difficult to be oxidized. Figure 9.6 shows the response of polymer to different ions under fixed voltage. It can be found that ①small amount of Li$^+$ causes sharp change of current, ②the influence of Na$^+$ to current is moderate, and ③ response from K$^+$ is the slowest. Thus, this polymer is the most sensitive to Li$^+$ and the least sensitive to K$^{+\ 12}$.

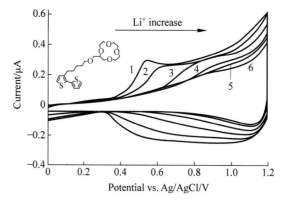

Figure 9.5 Cyclic voltammetry curve of polymer films in Li$^+$ solution with different concentrations

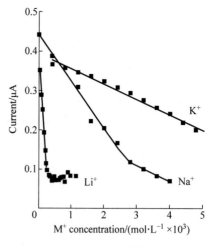

Figure 9.6 The response of polymer to different ions under fixed voltage. For Li$^+$, Na$^+$, and K$^+$, the applied voltage is 0.53 V, 0.59 V, 0.57 V, respectively

9.2.5 Fluorescence Sensors

Fluorescence sensor is one kind of optical sensors that use analyte-sensitive organic materials to detect changes in luminescence, optical absorption, refractive and reflectance, etc. A typical optical sensor often consists of an optical fiber with a conducting polymer immobilized on the fiber tip. By immobilization of different polymer layers onto the fiber tip, the sensor can be tuned to distinguish a particular analyte[13].

Among optical techniques, fluorescence is a very powerful transduction mechanism that allows for detection of even trace element based on a different outcome of fluorescence such as lifetime, wavelength (excitation and emission), anisotropy or intensity of the emission of fluorescent probes and energy transfer. The sensors can obtain the qualitative and quantitative identification of alien species.

The absorption of photons in the fluorescent organic semiconductors creates excitons, giving rise to strong fluorescence. The interaction of analytes with these fluorescent molecular can disrupt the extent of delocalization and result in fluorescence changes. When fluorescent sensor molecules contact with analyte, the mechanisms for fluorescence changes of the sensor molecules include: photoinduced electron transfer (PET), energy transfer and conformation change[14].

1. Photoinduced Electron Transfer

Figure 9.7 shows the processes of photoinduced electron transfer (PET). As can be seen, the electron in analyte orbital can jump to the HOMO of excited molecule (Figure 9.7(a)) or the excited molecule can transfer its electron in LUMO to the empty orbital of analyte (Figure 9.7(b)). Both processes will induce the decay of excited molecule to its ground state without light emission. Thus, PET process results in the fluorescence quenching of sensors.

13. 荧光传感器是光学传感器的一种,光学传感器是通过对待测物敏感的有机材料来探测分析其在光学发光、吸收、折射率和反射率等方面的变化。典型的光学传感器通常由一根光纤组成,其中导电聚合物固定在光纤的尖端。通过将不同的聚合物层固定到光纤的尖端上,可以调节光学传感器,以便区别特定的待测样品。

14. 分析物与这些荧光分子的相互作用,可以改变(电子)离域程度并导致荧光变化。当荧光传感分子与分析物接触时,传感分子的荧光变化机制包括光诱导电子转移(PET)、能量转移和构象变化。

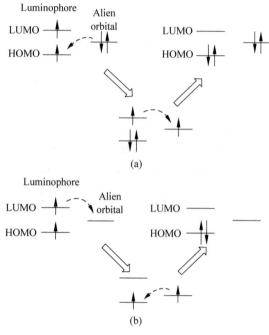

Figure 9.7 Scheme of photoinduced electron transfer process. (a) Alien orbits are electron filled, and (b) alien orbits are empty

2. Conformation Change

The luminescence of fluorescent molecules may be changed by the molecular configuration change with the external influence. These changes include the generation or reduction of excimer luminescence, the enhancement of luminescence caused by molecular rigidity, etc[15]. As shown in the Figure 9.8(a), A^{n-} decreases the distance between the two emitting units, and changes the molecular configuration, resulting in excimer emission. For contrast, Figure 9.8(b) shows decrease/disappearance of excimer emission upon analyte adding. The molecule is a pyrene derivative, and the two pyrene moieties tend to stack face to face and excimer emission can be clearly observed when no external species present. Adding I^-, the attraction between N^+ and I^- breaks the face to face stacking of pyrene parts and excimer emission decreases, or even disappears.

15. 荧光分子的发光可能会因为外部影响导致的分子构型变化而改变。这些变化包括激基二聚物发光的产生或减少、分子刚性引起的发光增强等。

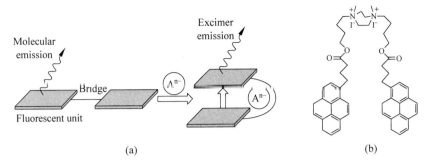

Figure 9.8　(a) Schematic occurrence of anionic induced excimer. (b) Sensing of I⁻

based on fluorescence changes of the excimer

3. Fluorescent Resonance Energy Transfer（FRET）

Another frequently employed transducing mechanism in organic fluorescent sensor is fluorescent resonance energy transfer，which refers that，the exciton in the fluorescent sensor transfers its energy to the analyte by resonance of dipole-dipole interaction， leading to the fluorescence quenching[16]. This mechanism is the same as Förster energy transfer，and is discussed in Chapter 3.

4. Amplifying Effect in Fluorescent Polymer

As shown in the Figure 9.9(a)，when sensing materials are luminescence materials of small molecules，an external quencher molecule can only quench a fluorescence molecule; when sensing materials are conjugated polymers (Figure 9.9(b))， an external quencher molecule can interact with any active site in conjugated polymer chains， hence rendering all excitons in the polymer chains being quenched via electronic or energy transfer， leading to the fluorescence quenching of the whole polymer chain， so the fluorescence quenching signal is magnified[17].

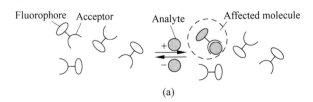

Figure 9.9　Amplifying effect in fluorescent polymer. (a) fluorescent small

molecule based sensor，(b) fluorescent polymer based sensor

16. 在有机荧光传感器中，另一个经常使用的传感机制是荧光能量共振转移，指的是荧光传感材料的激子将其能量通过共振的偶极-偶极相互作用传递给待分析物，导致荧光猝灭。

17. 当传感材料是小分子发光材料时，一个外部的猝灭分子只能猝灭一个荧光分子；当传感材料是共轭聚合物时（图9.9(b)），一个外部猝灭分子可以与共轭聚合物链中的任何活性位点相互作用，使得聚合物链中的所有激子通过电子或能量转移被猝灭，从而导致整个聚合物链的荧光猝灭，因此荧光猝灭信号被放大了。

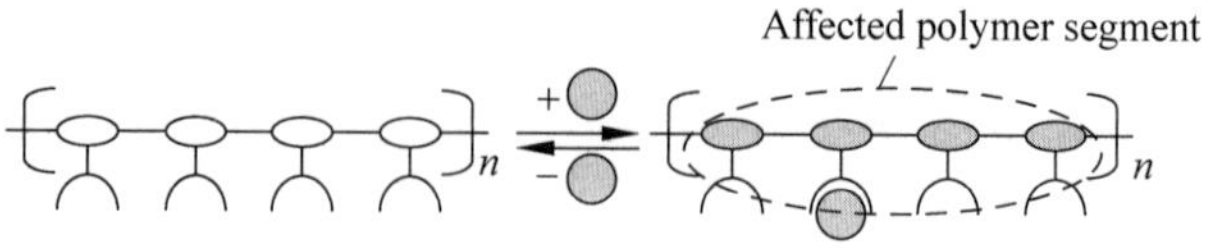

(b)

Figure 9.9 (Continuous)

9.2.6 Colorimetric Sensors

Colorimetric sensor is one of absorption-based sensing techniques, which depend on the conformational changes in organic molecules with interaction of the analyte; by such, the band gap and hence the absorption characteristics of the organic molecules exhibit a change that can be used as sensing[18]. In contrast to fluorescence, colorimetric techniques have advantages of low cost, requirement of inexpensive equipment and simplicity. The colorimetric sensors have been developed for both qualitative analyte identification and quantitative analysis. The colorimetric response (CR) has been used quite often to define the response of organic semiconductors to the analyte and is defined as follows:

$$CR = \frac{PB_0 - PB_1}{PB_0} \times 100\% \tag{9-5}$$

where $PB = A_{SW}/(A_{SW} + A_{LW})$, in which A_{SW} and A_{LW} is the absorbance at the short wavelength and the long wavelength component between isosbestic point, respectively, in the UV-Vis spectrum, and PB_0 and PB_1 are the values without and with analyte, respectively. The colorimetric sensors based on absorption property can be classified as thermochromism, solvatochromism, piezochromism, ionochromism and biochromism, etc.

1. Thermochromism

The dominant mechanism for thermochromism in organic materials is the existence of two molecular conformations with different optical gap, whose ratio changes with temperature. Thermochromism can also be observed in some solid state polymers, but the temperatures are higher than those in solution. Possible reason is that in solid state, both

18. 色度传感器是一种基于吸收的传感技术,它通过有机分子与待测样品之间的相互作用所产生的构象变化来传感。由此,改变了有机分子的带隙以及吸收特性,这可用来传感。

viscosity and interaction between chains are higher，making the conformation-change difficult. An interesting phenomenon observed is that the temperature range for thermochromic change in solid polymers decrease with the increase of side-chain length or flexibility[19].

Figure 9.10 shows the thermochromic effect curves of a thiophene derivative in water. The absorption peak is located at 544 nm before heating，while the absorption peak changes from 545 nm to 425 nm by heating. The temperature can be determined by the ratio of the absorbance at 545 nm and 425 nm through the calibrated temperature - ratio curve.

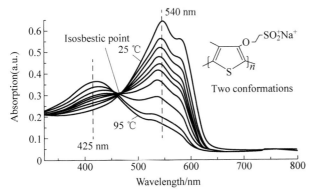

Figure 9. 10   Temperature-dependent UV-Vis absorption spectra of sodium poly2-(4-methyl-3-thiethyloxy)ethanesulfonate in water

In most polymers，the change of molecular framework between co-planar and non-planar is the dominant reason for thermochromism，as shown in Figure 9. 11. For irregular polymers (Figure 9.11(a))，a twist of one repeated unit can induce the twist of adjacent unit，leading to the continuous twist of the molecular chain and the breakdown of conjugation，hence the absorption spectrum blue-shifts. Figure 9.11(b) demonstrates a scheme of theromochromism for regular polymers. Due to the regular shape in a molecule，large molecular plane with good conjugation can be formed via self-assembly. Disturbance in side chain can change the regularity of molecule aggregation，hence breaking down the co-planarity and the absorption spectrum blue-shifts[20].

19. 有机材料热致变色的主要机制是存在两种具有不同光学带隙的分子构象,两种构象的比率随温度而变化。在一些固态聚合物中也可以观察到热致变色,但其温度高于溶液时的温度。可能的原因是,在固态下,黏度还有聚合物链之间的相互作用都较高,使得构象变化变得困难。一个观察到的有趣现象是,固态聚合物中热致变色变化的温度范围随着侧链长度或聚合物柔韧性的增加而减小。

20. 对于非规整型的聚合物(图9.11(a)),一个重复单元的扭曲可以诱发邻近单元的扭曲,造成分子链的连续扭曲及共轭的破坏,因此吸收光谱蓝移。图9.11(b)展示的是规整型聚合物的热致变色情形。由于分子的规整性,共轭性很好的大分子平面可以通过自组织形成。分子侧链的扰动改变了分子聚集的规整性,因而其平面性遭到破坏,吸收光谱蓝移。

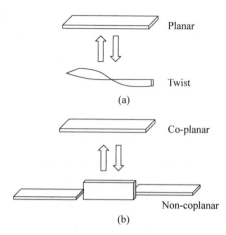

Figure 9.11　Conformational changes in polymers, twist structure (a), non-coplanar structure (b)

21. 材料 A 由于侧链的空间位阻太大,因而没有平面构型;材料 B 和 C 由于缺少侧链的空间位阻效应,即使在高温下,分子也可以保持平面构型。因此材料 A、B 和 C 均不能够在加热下形成另外一个构型,观测不到热致变色现象。图 9.12(a)给出了 C 类材料的吸收光谱随温度变化的曲线,没有观测到热致变色效应。将图 9.12(没有空间位阻侧链)与图 9.10(甲基作为空间位阻侧链)进行比较,侧链取代不同而导致的不同热致变色效应,就可以清楚看见。

Although thermochromism in polymers are mainly attributed to the conformational changes or the interaction in the polymer backbone, the position and the steric hinderence from side chains have significant influence. As shown in Figure 9.12(a), the three polymers, A, B, and C, exhibit no obvious thermochromism: Material A possesses a side chain with large steric hinderence, hence there is no planar conformation for A; in material B and C, due to the lack of steric hinderence from side chains, even at high temperature, the molecule can retain planar conformations. Therefore, material A, B and C can not change to another conformation by heating, and hence no observable thermochromatic phenomena. Figure 9.12(b) gives an example for C type material, e.g., no thermochromism. Comparison between Figure 9.12 (no side-chain for steric hinerence) and Figure 9.10 (methyl group is side chain for steric hinderence) well demonstrates the side chain effect on thermochromism[21].

2. Ionochromism

22. 在含有醚或冠醚取代基的共轭聚合物中,当这些醚或冠醚与碱金属配位时,侧链的空间排列可以得到修饰,从而影响聚合物主链的构型及其吸收光谱。

In conjugated polymers containing ether or crown ether substituents, the spatial arrangement of side chains can be modified when the ether or crown ether is coordinated with alkali metals, thus affecting the main chain configuration and absorption spectrum of the polymer[22].

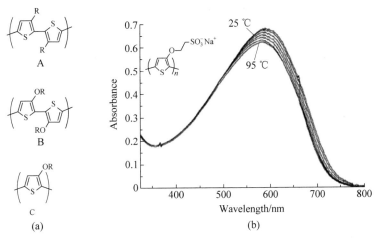

Figure 9.12　（a）Three molecules with no obvious thermochromic effects，

　　　　　　（b）temperature dependent absorption spectra of sodium poly

　　　　　　（2-（3-thiethyloxy）ethanesulfonate）in water

In Figure 9.13，the coordination between Na$^+$ and crown ether leads to rigid self-assembly of the macromolecules，resulting in increase of planarity and effective conjugation，absorption peak change from 440 nm to 545 nm[23].

23. 图 9.13 中，Na$^+$ 和冠醚之间的配位，导致了大分子的刚性自主装，结果增加了平面性及有效共轭，吸收峰位从 440 nm 变化到 545 nm。

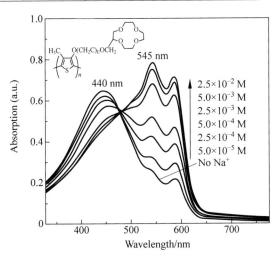

Figure 9.13　The absorption spectra of crown ether polymers at different concentrations of sodium ions

3. Solvatochromism and Piezochromism

The change of solvent can lead to different dipole-dipole interaction between the solute and solvent，inducing intramolecular or intermolecular deformation，which leads to

24. 溶剂的改变可以导致溶剂与溶质之间的偶极-偶极相互作用的不同,诱发了分子内或者分子间的形变,导致吸收光谱的变化。这称为溶致变色。例如,在良性溶剂下,聚 3-烷氧基-4-甲基噻吩的吸收峰位在 425 nm;当加入非良性溶剂时,吸收峰位红移至 545 nm。这个变化与热致变色效应类似,即都是由于主链的构型变化所致。另外,有机物的构型可以随着压力改变,从而产生吸收光谱的变化。此现象称为压致变色效应。如随着压力的增大,非平面型的(或高温型)的聚 3-烷基噻吩变为平面型结构,其吸收光谱会产生红移。

the change of absorption spectrum. This is termed as solvatochromism. For instance, in good solvent, poly(3-alkoxy-4-methylthiophen) exhibits absorption peak at 425 nm; adding bad solvent, the absorption peak changes to 545 nm. This change is similar to thermochromism, that is, being attributed to the conformational change of the backbone in the polymer. On the other hand, the conformation of organic materials can change with pressure, leading to changes in absorption spectrum. This is called piezochromism. For example, with the increase of pressure, the non-planar (or the high temperature) conformation of poly(3-alkylthiophene) changes to planar structure, leading to the red shift of the absorption spectrum[24].

4. Biochromism

The complexation between biological group and some functional groups of the polymer can lead to the color change of polymer. This phenomenon can be used to detect biological molecules, which is called biochromism. As shown in Figure 9.14, biospecies interact with the receptor unit of the sensor, which results in disorder of the polyacetylene based LB film, thus changes the sensor color.

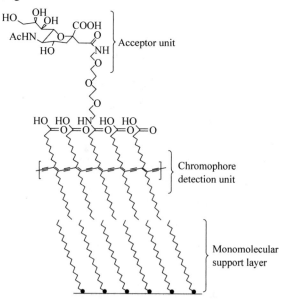

Figure 9.14　An example of the identification of hemagglutinin using a Languir-Blodgett membrane containing a polydiacetylene derivative of a sialic acid recognition unit

9.3 Organic Photodetectors

The photon detector is one of the cornerstones of modern technology, and it plays a critical role in numerous applications including imaging, communications, data retrieval, proximity and motion detection, environmental monitoring, and chemical analysis, to name but a few. In crude terms, photon detection can be divided into two broad categories: data communications, where light is used to carry an encoded signal; and sensing, where it is the properties of the light itself that are of interest. The frequency, intensity, and spectral characteristics of the optical signal vary tremendously from one application to the next[25]. In fiber-optic communication systems, photodetectors are used to receive infrared (IR) signals at rates of up to 100 GHz. In optical disc drives, they retrieve data at up to 200 MHz by detecting visible laser light reflected from the pits of a spinning disc. In remote controls for electronic equipment, optical "trip switches" for home security systems, proximity detectors for motor vehicles, and orientation/position sensors in optical-mouse devices, lower bandwidth is requited.

In choosing which detector to use for a given application, it is important to take into account both the optical demands of that application and wider issues relating to the circumstances in which the detector will be used. In data communication, the intensity of the optical signal is usually quite high, and the speed of response tends to be the overriding consideration. In sensing applications, other issues such as sensitivity, linearity, dynamic range, and spectral range are often more important. In some cases, tolerance to harsh operating conditions may be required, including resilience to corrosive chemicals or extremes of temperature. In others, issues such as the size, weight, power consumption, or cost of the photodetectors may be important. It is rarely the case that a single technology will meet all of an engineer's design criteria perfectly, and selecting a suitable detector frequently comes down to finding an acceptable compromise between many competing

25. 粗略地讲,光子检测可以分为两个广类:数据通信和传感,数据通信中光用于携带编码信号;传感中感兴趣的是光本身的性质。光的频率、强度和光谱特性从一种应用到下一种应用存在巨大的变化。

26. 很少有一种技术能够完美地满足工程师的所有设计要求,选择合适的探测器往往归结为在许多冲突的要求之间找到可接受的折中方案。

27. 绝大多数有机光电探测器可以归类为光电二极管,在于它们①即使在没有施加电压的情况下也能产生光电流,并且②没有内部增益(即每个吸收的光子最多可产生一个被外电路提取的电子)。有机光检测器在灵敏度方面无法与当前使用的光电倍增管(PMT)和雪崩光电二极管(APD)竞争。它们是基于低载流子迁移率材料的高电容器件,因此与 PMTs、APDs 和传统无机光电二极管相比,响应时间较慢。

criteria[26].

It is for the above reasons that the emergence of a new detection technology is important. Organic photodetectors (OPDs) will offer their own mix of advantages and disadvantages, making them superior for some applications and inferior for others. It is likely that new applications will arise for which conventional detectors are either technically or commercially unsuited, opening up completely new technological opportunities. Huge research efforts have been invested in OLEDs, OPVs, and OFETs, but OPDs have received far less attention to date. Although they are structurally similar to organic solar cells, their performance criteria are quite different. Material systems and device geometries that work well for solar energy applications may perform poorly when applied to photodetectors, yielding poor spectral characteristics, excessive noise, or sluggish response. OPDs should not therefore be viewed as a simple offshoot of existing solar cell research, but as technological devices in their own right with their own unique set of technical challenges.

The vast majority of organic photodetectors can be classified as photodiodes, in that they ① generate a photocurrent even in the absence of an applied bias and ②exhibit no internal gain (i. e., at most one electron is extracted into the external circuit for each absorbed photon). Organic photodetectors are unable to compete with currently used photomultiplier tubes (PMTs) and avalanche photodiodes (APDs) in terms of sensitivity. They are also high-capacitance devices based on low-mobility materials and consequently exhibit slow response times compared to PMTs, APDs, and conventional inorganic photodiodes[27]. On the other hand, their favorable processing characteristics may open up new applications that are currently unviable, e. g., in large-area light detection. The largest single-element photodetectors are 20 inch PMTs, and the largest imaging arrays are ~ 1600 cm^2 amorphous silicon panels. Using printing techniques, it should be possible to create organic photovoltaic (OPV) devices of much larger area at greatly reduced cost. In addition, by fabricating OPV devices on flexible substrates, conformal photodetectors can be created that adapt to the shape of the surface on which they are mounted. A critical area where OPD devices may in the future have an advantage over other technologies is in spectral response range. OPDs with a spectral range 350 nm to 1000 nm have already been reported, which is better than typical multialkali-based PMTs (300 nm to 850 nm) and approaches that of UV-enhanced silicon photodiodes (190 nm to 1100 nm)[28]. We are confident that OPDs with greatly enhanced spectral ranges will be developed in due course. Finally—and this is a point that we will consider—organic photodiodes may one day become the solution of choice for low bandwidth low light-level optical detection, offering comparable performance to the silicon photodiode at a price point close to that of the photoresistor.

9.3.1 Device Architectures

Same as the OPVs, there are two main device architectures for organic photodiodes: the planar heterojunction and the bulk heterojunction. The simplest planar heterojunction devices consist of a bilayer of a hole transporting material (the donor) and an electron transporting material (the acceptor) sandwiched between two electrodes. The simplest bulk heterojunction devices consist of a single blended layer of the donor and acceptor, again sandwiched between two electrodes. The key requirement in both cases is that the frontier orbitals of the two organic materials be substantially offset to encourage electron transfer from the donor to the acceptor or hole transfer in the opposite direction. The result is a partitioning of the holes and electrons into the donor and acceptor phases, respectively, from where they can be transported to the relevant electrodes[29]. The detail work principle can refer to Chapter 7.

9.3.2 Device Characteristics

1. Spectral Response

The spectral response of a photodetector is normally

28. OPD 器件未来可能比其他光探测技术具有优势的一个关键性能,是 OPD 的光谱响应范围。光谱响应范围为 350 nm 至 1000 nm 的 OPD 已经有报道,这优于典型的基于多碱金属混合的 PMT(300 nm 至 850 nm),并且接近紫外增强硅光电二极管(190 nm 至 1100 nm)。

29. 这两种情况下的关键要求是,两种有机材料的前沿轨道具有足够的偏移,以利于电子从给体转移到受体或相反方向进行空穴转移。结果是空穴和电子分别进入给体相和受体相,从各自的相中,它们可以被输送到相关的电极。

characterized in terms of its "photosensitivity", which is ordinarily measured under short-circuit conditions. The photosensitivity $S(\lambda)$ at an illumination wavelength λ is defined as the ratio of the photocurrent $I(\lambda)$ to the incident power $P(\lambda)$:

$$S(\lambda) = \frac{I(\lambda)}{P(\lambda)} \tag{9-6}$$

In situations where the photodiode is illuminated by a broadband excitation source, the resultant photocurrent I is given by

$$I = \int_{\lambda_{\min}}^{\lambda_{\max}} P(\lambda) S(\lambda) \mathrm{d}\lambda \tag{9-7}$$

The spectral response characteristics of a typical ITO/PEDOT:PSS/P3HT:PCBM/Al device is shown in Figure 9.15 in terms of photosensitivity. It also shows the spectral response characteristics of a typical Si photodiode (OSRAM, SFH2430) for comparison. The peak efficiencies of the organic device (0.24 A · W^{-1}) are comparable to those of the silicon device (0.18 A · W^{-1}), although the active range is much narrower due to the narrow absorption range of the P3HT:PCBM system. Note that although the silicon photodiode we have chosen is fairly typical, superior devices with substantially higher photosensitivity are available; likewise, superior organic devices can also be fabricated[30].

30. 尽管 P3HT:PCBM 体系较窄的吸收范围导致了相当窄的有效范围,有机器件的峰值光敏效率(0.24 A · W^{-1})与硅器件的峰值效率(0.18 A · W^{-1})相当。请注意,尽管我们选择的是相当典型的硅光电二极管,具有更高光敏性的高级器件是存在的;同样,也可以制造出更好的有机器件。

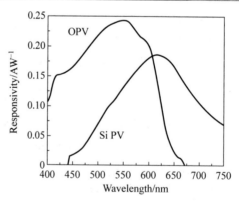

Figure 9.15　Photosensitivity spectral response curves for a Si based inorganic device and an ITO/PEDOT:PSS/P3HT:PCBM/Al organic device

2. Rise Time and Cutoff Frequency

The speed of response is a critical consideration in data

communications and time-resolved sensing applications，and it is usually characterized in terms of either the rise time or cutoff frequency defined as the reciprocal of rise time. The rise time of a photodiode is conventionally defined as the time required for the output to change from 10% to 90% of its final level in response to a step increase in the intensity (Figure 9. 16). The fall time is similarly defined as the time required for the output to change from 90% to 10% of its initial level in response to a step decrease in the intensity. The rise and fall times depend on the wavelength of the incident light (which affects the carrier generation profile) and the value of the load resistance R_L (which is conventionally chosen to be 50 Ω); higher load resistances result in longer response times due to an increased RC time constant (where the relevant capacitance C is the photodiode capacitance)[31].

31. 上升和下降时间与入射光波长（影响载流子产生轮廓）和负载电阻 R_L（通常选取的值是 50 Ω）有关；负载电阻越高,由于 RC 时间常数增加（其中相关电容 C 是光电二极管电容）,响应时间越长。

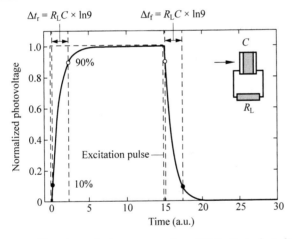

Figure 9. 16　The theoretical response of an RC-limited photodiode to a square wave light source. The rise and fall times are related to the RC time constant by $\Delta t = R_L C \times \ln 9$

3. Intrinsic Photodiode Noise Characteristics

1) Thermal Noise

The photodiodes can be modeled as a current generator in parallel with a diode，a shunt resistor，and a capacitor. The electrons inside any resistor undergo random motion as they are constantly buffeted backward and forward by atomic collisions，leading (in short circuit) to an overall current of zero. This motion results in a constant change of charge

32. 任何电阻器内的电子都会随机运动,因为它们不断被材料原子碰撞而前后运动,导致(在短路时)总电流为零。这种运动导致电荷分布的不断变化,因此在电阻器的端子上感应出随机波动的电压,称为热噪声或约翰逊噪声。随机时间的过程在频域中是均匀分布的。因此,热噪声在其特性上是"白色"的;即每单位频率的时间平均功率,称为功率谱密度,在所有频率下都是相同的。

33. 光电二极管在特定情况下,可以认为是一个分流电阻与二极管并联,并且两个电子元件都会产生热噪声(其方差以平方和的方式组合在一起)。在短路情况下,二极管内阻远高于分流电阻,因此分流电阻是热电流噪声的主要来源;在施加足够高的电压情况下,二极管内阻大幅下降,使其成为热噪声的主要来源。

34. 噪声等效功率(NEP)是表征探测器灵敏度的有用方法。NEP是最小可检测功率,形式上被定义为实现信噪比1时所需的光入射功率。

distribution and so induces a randomly fluctuating voltage across the terminals of the resistor, known as thermal or Johnson noise. A process that is random in time is evenly distributed in the frequency domain. It follows that thermal noise is "white" in its characteristics; i.e., the time-averaged power per unit frequency, known as the power spectral density, is the same at all frequencies[32].

The current noise is therefore smallest for large values of the shunt resistance. Importantly, biasing the resistor has virtually no effect on the voltage and current fluctuations since it has the minor effect of superimposing a very small constant drift velocity on the (much larger) instantaneous velocities of the charge carriers. The thermal noise of a discrete resistor is therefore largely independent of the applied bias. In the specific case of a photodiode, the shunt resistor is in parallel with a diode, and both "components" contribute thermal noise (whose variances combine in a sum of squares manner). In short circuit, the diode resistance is much higher than the shunt resistance, and the shunt resistance is therefore the dominant source of thermal current noise; at sufficiently high applied biases, the diode resistance drops substantially, causing it to become the dominant source of thermal noise[33].

2) Shot Noise

The second major source of noise is shot noise, which is due to statistical fluctuations in the flowing current caused by the discrete nature of the electrons. The signal-to-noise ratio (SNR) for a signal dominated by shot noise is given by

$$SNR \propto \sqrt{I} \qquad (9\text{-}8)$$

Shot noise is therefore most important at low photocurrents where the statistical fluctuations in the flowing electrons are most evident.

3) Noise Equivalent Power

The noise equivalent power (NEP) is a useful means of characterizing the sensitivity of a detector. The NEP is the minimum detectable power and is formally defined as the incident power required achieving a signal-to-noise ratio of 1[34].

本章小结

1. 内容概要：

本章在对传感器进行背景介绍的基础上，讨论了传感的主要理论模型，以及基于有机半导体材料的各类传感机制，包括化学电阻传感、电化学传感、荧光传感、色度传感（热致变色、离子变色和生物变色）。最后，对有机光电传感（光检测器）的检测特征进行了比较详细的介绍。

2. 基本概念：
传感、待测物、传感原理、灵敏度、选择性、化学电阻传感器、电位式传感器、荧光传感器、外来轨道、放大效应、光诱导电子转移、色度传感器、热致变色、离子变色、溶致变色、压致变色、生物变色、构型变化、等吸光点、噪声等效功率、光检测器、散粒噪声、热噪声。

3. 主要公式：

（1）Langmuir 等热吸附：$\dfrac{P}{\theta} = \dfrac{1}{\chi} + P$，$\chi = \dfrac{\tau_a}{n_0} \times \dfrac{I}{P}$

（2）BET 吸附：$V = \dfrac{YV_{max}\dfrac{P}{P_V}}{\left(1-\dfrac{P}{P_V}\right)\left[1+\dfrac{(Y-1)P}{P_V}\right]}$，$Y \approx \exp\left[(\Delta H_{des} - \Delta H_{vap})/(RT)\right]$

（3）Freundlich 吸附：$\dfrac{\chi}{m} = KP^{\frac{1}{n}}$

（4）电位与待测物浓度关系：$E = E_0 + 2.303\dfrac{RT}{zF}\lg a_{H^+}$

（5）比色响应：$CR = \dfrac{PB_0 - PB_1}{PB_0} \times 100\%$

（6）光灵敏度：$S(\lambda) = \dfrac{I(\lambda)}{P(\lambda)}$

（7）光电流：$I = \displaystyle\int_{\lambda_{min}}^{\lambda_{max}} P(\lambda)S(\lambda)\mathrm{d}\lambda$

（8）散粒噪声：$SNR \propto \sqrt{I}$

Glossary

Active material	活性材料
Aggregation emission	聚集发光
Alien orbital	外来轨道
Amplifying effect	放大效应
Analyte	被检测物，待测物
Biochromism	生物变色效应
Biosensor	生物传感器

425

Charge transfer	电荷转移
Chemical sensor	化学传感器
Colorimetric sensor	色度传感
Conformation change	构型变化
Conjugated polymer	共轭聚合物
Coordination	配位
Detection	侦查，探测
Effective conjugation	有效共轭
Emission	发射
Excimer emission	基激二聚物发射
Fluorescent sensor	荧光传感器
Fluorescent resonance energy transfer（FRET）	荧光共振能量转移
Gas sensor	气体传感器
Humidity sensor	湿度传感器
Information system	信息系统
Inorganic sensor	无机传感器
Intramolecular charge transfer material	分子内电荷转移化合物
Ion sensor	离子传感器
Ionochromism	离子变色效应
Isosbestic point	等吸光度点
Noise equivalent power	噪声等效功率
Organic sensor	有机传感器
Photo absorption	光吸收
Photodetector	光检测器
Photoinduced electron transfer，PET	光诱导电子转移
Physical sensor	物理传感器
Piezochromism	压致变色效应
Potentiometric sensor	电位式传感器
Resistor	电阻器
Response	反应
Sensitive	敏感的，灵敏的
Sensor	传感器
Shot noise	散粒噪声
Signal processing	信号处理
Solvatochromism	溶致变色效应
Thermochromism	热致变色效应
Thermosensor	热敏元件
Thermal noise	热噪声
Transducing principle	传感原理，转换原理

Problems

1. True or false questions

(1) Liquid/gas chromatography and sensor device are both tools for the detection of analytes or properties from analytes; compared to the former, the latter is more convenient and costless.

(2) One of the advantage of organic sensor over the inorganic counterpart is its high speed.

(3) In terms of adsorption model, Langmuir Model describes the adsorption of gas molecules on a surface; there exists interaction between the adsorbate molecules.

(4) Luminescence based organic sensor doesn't belong to optical sensor.

(5) In organic conductometric sensor, the change of conductivity of the organic material is account from the interaction between the analyte and the molecule in the organic material.

(6) The luminescence quenching is often severe in a small molecule compared to a polymer.

(7) The device structure for organic photodetector is very similar to that for organic solar cell.

2. Please enumerate the advantages and disadvantages of the sensors based on organic materials

3. Please enumerate the mechanisms of organic sensors.

4. A tablet of a polyamine derivative was set between two electrodes, and exposed to vapors of organic solvent. Figure 9.2(b) shows the voltage change before and after exposed to the solvent vapor of: (1) CH_2Cl_2, (2) $CHCl_3$, (3) CHCN, (4) MeOH, (5) DMSO, (6) NMP. Please point out which one is the most, and the least sensitive solvents to the organic sensor, respectively.

5. For colorimetric type sensor based on absorption property, please emulate the main mechanism leading to the change of colorimetric property in a material?

6. Figure 9.10 shows the changes of material absorption with heating, please point out the phenomenon with heating, and explain possible reasons.

7. According to Figure 9.13, please point out the effect of Na^+ additive, and explain it in terms of working principle.

8. What is PET process in a fluorescent sensor? What are the two common types of PET? Please, with the help of drawing, state one PET process in detail.

参 考 文 献

[1] 罗勤慧.配位化学［M］.北京：科学出版社,2012.

[2] 黄维,密保秀,高志强.有机电子学［M］.北京：科学出版社,2011.

[3] 樊美公,姚建年,佟振合,等.分子光化学与光功能材料科学［M］.北京：科学出版社,2008.

[4] NAITO H. Organic semiconductors for optoelectronics ［M］. West Sussex：John Wiley & Sons Inc.,2021.

[5] NGUYEN T-P. Organic electronics 1：materials and physical processes ［M］.London：ISTE Ltd and NJ：John Wiley& Sons Inc.,2021.

[6] BISQUERT J. The physics of solar energy conversion ［M］.Boca Raton：CRC Press,2020.

[7] STEPHEN R F. Organic electronics：foundations to applications ［M］.Oxford：Oxford University Press,2020.

[8] SOROUSH M,LAU K K S. Dye-sensitized solar cells：mathematical modeling,and materials design and optimization ［M］.London：Academic Press,2019.

[9] YERSIN H. Highly efficient OLEDs：materials based on thermally activated delayed fluorescence ［M］.Weinheim：Wiley-VCH Verlag GmbH & Co. KGaA,2019.

[10] MÄNTELE W, DENIZ E. UV-Vis absorption spectroscopy：Lambert-Beer reloaded ［J］. Spectrochimica Acta Part A：Molecular and Biomolecular Spectroscopy,2017,173：965-968.

[11] SUN S-S,DALTON L R. Introduction to organic electronic and optoelectronic materials and devices ［M］.2nd ed.London：CRC Press,2017.

[12] GARCIA-BREIJO E,PÉREZ B G-L,COSSEDDU P. Organic sensors：materials and applications ［M］.London：The Institution of Engineering and Technology,2016.

[13] OSTROVERKHOVA O. Organic optoelectronic materials：mechanisms and applications ［J］. Chemical Reviews,2016,116：13279-13412.

[14] VALIZADEH P. Field effect transistors,a comprehensive overview：from basic concepts to novel technologies ［M］.New Jersey：John Wiley & Sons,Inc.,2016.

[15] GASPAR J D,POLIKARPOV E. OLED fundamentals：materials,devices,and processing of organic light-emitting diodes ［M］.Boca Raton：CRC Press,2015.

[16] KÖHLER A,BÄSSLER H. Electronic processes in organic semiconductors：an introduction ［M］. Weinheim：Wiley-VCH Verlag GmbH & Co. KGaA,2015.

[17] MIKHNENKO O V,BLOM P W M,QNGUYEN T. Exciton diffusion in organic semiconductors ［J］.Energy & Environmental Science,2015,8：1867-1888.

[18] SHRIVER D,WELLER M,OVERTON T,et al. Inorganic chemistry ［M］.6th ed.Oxford：Oxford University Press,2014.

[19] STONE A J. The theory of intermolecular forces ［M］.2nd ed.Oxford：Oxford University Press,2013.

[20] LI M F,NATHAN A,WU Y,et al. Organic thin film transistor integration：a hybrid approach ［M］.Weinheim：Wiley-VCH Verlag GmbH & Co. KGaA,2011.

[21] KALYANASUNDARAM K. Dye-sensitized solar cells ［M］.Lausanne：EPFL Press,2010.

[22] KAMPEN T U. Low molecular weight organic semiconductors ［M］.Weinheim：WILEY-VCH Verlag GmbH & Co. KGaA,2010.

[23] KYMISSIS I. Organic field effect transistors：theory,fabrication and characterization ［M］.New York：Springer Science + Business Media,LLC,2009.

[24] MCINTIRE R,DONNELL P. Integrated circuits,photodiodes and organic field effect transistors

［M］. New York：Nova Science Publishers Inc. ，2009.

［25］ STALLINGA P. Electrical characterization of organic electronic materials and devices ［M］. West Sussex：John Wiley & Sons Ltd. ，2009.

［26］ WÖLL C. Physical and chemical aspects of organic electronics：from fundamentals to functioning devices ［M］. Weinheim：WILEY-VCH Verlag GmbH & Co. KGaA，2009.

［27］ BRABEC C，DYAKONOV V，SCHERF U. Organic photovoltaics：materials，device physics，and manufacturing technologies ［M］. Weinheim：WILEY-VCH Verlag GmbH & Co. KGaA，2008.

［28］ BERNARDS D A，OWENS R M，MALLIARAS G G. Organic semiconductors in sensor applications ［M］. Berlin Heidelberg：Springer-Verlag，2008.

［29］ SCHWOERER M，WOLF H C. Organic molecular solids ［M］. Weinheim：John Wiley & Sons Inc. ，2007.

［30］ KLAUK H. Organic electronics：material，manufacturing and applications ［M］. Weinheim：WILEY-VCH Verlag GmbH & Co. KGaA，2006.

［31］ BRÜTTING W. Physics of organic semiconductors ［M］. Weinheim：Wiley-VCH Verlag & Co. KGaA，2005.

［32］ KALINOWSKI J. Organic light-emitting diodes：principles，characteristics，and processes ［M］. New York：Marcel Dekker，2005.

［33］ SUN S-S，SARICIFTCI N S. Organic photovoltaics：mechanisms，materials，and Devices ［M］. Boca Raton：Taylor & Francis Group，2005.

［34］ PALSSON L-O，MONKMAN A P. Measurements of solid-state photoluminescence quantum yields of films using a fluorimeter ［J］. Adv. Mater. ，2002，14：757-758.

［35］ KALINOWSKI J. Electroluminescence in organics ［J］. J. Phys. D：Appl. Phys. ，1999，32：R179-R250.

［36］ POPE M，SWENBERG C E. Electronic processes in organic crystals and polymers ［M］. 2nd ed. Oxford and New York：Oxford University Press，1999.

［37］ KLESSINGER M，MICHL J. Excited states and photochemistry of organic molecules ［M］. New York：VCH Publishers，1995.